INTRODUCTION TO PHASE
EQUILIBRIA IN CERAMIC SYSTEMS

CRC Press
Taylor & Francis Group
Boca Raton London New York

CRC Press is an imprint of the
Taylor & Francis Group, an **informa** business

First published 1984 by Marcel Dekker, Inc.

Published 2019 by CRC Press
Taylor & Francis Group
6000 Broken Sound Parkway NW, Suite 300
Boca Raton, FL 33487-2742

© 1984 by Taylor & Francis Group, LLC
CRC Press is an imprint of Taylor & Francis Group, an Informa business

First issued in paperback 2019

No claim to original U.S. Government works

ISBN-13: 978-0-367-45183-7 (pbk)
ISBN-13: 978-0-8247-7152-2 (hbk)

Visit the Taylor & Francis Web site at
http://www.taylorandfrancis.com

and the CRC Press Web site at
http://www.crcpress.com

Hummel, Floyd A., [date]
    Introduction to phase equilibria in ceramic systems

    Includes bibliographies and index.
        1. Ceramics.   2. Phase rule and equilibrium.
I. Title.
TP810.H86   1984   666   84-5004
ISBN: 0–8247–7152–4

Current printing (last digit):
11

PREFACE

As W. Eitel states in his Preface, "There is no lack of textbooks on hetero-geneous equilibria." Among the early classics were "The Phase Rule" by Alex-ander Findlay and "The Phase Rule and Phase Reactions" by S. T. Bowden and two books which are now out-of-print, "The Phase Rule" by A. C. D. Rivett and "Principles of Phase Theory" by D. A. Clibbens. More recently, the very fine books by Eitel, Ricci, and Rhines have furnished rather comprehensive geo-metrical approaches to heterogeneous equilibria in silicate, aqueous salt, and metal systems. The books by Prince, Gordon, and Reisman are some of the best examples of the use of the thermodynamic approach with the phase diagram. Each of these books must be regarded as excellent and having its own special strength.

However, none of these texts are completely adequate for use in teaching phase equilibria to undergraduates in Ceramic Science and Engineering. In Gordon's Preface he lists five comments on other texts: "(1) they are heavily oriented toward the viewpoint of the chemist, and therefore, though they treat the systems involved thermodynamically, they give the area of materials systems little attention; (2) they are highly mathematical, with a language and symbolism outside the realm of easy familiarity to the materials engineer; (3) they are treatises rather than texts; (4) they deal only briefly with phase diagrams as one of many subjects in a book of a more general nature; or (5) they give only what might be called the geometrical and phenomenological approach to phase dia-grams, with little or no discussion of the thermodynamic foundations."

Comment (1) applies to all of the early classics such as Findlay, comment (2) applies to certain first-rate, top-level books such as Ricci where the language and symbolism are outside the realm of easy familiarity to the materials scientist or engineer, and comment (3) applies to many modern textbooks because they lack examples and problems. The many excellent books on calculus and other subjects in mathematics are excellent because abundant sets of problems have been provided for use by the experienced or inexperienced instructor. Comment (4) is discussed in Sec. II. A. of Chapter 1. Many textbooks in ceramics contain a single chapter on phase equilibria and the book on "The Defect Solid State" with the excellent chapter by D. E. Rase is a prime example of Gordon's #4 comment. With respect to comment (5), it is simply necessary for a student to know the basic geometry *before* working on the thermodynamics of phase diagrams. The student usually becomes acquainted with general thermodynamics while taking physical chemistry and other courses in his major discipline. Later, it may be necessary to study the specific thermodynamics of phase diagrams.

Other comments can be applied. The very fine textbook of F. N. Rhines, "Phase Diagrams in Metallurgy," is obviously written for metallurgists and therefore stresses metal systems, which, in general, react more rapidly and contain phases of higher crystallographic symmetry than ceramic systems. Solid solubility is emphasized, whereas the intermetallic compound is not; this has been traditional for the metallurgist, whereas the ceramist has long been acutely aware of the intermediate compound but only more recently of the tremendous importance and effects of solid solution on phase transitions and the chemical and physical properties of compounds. Rhines very appropriately emphasizes the importance of the *application* of phase diagram information and the need to develop space perception. Eitel very appropriately emphasizes the great importance of *non-equilibrium* in silicate systems. in more recent years, this latter notion has been extended to oxide systems of all types including simple oxides, borates, silicates, phosphates, titanates, and other "ceramic" systems.

Ceramics, metallurgy, and polymers have always had a few close relationships, but since ceramic, metallurgical, and polymer science and engineering are now combined as materials science and engineering in many universities, certain phase-related aspects of the three disciplines are becoming closer and closer as time continues, similar to crystal structure and crystal chemistry relations.

For example, at one time "powder" and "glass" were not very important aspects of metallurgy, but now "powder metallurgy" and "metallic glasses" are becoming as important as conventional fusion techniques, at least to the basic science of metals. In contrast, ceramics is now deeply involved in the production of single crystals and polycrystalline materials by fusion, relative to its historical powder methods of producing sintered porous or vitreous products and relative to its conventional methods of producing (noncrystalline) glasses, glazes, and enamels by fusion. With respect to phase equilibria, current technology increases the need for knowing the liquid-solid relationships of one atmosphere phase diagrams, but very often the vapor phase must also be considered or not neglected. For example, chemical vapor deposition (CVD) is an established method of producing solid (crystalline or noncrystalline) films. Eventually, each of the three disciplines will require a knowledge of phase relationships which involve not only the liquid and "solid state," but like all basic phase equilibria, the solid, liquid, and vapor state relationships' (p-T-X) diagrams. The pressure variable is becoming more important in "ceramic" systems as time goes on and eventually two, three, and higher order systems will have to be treated in terms of the (p-T-X) type of diagram. However, at the present time, most ceramic operations are carried out at or near one atmosphere and it is only necessary to discuss the influence of pressure in one-component systems and at rare intervals in later chapters.

At this point, one concludes that there is still a need for a textbook for undergraduates in Ceramic Science and Engineering which will contain sufficient geometrical analyses to enable the student to move into any of the more highly specialized textbooks or treatises with great ease. For example, the examination of ceramic systems under controlled oxygen pressure is now very important in the fields of slags, pigments, ferroelectrics, and ferromagnetics, and a knowledge of the principles set out in the Muan and Osborn book, "Phase Equilibria Among Oxides in Steelmaking," is an invaluable supplement in these areas.

Phase equilibria is interwoven (sometimes in a very intricate way) with crystal chemistry and at appropriate points in the text, this interrelationship will be illustrated through the use of examples of real systems, following the practice of Eitel and Rhines and others.

The author wishes to acknowledge the good fortune to have studied physical chemistry under Professor T. R. Briggs of Cornell, who was an associate of Professor Wilder D. Bancroft, founder of the "Journal of Physical Chemistry." Professor Bancroft was, in turn, one of the persons responsible for communicating the papers of Roozeboom, Schreinemakers, Buchner, and Aten, who were the interpreters of the great works of Gibbs. This chain was in a large measure responsible for placing the Phase Rule and all of its ramifications in a position where the American chemist, physicist, geologist, geochemist, mineralogist, metallurgist, and ceramist could apply it to real systems.

Finally, I wish to thank Dr. J. J. Brown, Professor of Materials Engineering at Virginia Polytechnic Institute and State University, for contributions to Chapters 1 and 2 which resulted in better organization and better discussion of the Phase Rule, Mrs. Nancy Bierly of Aaronsburg, Pa., for typing assistance and Mr. Clifford Warner of Cepiad Associates, Boalsburg, Pa., for drafting figures in chapters 2, 3, 5-8.

Floyd A. Hummel

CONTENTS

1

INTRODUCTION AND DEFINITIONS

I. HISTORICAL

A. General

The entire field of chemical equilibria involving the application of the simple equation $F = C - P + 2$ stems from the original thermodynamics of J. Willard Gibbs (1), most particularly his paper on the "Equilibrium of Heterogeneous Substances" (2). The student is urged to read the account of the life and work of this outstanding genius as recorded by one of his students (3).

The basic mathematical and thermochemical treatment of Gibbs remained obscure and little used until the interpretations of Roozeboom (4), Shreinemakers (5), and Bancroft (6) appeared. Through the efforts of these writers, the Phase Rule was placed in position for the widespread application to all types of chemical systems. The later texts of Tamman (7) and Vogel (8) deserve special mention. Since the early work, many excellent textbooks have been written, most of them in the field of metallurgy or aqueous solution chemistry.

In more recent times, the work of the Geophysical Laboratory of the Carnegie Institution of Washington, D.C. has been of special significance and considerable use to ceramists and others working in the field of high temperature silicate and oxide chemistry. An account of the history, methods, and achievements of this Institution is available (9), as well as a list of specific systems which have been studied (10). The names of Adams, Day, Bowen, Morey, Shepherd, Rankin, Wright, Merwin, Greig, and Kracek are associated with most of the early work of this laboratory, while those of Schairer, Osborn, Tuttle, Keith, Yoder, Roedder, and Schreyer are prominent in later publications. Morey

(11,12) wrote two papers on the analysis and interpretation of phase diagrams, especially for glass technologists, high temperature oxide chemists and ceramists.

The equilibrium work of Insley, McMurdie, Geller, Bunting, Lang, Levin, Roth, Waring, Robbins, and Schneider at the National Bureau of Standards deserves special mention, since the systems studied have direct application to optical glass, glazes, special refractory bodies and ceramic dielectric wares. Apparently no tabulation of the systems studied has been published, but most of them have been incorporated in the Levin, Robbins, and McMurdie compilations (13a,b).

B. Significance to Ceramic Science and Engineering (The Orton Lectures)

It is noteworthy that the first Orton Memorial Lecture was delivered to the American Ceramic Society in 1933 at Pittsburgh, Pennsylvania, by Edward W. Washburn on "The Phase Rule in Ceramics". Several subsequent Orton Lectures on this topic or on closely related topics such as "Natural and Artificial Ceramic Products" (Arthur L. Day, 1934), "Petrology and Silicate Technology" (Norman L. Bowen, 1943), and "Spark Plug Insulation" (Frank H. Riddle, 1959) attest to the general significance of the Phase Rule in ceramic science and engineering.

The first collection of phase diagrams, consisting of 178 diagrams, was published by F. P. Hall and Herbert Insley as the October, 1933 issue of the Journal of the American Ceramic Society. They received the first Ross Coffin Purdy Award in 1949 for this outstanding contribution to ceramic literature and for laying the cornerstone for all subsequent collections.

Subsequent compilations include the following:

Year	No. of Diagrams	Collectors
April, 1938*	154 (Supplement to 1933)	Hall and Insley
Nov., 1947*	507 (Part II of Nov. Jour)	Hall and Insley
Dec., 1949*	25 (Supplement to 1947)	McMurdie and Hall
1956	811	Levin, McMurdie and Hall
1959	462 (Supplement to 1956)	Levin and McMurdie
1964	2066	Levin, Robbins and McMurdie
1969	2082	Levin, Robbins and McMurdie
1975	850	Levin and McMurdie
1981	590	Roth, Cook and Cleek

*Part of regular issues of the Journal of the American Ceramic Society. Since 1956 the collections of diagrams have appeared as special bound publications.

Originally, the collection was confined to oxide and silicate systems, but starting in 1947, sulfide, fluoride and many other types of systems important to ceramic technology have been included.

At the present time, the significance of phase studies in ceramic systems no longer has to be justified. Every year, a certain amount of equilibrium studies and new phase diagrams appear in a great variety of journals throughout the world, and there are long-range plans for keeping the collection of "Phase Diagrams For Ceramists" up to date.

II. LITERATURE SOURCES

A. Textbooks in Ceramics

Various general and specialized textbooks have recognized the contributions of the equilibrium diagram and have devoted one or more chapters to a discussion of the application to a particular subject. Examples are Chapter 4, "Modern Glass Practice" by S. R. Scholes; Chapter 13, "Elements of Ceramics" by F. H. Norton; Chapter 13, "Refractories" by F. H. Norton; Chapter 3, "Enamels" by A. I. Andrews; Chapter 2, "Ceramic Whitewares" by R. Newcomb, Jr., several chapters in "Steelplant Refractories" by J. H. Chesters; Chapters 9 and 11, "Introduction to Ceramics" by W. D. Kingery; and Chapter 10, "Chemistry and Physics of Clays and Other Ceramic Materials" by R. W. Grimshaw.

It is interesting to quote from the extended Chapter 7 on Melting, Fusion, and Crystallization in "Ceramics: Clay Technology" written by Hewitt Wilson in 1927:

> Raw and dried clay is largely composed of colloidal and amorphous material. High temperatures tend to eliminate the colloidal phases and to replace them with the corresponding silicate glasses, together with small amounts of the more easily and quickly formed crystalline mullite which has been found in low-temperature paving brick and high-temperature porcelain bodies. Although as in chinaware bodies, the crystalline material such as feldspar and quartz may amount to 50 percent of the composition, yet but little of this is obtained as crystalline matter after heating, because the rate of cooling in ceramic furnaces and kilns is too fast for the crystalline growth of most silicates. THE CERAMIC INDUSTRY AS A WHOLE IS PRIMARILY INTERESTED IN THE PRODUCTION OF AMORPHOUS MATERIALS. The structural pottery industries wish to produce ceramic bodies which are hard and strong, that is, with a large content of glass bond, but which have sufficient rigidity at the maximum temperature of the kiln to resist deformation or loss of shape. On the other hand, these viscous bodies afford very little opportunity for the development of crystals, and mullite appears to be the only one active enough to develop in the ordinary clay mixture.

At present, however, interest is being displayed in those crystalline forms which are useful in refractory work because of their rigidity under pressure at temperatures close to their melting points and because of the greater inertness of crystalline materials to the action of active agents such as slags and glasses. The electrical-insulator industry has recently introduced a high content of mullite into spark-plug bodies because of the better resistance to a combination of electrical and thermal stresses. The Corning Glass Company* and others are engaged in commercial experiments of a revolutionary character. Mullite is melted in an electric furnace and cast directly in sand molds to produce glass tank shapes. These are carefully annealed to prevent strain and thus produce a solid mass of interlocking crystals, with the maximum resistance to load and slag or glass action at high temperatures. Heretofore, mullite, corundum, and other crystalline masses have been crushed, bonded with plastic fire clays, milk of lime, etc., and refired in ordinary combustion kilns. In some cases, as in silica and special mullite brick, a second interlocking crystallization could be produced. In most cases, however, the bond was amorphous and the character of the product very largely depended on this bond. Recrystallized silicon carbide refractory products have been on the market for many years. The second growth of crystals is produced in electric resistance furnaces without a silicate bond. Silicon carbide has no melting point but rather a dissociation temperature at 2240°C.

Other crystalline ceramic forms include corundum, Al_2O_3, in aluminous bauxite and diaspor, calcium silicates and aluminates in different cements, periclase in magnesia brick, and spinel, $MgO \cdot Al_2O_3$, in the "Diamel" brick of the Vitrefrax Company. In selecting such high-temperature refractories for casting wares, it is necessary to obtain fluidity with comparatively low vapor pressure over a workable temperature range, together with rapid crystallization. Magnesia and lime vaporize very fast, and, while quartz melts at comparatively low temperatures, it crystallizes too slowly from its melt. Fused silica finds its best service in the production of amorphous silica glass wares. With the recent rapid commercial development of the electric furnace, we will soon have more data on the behavior of these high-temperature compounds. The delayed solution of the lime problem may be found in the production of some high-temperature compound of this cheap, abundant, very refractory, but as yet, far too unstable and active material for high-temperature service.

The influence of clay and the technology of its heat treatment is already evident in the title of the book, but the three paragraphs quoted above emphasize how few materials besides clay (and mullite) were thought to be of interest at that time. After mullite, only a small group of materials are mentioned, name-

---

*Anonymous: "Tank Block Problem Solved at Corning", Glass Ind., 7, 257 (1926).

ly silica, silicon carbide, corundum, spinel, MgO, calcium aluminates, and calcium silicates. Moreover, it was said that "THE CERAMIC INDUSTRY AS A WHOLE IS PRIMARILY INTERESTED IN THE PRODUCTION OF AMOR-PHOUS GLASSY MATERIALS," which even at that time was not true.

These statements form a striking contrast with our present wide interests, in crystalline or glassy borates, aluminates, silicates, titanates, zirconates, stannates, vanadates, phosphates, niobates, chromates, molybdates, tungstates, fluorides, and sulfides. Regardless of the form in which these compositions are desired, such as single crystals, fibers, foams, as finely divided sintered aggregates, or as finely divided aggregates in a glassy matrix, it demonstrates the tremendous effort which is being put forth to discover new phases, determine their equilibrium and non-equilibrium relationships in various systems, and to measure a wide variety of their chemical and physical properties.

The whole movement represents a transition period (started around 1943 during WWII) in Ceramics with less and less emphasis on easily formed clay products which can meet only limited general technological requirements and more and more emphasis on a great variety of difficult-to-form (in general) special compositions which meet severe and highly specific scientific and technological requirements.

B. Textbooks on Phase Equilibria

The general textbooks in Ceramics mentioned in Section B1 above cannot adequately cover the subject of phase equilibria in oxide systems in one or two chapters. The reading list on phase equilibria given at the end of this chapter includes most of the easily accessible books written in English. The instructor may wish to supplement the reading in this book with some specialized material in one of the fourteen texts listed. The most recent book of Ehlers contains some very useful material for ceramists, although it is directed toward geologists, as the title implies.

C. General Articles

A number of articles have appeared which suggest the increased use of the quantitative calculations which can be made from the equilibrium diagram. Of general interest in this connection are the papers of Dahl (14, 15, 16, 17) and Foster (18, 19). Dahl illustrated his calculations with applications in the cement industry while Foster stressed the proper use of the diagram in the heat treatment of complex mineral assemblages, and non-equilibrium situations. He also stressed the importance of the solid state reaction in ceramics. Apler (20, 21) has edited two series of books dealing with the application of phase diagrams to various fields of ceramic science and engineering.

D. Books on Techniques of Construction and Application
of Phase Diagrams

The compilation of "Phase Diagrams for Ceramists" by Levin, Robbins, and McMurdie (13a) contains a very good list of the books and papers which are available on techniques of construction of phase diagrams (pages 32-36).

III. STATEMENT OF THE PHASE RULE AND DEFINITIONS

In the processing and manufacture of ceramic products, the reactions which occur are understood more clearly if the phase relations under equilibrium conditions are known. The chemical and physical properties of ceramic products are related to the number, composition, and distribution (microstructure) of the phases present. Temperature, pressure, and concentration are the principal variables which determine the kinds and the amounts of the phases present.

Construction of phase diagrams is based on the assumption that the system under consideration is at equilibrium. In the development of reliable information on phase relations, this condition must be satisfied. In a practical sense, however, as in the manufacture or use of a ceramic product, circumstances may not permit a condition of equilibrium to be established. In many cases, it is known that the system is driving toward or approaching equilibrium, and the knowledge of the direction in which the reaction is progressing or the direction and amount by which it deviates from equilibrium can be of great value. This gives rise to the large areas of research involving the mechanism and kinetics of reactions which are so closely related to phase equilibrium.

The phase rule developed by J. Willard Gibbs (1, 2) was derived from the first and second laws of thermodynamics. If sufficient thermodynamic data were available, equilibrium relations of the phases could be calculated. Usually, such data have not been generated. However, the understanding of the thermodynamic basis for the phase rule and the manner by which phase relations can be represented in temperature-pressure-composition diagrams is extremely helpful to ceramists, even though most phase diagrams are determined by experimental laboratory techniques.

There are numerous definitions whose meaning one must understand and which must be used repeatedly in discussing one, two, three, and four component systems. In this chapter it is the intention to introduce those which are involved in the Phase Rule itself and a few others which are needed to describe some common events which occur in simple one-component systems.

A. Equilibrium

The fundamental concept upon which the phase rule, and the phase diagram is based, is the state of thermodynamic equilibrium. Generally equilibrium in a system can be recognized by applying the following criteria: (1) the properties of the system do not change with the passage of time, (2) the state in question can be obtained by approaching the condition from more than one direction, i.e. by temperature, pressure, or, concentration: It should be noted that there is no limit to the "time" parameter in the definition of equilibrium. Cases are known when minutes, hours, years, and even thousands or millions of years are insufficient for attainment of the equilibrium state. It is this equilibrium end-point, and only this, that is represented by the phase equilibrium diagram, drawn to conform to the phase rule.

The thermodynamic definition of equilibrium can be obtained from the Gibbs free energy equation which can be stated as

$$G = PV - TS + \mu_1 X_1 + \ldots \mu_i X_i \qquad (1.1)$$

where

G = Gibbs free energy
P = pressure of the system
V = volume of the system
T = absolute temperature
S = entropy
μ_i = chemical potential of ith component
X_i = mole fraction of ith component

The equilibrium state is defined as that condition when G is a minimum for a particular set of external conditions. Of course, if the external conditions are changed (i.e. temperature, pressure, etc.) the equilibrium state will change and G will strive to assume a new minimum value.

The Gibbs free energy is a function of the independent variables T, P, and X_i. The equation can be rewritten in the differential form as

$$dG = VdP - SdT + \sum_i \mu_i dX_i \qquad (1.2)$$

wherein i has a maximum value of C-1 if C refers to the minimum number of components necessary to define the system. In the special case when C = 1, the third term in equation (1.2) vanishes and the equation, as it is most commonly used, becomes

$$dG = VdP - SdT \qquad (1.3)$$

Because the Gibbs free energy is a measure of the available energy, it there-fore represents the driving force for a reaction. For a reaction to occur spontane-

ously, the free energy change must be negative (dG < 0). When the free energy
change is zero (dG = 0), the system is in a state of equilibrium and no further
changes will occur.

As will be shown in Chapter 2, the phase diagrams can be calculated from
the Gibbs free energy equation. This, however, can be done only with great difficulty and consequently most phase diagrams are a result of experimental
determinations.

It should be noted that there are other thermodynamic potentials besides
G such as the enthalpy, H, the Helmholtz free energy, F, the internal energy, E,
and others but these are functions of other combinations of dependent variables.
Phase diagrams can be obtained from each of these potentials but it is most convenient and reasonable from an experimental standpoint to construct a phase
diagram using the variables P, T, and X. It is for this reason that when we refer
to the phase rule or free energy while discussing phase equilibria it is universally
understood that we are referring to the Gibbs version.

B. System

A system is any portion of the material universe which can be isolated completely and arbitrarily for consideration of the changes which may occur within it
under varying conditions. In ceramic and metal systems it is often difficult to experimentally isolate them due to reaction with the containing vessel at the high
or extra-high temperatures involved. Liquids become extremely reactive at
$1700°C$ - $2000°C$ and even solid-solid reactions become vigorous and rapid.
Often a system may be considered as composed of smaller systems, which together make up the larger system. For example, consider the reactions between
the two common ceramic oxides SiO_2 and Al_2O_3. These two materials constitute a system which is called the system Al_2O_3-SiO_2. We could deal with
smaller systems within this system, such as the system SiO_2 or the system Al_2O_3,
or even a small compositional range, and analyze its behavior with respect to
variations in temperature, pressure, and composition. To carry this notion still
further, the system SiO_2 or Al_2O_3 could be regarded, under certain conditions,
as only part of the system Si - O_2 or Al - O_2. In fact, it is appropriate at this
time to point out that the collection of "Phase Diagrams for Ceramists" begins
with systems involving one metal and oxygen, two metals and oxygen, three
metals and oxygen, etc., a very logical classification based on the fundamental
notion of a system. Metallic systems when combined with oxygen lead to the
ceramic systems.

C. Phase

A phase is any portion of a system which is physically homogeneous within itself and bounded by a surface so that it is mechanically separable from any other

portions. A separable portion need not form a continuous body, as for example, one liquid dispersed in another. A system may contain one phase or many phases. If it is a one phase system, homogeneous equilibria is involved; if the system is composed of two or more phases, heterogeneous equilibria is involved as in the case for many metal and ceramic systems. Phases are distinguished by their different physical character. The physical character of the physical states of matter (i.e. gases, liquids, and crystals) is different. These states of matter, then, are physically distinct and represent different phases. Although gases at ordinary pressure are completely soluble (miscible) in one another and thus represent only one phase, liquids or crystals may be found to exist in several different phases, each phase physically distinct from the others. Water and mercury, for example, are both liquids each representing a different phase. Crystalline silica, SiO_2, may exist in several different crystalline structures, each consisting of a different phase. As exemplified by solutions, either liquid or crystalline, the homogeneous character of the phase is not confined to a rigid chemical composition, since the existence of a variation in chemical composition of liquid or crystalline phases often does not alter the homogeneous character nor cause any abrupt or distinctive change in the physical structure of the phase. As solid phases are considered to be distinguished mainly by their physical character, such physical characteristics as density, crystal structure, or optical properties are used to aid in the identification and distinction of phases in a system.

Normally only one gas phase is considered to exist in a system, since all gases under ordinary conditions are found to be completely miscible. At very high pressures there may exist limited mutual solubility of two gases, but this is a very special case since the densities have then been increased to the point where their behavior approaches that of liquids.

There are certain fine details concerning the definition of a phase which need some elaboration. Quoting from Ricci, pp 3-4.

> Variations on a very small, or micro, scale and distributed according to statistical laws of probability are possible in every homogeneous phase: variations caused by the isotopy of the elements present, random defects and random arrangements in certain crystalline phases, and the statistical variation of density, and hence of all properties, in liquid and gas phases which become so pronounced as to cause visible effects near critical points (critical opalescence). These statistical variations are reproducible functions of the variables considered in the Phase Rule and persist when the system is in final equilibrium. Phases involving only such variations may be called not only homogeneous but uniform. Another non-uniformity possible in a homogeneous phase of an isolated equilibrium system free of the forces of gravitational and other such fields seem to be that of surface energy, if the phase is a subdivided one. The subdivided phase in a 2-phase colloidal system, for example, may not have the same surface development

in all its pieces. But if there is such a thing as a reproducibly stable colloidal system, with an equilibrium state which is a function of T, P, and composition alone, independent of time and of the relative amounts of the phases, then this non-uniformity must be a regular one, following some statistical distribution fixed solely by these variables. If the colloidal system, then, is stable and in reversible equilibrium, the distribution of its surface energy must be assumed to be either uniform or a reproducible function of the stated variables. If it is not stable and not at equilibrium, it cannot be discussed in terms of the Phase Rule or represented in the terminology and by means of the graphical methods used in the Phase Rule.

Finally, it must be pointed out that, from the point of view of the Phase Rule, even homogeneity is secondary in the definition of a phase, which may also be heterogeneous. If two or more homogeneous parts of a heterogeneous system always occur in the same proportions, giving a total mass of constant composition, they constitute together a single phase, even though obviously and macroscopically heterogeneous and mechanically separable into different parts. This is the case, for example, when a solid exists in two optically enantiomorphic forms but is instantaneously and completely racemized in passing through the liquid or dissolved state. Thus solid sodium chlorate or solid quartz, in equilibrium with a melt or a solution, will consist of a 1:1 mixture of d- and l- crystals. Such mixtures may be mechanically separable into two different substances, two species with different properties, but the behavior in the heterogeneous system is that of a single phase. This is so not simply because the thermodynamic properties of the two forms are the same or even because the forms are equally probable upon crystallization, but because their ratio when in dynamic equilibrium with a melt or a solution is constant. They would still constitute a single phase if they occurred in any fixed proportion other than 1:1.

In addition to the type of phases discussed above, there is the interesting case of silicate glasses. Such compositions have always been considered single phases (excepting opal or other opacified glasses), but recent experimentation with the electron microscope has proven that the structure of these materials may not be the purely random distribution of atoms or ions as first supposed but may rather be composed of "regions" of different degree of order.

To summarize regarding phases, from the standpoint of the Phase Rule, colloidal phases, optical isomers, and glasses are at present to be considered as consisting of single homogeneous phases which are mechanically separable, even though an actual experimental mechanical separation of the phases may be difficult or impossible, as in the case of separating NaF or CaF_2 crystals from an opal glass, or when a glass in glass separation is involved which requires a magnification of 100,000 in the electron microscope to detect it. However, this fact does not invalidate the definition since the limits of practicability are understood.

D. Components

The *components* of a system are the smallest number of independently variable chemical constituents necessary and sufficient to express the composition of each phase present in any state of equilibrium. In the alumina-silica system, Al_2O_3 and SiO_2 are the components of the system since all phases and reactions can be described using only these two materials. Al, Si, and O would not be the components for they are not the *least* number of chemical substances by which the system can be quantitatively expressed. Consider the reaction:

$$CaCO_3 \rightleftarrows CaO + CO_2$$

At equilibrium, all three chemical constituents are present but they are not all components because they are not all independently variable. Choosing any two of the three phases fixes the composition of the remaining phase.

In selecting the components of this system, it is helpful to arrange the information in tabular form (as shown below) where the composition of each phase is expressed in terms of positive, negative, or zero quantities of the chemical constituents chosen as components.

Composition of Phase in Terms of Chemical Constituents

Phase	$CaCO_3$ and CaO		CaO and CO_2		$CaCO_3$ and CO_2	
$CaCO_3$	+	0	+	+	+	0
CaO	0	+	+	0	+	−
CO_2	+	−	0	+	0	+

From an examination of the above table, it is apparent that any two of the chemical constituents could be selected as the components of the system because each selection resulted in a suitable expression for the composition of each phase present at equilibrium. Note, however, that only one of the selections, CaO and CO_2, resulted in an expression involving positive quantities only. In selecting the components of a system it is preferable to use only positive quantities if possible.

In ceramic systems, special mention should be made of the influence of variable valence elements on the selection of the components of a system. We are especially referring to the first transition series of elements, but the notion applies to the rare earth series or lanthanides and to any element which can commonly exist in more than one valence state.

For example, the more common valence states of titanium are Ti^{2+}, Ti^{3+}, and Ti^{4+}, and the oxides TiO and Ti_2O_3 can be produced starting with the most common oxide TiO_2. Vanadium exists as V^{2+}, V^{3+}, V^{4+}, and V^{5+} and each of the corresponding oxides are available for use as components of ceramic systems, even though V_2O_5 is by far the most common. Manganese exists in Mn^{2+}, Mn^{3+}, Mn^{4+}, Mn^{5+}, and Mn^{7+} states, but only the oxides MnO and MnO_2 are readily available for use as raw materials. Finally the two common valence states of iron are Fe^{2+} and Fe^{3+} and the three oxides FeO, Fe_2O_3, and Fe_3O_4 (magnetite) are well known and readily available. If one wishes to investigate systems containing these oxides, it is absolutely necessary to control the oxygen pressure in the system if a "clean-cut" system containing only *one* oxidation state of the transition metal element is desired. The best example of such control is the work of Osborn and Muan as recorded in separate publications and in "Phase Equilibria Among Oxides in Steelmaking" where great care is taken to control the $Fe^{2+} \rightleftarrows Fe^{3+}$ equilibrium.

As a final note, it can be emphasized that the binary systems $TiO-SiO_2$, $Ti_2O_3-SiO_2$, and TiO_2-SiO_2 will have vastly different configurations as far as the phase diagram is concerned. Only the latter system has been experimentally determined with any degree of completeness. Work in the other two systems would require control of the oxygen pressure to assure the existence of only TiO or Ti_2O_3.

E. Variance (or Degrees of Freedom)

Variance is the number of intensive variables such as temperature, pressure, and/or concentration of the components in a phase which must be arbitrarily fixed in order that the condition of the system may be perfectly defined. A system is described as invariant, monovariant, bivariant, etc., according to whether it possesses respectively, zero, one, two, etc., degrees of freedom. Ordinarily, it is assumed that temperature, pressure and concentration of the components are the only externally controllable variables which will affect the phase relationships in chemical or ceramic systems. If electrostatic, magnetic or gravitational fields, or surface forces become of such magnitude as to influence the system, these external physical variables would have to be considered and the constant (2) in the Phase Rule would have to be increased.

F. The Phase Rule

Originally deduced by J. Willard Gibbs (1, 2) and brought to general use by Roozeboom (4), the phase rule serves to define the conditions of equilibrium in terms of a relationship between the number of phases and the components of a system:

$$F = C - P + 2 \text{ (temperature and pressure)} \tag{1.3}$$

where

> F = the variance or degrees of freedom
> C = th number of components
> P = the number of phases.

This rule serves to limit the possible complexity of phase diagrams by limiting the maximum number of phases that can coexist at equilibrium. For example, no more than 3 phases can coexist in a 1 component system, 4 in a 2 component system, 5 in a 3 component system, etc.

The phase rule is a direct consequence of the Gibbs free energy relationship. Consider a one component system that contains two phases, α and β, which are in equilibrium. The total change in free energy is zero; therefore:

$$
\begin{aligned}
dG &= 0 \\
&= dG^\alpha - dG^\beta \\
&= (-SdT + VdP + \mu_1^\alpha dX_1) - (-SdT + VdP + \mu_1^\beta dX_1) \\
&= \mu_1^\alpha dX_1 - \mu_1^\beta dX_1
\end{aligned} \tag{1.4}
$$

This shows that at equilibrium any energy gain by phase α must equal the energy loss by phase β. The chemical potentials of the various phases then are *equal* in the equilibrium state.

In a two component system, coexistance of two phases would give rise to two equations:

$$
\mu_1^\alpha(T,P,X_1^\alpha) = \mu_1^\beta(T,P,X_1^\beta) \tag{1.5}
$$

$$
\mu_2^\alpha(T,P,X_2^\alpha) = \mu_2^\beta(T,P,X_2^\beta) \tag{1.6}
$$

It can be recalled that the independent variables involved in these equations are temperature (T), pressure (P), mole fraction of component one in each phase (X_1^α and X_1^β), and mole fraction of component two in each phase (X_2^α and X_2^β). However, the mole fraction of component 1 in a phase is

$$
X_1^\alpha = \frac{n_1}{n_1 + n_2} \tag{1.7}
$$

where

> n_1 = moles of component 1 in phase α
> n_2 = moles of component 2 in phase α

Table 1.1   Chemical Potential Equations for a System Containing C
Components and P Phases

$$\mu_1^\alpha(T,P,X_1^\alpha) \quad = \quad \mu_1^\beta(T,P,X_1^\beta) \quad = \quad \ldots \mu_1^P(T,P,N_1^P)$$

$$\mu_2^\alpha(T,P,X_2^\alpha) \quad = \quad \mu_2^\beta(T,P,X_2^\beta) \quad = \quad \ldots \mu_2^P(T,P,N_2^P)$$

$$\mu_{C-1}^\alpha(T,P,X_{C-1}^\alpha) = \mu_{C-1}^\beta(T,P,X_{C-1}^\beta) = \ldots \mu_{C-1}(T,P,N_{C-1}^P)$$

and, for each phase, the sum of the mole fractions equals unity:

$$X_1^\alpha + X_2^\alpha = 1 \tag{1.8}$$

If X_1^α is known, then X_2^α must equal $1 - X_1^\alpha$. In the general case, if all but one
fraction is known, the remaining one can be calculated. In the above example,
then, there are a total of only four independent variables, T, P, X_1^α and X_1^β. The
variance of the system is determined by comparing the number of independent
variables with the number of equations available. In this case there are two equations (1.5) and (1.6) and four independent variables. Two equations allow us to
solve for a maximum of two unknowns, therefore two of our variables are not
fixed by the system and can be changed without changing the number of phases
in equilibrium. The system is said to have a variance of two, or, in other words,
two degrees of freedom. The exact variables free to change are not specified,
but, of course, must be pressure, temperature, or the composition of the phases.

   In the completely general case as shown in Table 1.1 where there are C
components and P phases, there will be (1) P-1 equations for each component
or a total of C(P-1) equations, and (2) C-1 variables for each phase plus 2 for
temperature and pressure or a total of P(C-1) + 2 independent variables. The
variance (F) must equal the difference between the number of variables and
number of available equations or:

$$\begin{aligned}
F &= \text{variables} - \text{equations} \\
&= [P(C-1) + 2] - [C(P-1)] \\
&= PC - P + 2 - PC + C \\
&= C - P + 2
\end{aligned} \tag{1.9}$$

   It is now easy to see that the constant two in the phase rule comes from
the variables temperature and pressure. If either of these is fixed, then this constant must be reduced accordingly.

It is interesting to note that Bancroft (6) wrote about the Phase Rule in 1897 and again in 1948 (22). One must infer that his enthusiasm and desire to explain the usefulness of the simple equation had increased over a fifty-one year interval.

PROBLEMS

1. Carefully consider the conditions under which the following common materials are either one or two component systems:

1. NH_4Cl	6. H_2O, D_2O
2. $CaCO_3$	7. PbO
3. H_2O	8. V_2O_5
4. SiO_2	9. NH_4NO_3, H_2O
5. Oxygen, ozone	10. Cr_2O_3

2. Read some of the classical textbooks such as Findlay, Bowden, or Ricci on the definition of "phase" and "components".
3. Explain what is meant by the two parts of the definition of equilibrium; a) that there is no change in the system with the passage of time and b) that the same state can be arrived at from different directions (the reversibility criterion).
4. Consider the Solvay Process reaction,

$$NaCl + NH_4HCO_3 \rightleftarrows NaHCO_3 + NH_4Cl$$

as carried out in water solution. (Reciprocal salt pairs)
In the presence of H_2O, how many components would be needed to represent this system? Is it likely that you could step into the laboratory and saturate H_2O at $25°C$ with all four salts?? $F = C - P + 2$.
5. Write a short paragraph on the meaning and definition of F (degrees of freedom).

REFERENCES

1. J. W. Gibbs, The Collected Works of J. Willard Gibbs, 2 Volumes, Longmans. Green and Company, New York, (1928); from Trans. Connecticut Acad. of Arts and Sciences, 1874-1878.
2. J. W. Gibbs, Equilibrium of Heterogeneous Substances, Trans. Conn. Acad. Sci. 3, 108-248, 343-524, 1874-1878.
3. Lynde Phelps Wheeler, Josiah Willard Gibbs, Yale University Press, 1951.

4. H. W. Bakhuis–Roozeboom, Die Heterogenen Gleichgewichte vom Stand-
   punkte der Phasenlehre (The Heterogeneous Equilibria from the Standpoint of the Phase Theory) 6 parts, F. Vieweg, Braunschweig, 1901; con-
   tinued after 1904 by A. H. W. Aten, E. H. Büchner, F. A. H. Schreine-
   makers. Vol. I: H. W. B. Roozeboom, Die Phasenlehre–Systeme aus einer
   Komponente, 1901. Vol. II: Systeme aus zwei Komponente. Part 1:
   H. W. B. Roozeboom, Systeme aus zwei Komponente, 1904. Part 2:
   E. H. Büchner, Systeme mit zwei flüssigen Phasen, 1918. Part 3: A. H. W.
   Aten, Pseudobinäre Systeme, 1918. Vol. III: Die Ternäre Gleichgewichte.
   Part 1: F. A. H. Schreinemakers, Systeme mit nur einer Flüssigkeit, ohne
   Mischkristalle and ohne Dampf, 1911. Part 2: F. A. H. Schreinemakers,
   Systeme mit zwei and mehr Flüssigkeiten, ohne Mischkristalle, and ohne
   Dampf, 1913.
5. F. A. H. Schreinemakers, Mischkristalle in Systemen drierer Stoffe, Z.
   physik. Chem. 50(2) 169-99; 51(5) 547-76; 52(5) 513-50 (1905).
6. W. D. Bancroft, The Phase Rule, J. Phys. Chem. 1, 255 (1897).
7. G. Tamman, Lehrbuch der Heterogenen Gleichgewichte (Textbook of
   Heterogeneous Equilibria), Vieweg, Braunschweig, 1924.
8. R. Vogel, Die Heterogenen Gleichgewichte, Vol. II in Handbuch der Metall
   physik, published by G. Masing, Leipzig, 1937.
9. G. W. Morey, The Studies in Silicate Chemistry of the Geophysical Laboratory of the Carnegie Institution of Washington, Journal Soc. Glass. Tech.
   Trans., 20, 245-256 (1936).
10. Leason H. Adams, List of Systems Investigated at Geophysical Labora-
    tory, Bowen Volume–American Journal of Science, pp. 1-26 (1952).
11. George W. Morey, Analytical Methods in Phase Rule Problems, J. Phys.
    Chem. 34(8), 1745-1750, 1930.
12. George W. Morey, The Interpretation of Phase Equilibrium Diagrams,
    The Glass Industry 12(4), 69-80, 1931.
13a. Ernest M. Levin, Carl R. Robbins, and Howard F. McMurdie, Phase Dia-
    grams for Ceramists, edited and published by the American Ceramic
    Society, 1964.
13b. Ernest M. Levin, Carl R. Robbins, and Howard F. McMurdie, Phase Dia-
    grams for Ceramists, 1969 Supplement.
14. L. A. Dahl, Interpretation of Phase Diagrams of Ternary Systems, Journal
    Phys. Chem., 50, 96-119 (1946).
15. L. A. Dahl, Analytical Treatment of Multicomponent Systems, J. Phys.
    and Colloid Chem., 52, 698-729 (1948).
16. L. A. Dahl, Equilibrium in Heterogeneous Systems of Two or More
    Components, J. Chem. Education, 26, 411-21 (1949).
17. L. A. Dahl, Parametric Equations in the Treatment of Multicomponent
    Systems, J. Phys. and Colloid Chem., 54, 547-64 (1950).

18. Wilfrid R. Foster, Contribution to the Interpretation of Phase Diagrams for Ceramists, Jour. Amer. Cer. Soc., 5, 151–160, 1951.
19a. Wilfrid R. Foster, Solid State Reactions in Phase Equilibrium Research, I, Bull. Amer. Cer. Soc., 8, 267–270, 1951.
19b. Wilfrid R. Foster, Solid State Reactions in Phase Equilibrium Research, II, Bull. Amer. Cer. Soc., 9, 291–296, 1951.
20. Allen M. Alper, Volume 5, High Temperature Oxides, Parts I-IV, Refractory Materials, A Series of Monographs, Academic Press, New York, 1971.
21. Allen M. Alper, Volume 6, Phase Diagrams: Materials Science and Technology, Volumes, I-V, Refractory Materials, A Series of Monographs, Academic Press, New York, 1970.
22. W. D. Bancroft, Components and Phases, Chemistry 21(6), 28–40, 1948.

READING LIST ON PHASE EQUILIBRIA

1. "Phase Diagrams for Ceramists", by Ernest M. Levin, Carl R. Robbins and Howard F. McMurdie, The American Ceramic Society, Inc., 1964. Compiled at the National Bureau of Standards.
2. "The Phase Rule and Phase Reactions", by S. T. Bowden, Macmillan Co., London, 1938.
3. "The Phase Rule", by Alexander Findley, A. N. Campbell, and N. O. Smith Longmans, Green and Co. Ninth Edition, Dover Publications, Inc., New York, 1951.
4. "Silicate Melt Equilibria", by Wilhelm Eitel, Rutgers University Press, New Brunswick, New Jersey, 1951.
5. "The Phase Rule and Heterogeneous Equilibrium", by John E. Ricci, D. Van Nostrand Company, Inc., New York, 1951.
6. "The Defect Solid State", Chapter IX on Phase Equilibria, by D. E. Rase, p. 321–373, Interscience Publishers, 1957.
7. "Phase Diagrams in Metallurgy", Fredrick N. Rhines, McGraw Hill Book Company, 1956.
8. "Ternary Systems", by G. Masing and B. A. Rogers, Reinhold Publishing Corp., New York, 1944. Paperback, Dover Publications, Inc., 1960.
9. "Ternary Equilibrium Diagrams", by D. R. F. West, MacMillan and Co., London, England, 1965.
10. "Phase Equilibria Among Oxides in Steelmaking", Arnulf Muan and E. F. Osborn, Addison-Wesley, 1965.
11. "Alloy Phase Equilibria" by A. Prince, American Elsevier Publishing Co., Inc., 52 Vanderbilt Avenue, New York, N.Y., 1966.
12. "Principles of Phase Diagrams in Materials Systems", by Paul Gordon, McGraw-Hill Book Company, Inc., 330 West 42nd St., New York, N.Y. 10036, 1968.

13. "The Interpretation of Geological Phase Diagrams", by Ernest G. Ehlers, W. H. Freeman and Co., San Francisco, 1972.
14. "Basalts and Phase Diagrams", by Stearns A. Morse, Springer-Verlag, 175 Fifth Avenue, New York, N.Y. 10010, 1980.
15. "Phase Equilibria" by Arnold Reisman, Academic Press, 111 Fifth Avenue, New York 10003, 1970.

2

THE ONE COMPONENT SYSTEM

I. THERMODYNAMIC BASIS OF THE PHASE DIAGRAM

The phase equilibrium diagram is a direct consequence of the Gibbs free energy equation. For a one component system the differential form of this equation is

$$dG = VdP - SdT \qquad (2.1)$$

and upon integration becomes

$$G(P_1,T_1) = (G(P_0,T_0) + \int_{P_0}^{P_1} V(P,T_0)dP - \int_{T_0}^{T_1} S(P_0,T)dT \qquad (2.2)$$

where the point $(G(P_0 T_0)$ is some arbitrary reference point on the free energy surface. Providing $V(P,T_0)$ and $S(P_0,T)$ are known, the equation can be solved and a free energy surface constructed. In actual practice, the variation of V as P changes can be experimentally determined for most phases quite easily and the integral $\int_{P_0}^{P_1} V(P,T_0)dP$ evaluated graphically. The second integral can be evaluated at constant pressure by noting that $dS = \frac{C_p}{T}dT$ where C_p is the heat capacity* of the material at constant pressure. Heat capacity increases with temperature and for most metal and ceramic materials can be represented over a limited temperature range by an equation of the type

$$C_p = a + bT \tag{2.3}$$

and over wider temperature ranges by

$$C_p = a + bT + cT^2 \tag{2.4}$$

or

$$C_p = a + bT + cT^{-1/2} \tag{2.5}$$

or

$$C_p = a + bT + cT^{-2} \tag{2.6}$$

where a, b, and c are experimentally determined constants. With the proper selection of equation, the second integral can be evaluated and the free energy surface constructed.

Figure 2.1 illustrates the free energy surfaces for these phases in a one component system. The three surfaces enclosed by the curves hgik, lief, and acki are drawn with curvatures and slopes to approximate the real free energy surfaces for solid, liquid, and vapor phases respectively. The intersection of these surfaces gives rise to the curves ij, jk, and jl. Recalling that the stable or equilibrium phase is always the one which possesses the lowest free energy, it can be seen that the intersections of these surfaces represent the transition from one phase to another. By convention, the free energy at equilibrium can be taken to be zero. When the intersections of the surfaces are projected onto this G=0 plane, a two dimensional diagram that shows the equilibrium phase relationships as a function of T and P results. The phase diagram for the system H_2O shown in Figure 2.2 serves as a real example.

_____

*To raise the temperature of a material, a certain amount of heat must be added. By definition, the heat required to raise the temperature of a given amount of material one degree centigrade is the heat capacity. The units chosen are usually calories per mole per degree. If the heat capacity has been determined at constant volume, the term C_V is used. These terms are not equal because C_p includes the work of expansion at constant pressure.

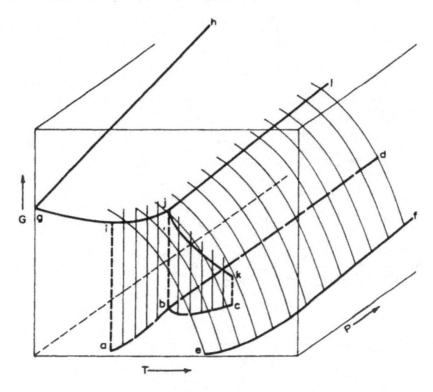

Figure 2.1. Schematic G–P–T Diagram Showing Origin of P–T Diagram

II. APPLICATION OF THE PHASE RULE TO THE SYSTEM
H₂O AT ORDINARY TEMPERATURE AND PRESSURE

The system H_2O illustrated in Figure 2.2 is very useful as an example of a simple one component phase diagram and of the application of the phase rule previously derived (1.9). In the system H_2O chemical composition is not a variable because only one component is being considered. In a one phase region, there are two degrees of freedom (obtained by substituting C = 1 and P = 1 in the equation F = C - P + 2). The two variables which must be specified to define the system are T and P.

Where two phases (P = 2) coexist in a one-component system, there is one degree of freedom, F = 3 - 2 = 1. Where three phases (P = 3) coexist in a one-component system, the degrees of freedom equal zero, F = 3 - 3 = 0. Such is the case of point E in Figure 2.2 where three phases, vapor (steam), liquid (water), and crystals (ice), coexist under equilibrium conditions. Point E is called an in-

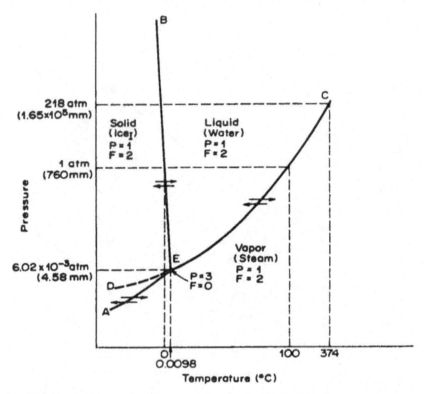

Figure 2.2. Schematic Diagram of the System H_2O at Ordinary Pressures

variant point, at which there are no degrees of freedom, or a triple point, since three phases coexist at this point.

On the boundary lines, AE, BE, and CE, there are two phases in equilibrium, and the system has only one degree of freedom. The one degree of freedom on the boundary line indicates that pressure and temperature are dependent on each other and the specification of one of these variables will automatically fix the other variable. In areas where only one phase exists, called the field of stability of the phase, pressure and temperature may be varied independently of each other. Thus, in one-component systems, the degrees of freedom may be zero, one, or two. To summarize:

3 phases coexisting: $F = 3 - 3 = 0$, invariant equilibrium (triple point)
2 phases coexisting: $F = 3 - 2 = 1$, univariant equilibrium (boundary lines)
1 phase: $F = 3 - 1 = 2$, bivariant equilibrium (field of stability)

The line AE is called the sublimation curve. It represents the conditions of temperature and pressure under which solid and vapor can coexist. The upper end of the sublimation curve is at the triple point where solid, liquid, and vapor coexist. The curve EC is called the vaporization curve and represents the temperature-pressure conditions under which liquid and vapor can coexist. In the absence of supercooling, its lower terminus is the triple point. Its upper terminus is the critical point (the temperature above which the gas cannot be liquified no matter how great the pressure).

The curve EB is the fusion curve or melting point curve. It represents the conditions of temperature and pressure at which solid and liquid are in equilibrium. Line EB represents the variation in melting point of the solid phase with change in pressure. Increased pressure generally causes an increase in the melting point, but there are several cases where the reverse is true. The slope of line EB also indicates the relative density difference between the crystals and the liquid. A solid is generally more dense than its liquid, but this is not the case for the system H$_2$O (Figure 2.2) where the liquid-crystal boundary line, EB, slopes with increasing pressure toward the left. Increase in pressure effects an increase in density, and from Fig. 2.2 it may be noted that an increase in pressure alone (temperature constant) can cause the solid to change to the more dense liquid phase.

It is especially important to note that there are some outstanding exceptions to the general rule that the solid is more dense than the liquid. In simple systems, H$_2$O and metallic bismuth or metallic antimony are notable examples and in ceramic systems lithium aluminosilicate and magnesium aluminosilicate glasses (undercooled liquids) are more dense than the crystalline phases β-spodumene or cordierite.

III. LE CHATELIER'S PRINCIPLE AND THE CLAUSIUS-CLAPEYRON EQUATION

In a qualitative way, one can predict the effect of changes on the equilibrium of a system by means of the principle of Le Chatelier, which can be stated as follows: "If an attempt is made to change the pressure, temperature or concentration of a system in equilibrium, then the equilibrium will shift in such a manner as to diminish the magnitude of the alteration in the factor which is varied.". Another way to state the principal is to say that, "If an attempt is made to change the pressure, temperature, or concentration of a system in equilibrium (place a constraint on it), the system reacts in such a way as to partially or totally nullify the effect of the change."

In the example cited previously (solid and liquid in equilibrium), if the volume of the system was held constant and heat was added to the system the expected result would be an increase in the temperature of the system. According to the Le Chatelier principle, that reaction will occur which tends to diminish the magnitude of the temperature rise. In this case, more of the liquid would solidify because melting absorbs heat (endothermic). The greater specific volume of the crystal would cause an increase in the pressure of the system and the equilibrium between solid and liquid would thus shift to a position of higher pressure and lower temperature (toward B on curve EB of Fig. 2.2).

If the heat content of the solid-liquid system were held constant and the pressure was decreased, a reaction would occur which would tend to diminish the effect of the pressure decrease, that is, the volume would increase. Solidification of more of the liquid would occur because solidification is accompanied by an increase in volume. The heat released by solidification causes the temperature of the system to increase. Thus the equilibrium between solid and liquid would shift to a position of higher temperature and lower pressure. (Toward point E on curve EB of Fig. 2.2).

Similar examples can be worked out for solid-vapor and for liquid-vapor systems at equilibrium.

A quantitative expression of the principle of Le Chatelier is given by the Clausius-Clapeyron equation which can be related to the slopes of the boundary lines in the pressure-temperature diagram.

$$dP/dT = \Delta H/T\Delta V \tag{2.7}$$

where

>   P = pressure on the system
>   T = the absolute temperature
> ΔH = the enthalpy change accompanying the phase change, i.e., solid to
>   liquid, liquid to vapor, etc., usually given in calories per mole.
> ΔV = the change in volume accompanying the phase change, usually given
>   in cc per mole.

The slopes of the boundary lines are usually positive because in going from a low to a higher temperature the enthalpy change is positive and the volume change is most often positive. There are several exceptions involving solid-liquid transformation in which the change in specific volume on heating is not positive and consequently the slopes of the melting point curves are negative.

It is instructive to use the Clausius-Clapeyron equation to determine the effect of pressure on the melting temperature of Ice_1 using the following data:

$$\frac{dT}{dP} = \frac{T(V_2 - V_1)}{Q}$$

assuming,

$$T = 273°K$$
$$V_2 = 1.0 \text{ cm}^3/\text{gm (liquid)}$$
$$V_1 = 1.1 \text{ cm}^3/\text{gm (Ice}_1)$$
$$Q = 80 \text{ cal} = 80 \text{ cal} \times 42,670 \text{ gm-cm/cal}$$
$$dP = 1 \text{ atm} = 1033.3 \text{ gm/cm}^2$$
$$dT = \frac{273 \times (-0.1 \text{ cm}^3/\text{gm}) \times 1033.3 \text{ gm/cm}^2}{80 \times 42,670 \text{ gm-cm}} = -0.00826°$$

or, in other words, a change in pressure of one atmosphere produces an 0.008°C change in temperature at 273°K. This calculation not only serves to illustrate the use of the equation, but also gives one example of the fact that Fig. 2.2 is not drawn to scale.

Further discussion of the simple system H_2O shown in Fig. 2.2 is in order. Such a common material as water is well known and its different phases, water— the liquid form, ice—the crystalline form, and water vapor—the gaseous form, are very familiar to us at atmospheric pressure. In Fig. 2.2, the changes at atmospheric pressure on heating ice crystals is noted on the constant pressure (isobaric) line across the diagram. Ice changes to water at 0°C.* Water boils at 100°C.

As pressure on the system is increased, the temperature at which liquid changes to vapor is increased; thus water boils at a lower temperature at high altitudes where pressure is lower. Decreasing pressure, while it lowers the boiling point, causes the melting point to increase. Generally, however, liquids are less dense than their crystalline form, and the melting points increase with increasing pressure.

The boundary line between liquid and vapor represents the temperature and pressure of boiling points of the liquid or the condensation points of the vapor. As pressure is increased, the boiling point is increased as is indicated by the line EC in Fig. 2.2. Liquid and vapor are in equilibrium at this boundary. At point C, called the critical point, the distinction of the two phases is lost; the vapor and liquid have become homogeneous at such high temperatures and pressures. This critical temperature for water is 374.0°C at a critical pressure of 217.7 atmospheres.

Although phase equilibrium diagrams express relationships of materials subjected to the variables, temperature, composition, and pressure, it should be recognized that the environment generally experienced by materials is the exposure to the atmosphere, and thus to atmospheric pressure. A system in which

---

*In a closed system and with pure water (i.e., water containing no air in solution), the freezing point of water is actually = 0.0099°C at a 760 mm pressure.

the materials are exposed to the atmosphere is called an "open" system, whereas a system in which the materials are exposed only to their own vapor pressures is called a "closed" system.

Some crystals have very high vapor pressures, and at atmospheric pressure, the crystals may evaporate (sublime) completely.

IV. METASTABILITY AND NON-EQUILIBRIUM

The fact that water may be supercooled illustrates a metastable condition (Fig. 2.2, ED). A metastable phase always has a higher vapor pressure than the stable phase in the same temperature range. Supercooled water is a metastable phase since this phase has a higher vapor pressure in its temperature range than that of ice (a more stable phase).

A metastable phase is fundamentally in a non-equilibrium state. However, it is frequently described as a state of "metastable-equilibrium" due to the extremely long time which is required to return to the stable equilibrium condition. One requirement for the stable equilibrium state is that the conditions of equilibrium can be approached in more than one manner. "Metastable-equilibrium" does not comply completely with the definition of equilibrium, for with metastable-equilibrium, the method of approach is usually through only one definite procedure. Supercooled water, for example, can only be attained by cooling water and not by heating ice. Still, supercooled water fulfills the other equilibrium requirement in that the phase experiences no further change with the passage of time.

The addition of a stable phase tends to effect the transformation of the metastable phase to the stable phase. The addition of ice, the stable phase, will cause the spontaneous transformation of the supercooled water to ice, which is called seeding (i.e., increasing the surface area of the forming (stable phase). External influences such as a mechanical disturbance (shaking) will also cause the metastable phase to revert to a more stable phase. Supercooled water is said to be extremely sensitive to a mechanical disturbance. In other words, it is an extremely unstable metastable phase! Supercooled water as a metastable phase has a certain region of temperature in which it may remain as water; that is, the degree of supercooling is limited to a certain minimum temperature (Point D, Fig. 2.2) below which supercooled water reverts to ice crystals.

The entire discussion given above infers that a clean-cut distinction should be made between the terms stable, metastable, and unstable phases as they are applied to all kinds of systems, but especially oxide and silicate systems.

The terms stable and metastable are most fundamental and are governed by equilibrium thermodynamics. That is, for the most part, phase diagrams attempt to show the true equilibrium phases which exist in a system under certain temperature and pressure conditions. It is also possible to construct phase non-equilibrium diagrams when the persistence of metastable phases is so great that they APPEAR to be stable phases. Very often the metastable, nonequilibrium relationships are superimposed on the diagram of the equilibrium relationships for the sake of clarity. At other times, a separate diagram is constructed showing the metastable nonequilibrium conditions.

The word "unstable" generally refers to the kinetics involved in the change from the metastable to the stable condition. For example, the change from undercooled liquid water to solid Ice_I takes place almost instantly when seeded or shaken mechanically. Undercooled liquid water is then referred to as an extreme example of a very unstable metastable phase. On the other hand, tridymite, cristobalite, and silica glass can exist indefinitely at room temperature (in some cases for thousands of years) without showing any tendency to revert to α-quartz which is the thermodynamically stable form of SiO_2 at room temperature and one atmosphere pressure.

Many, many oxide and silicate and some metal systems demonstrate these very slow kinetics (reversion from a metastable condition to a stable condition may take seconds, minutes, hours, days, years, or multiple of years). The phenomena is frequently closely related to the structure, particularly the crystal structure, of the material in question. The crystal structures of the metal systems are in general simpler than those of the ceramic systems and changes from metastable to stable conditions are in general faster. Some polymer systems are so complex structurally that their inversion, conversion, or crystallization kinetics are similar to silicate glasses, the prime examples of sluggishness in ceramic systems.

V. THE BRIDGMAN ICE DIAGRAM AND THE IMPORTANCE OF PRESSURE; POLYMORPHISM OF SOLIDS; PHASE DIAGRAM-PROPERTY DIAGRAM

It has been stated that the system H_2O shown in Figure 2.2 is for P–T conditions which are "ordinary"; that is, the pressures and temperatures shown are not high or extra high.

It is now very appropriate to show the Bridgman Ice Diagram (1) Figure 2.3 (P. W. Bridgman (2) of Harvard was one of the first persons to illustrate the importance of pressure on material systems). This diagram is schematic, not to scale, and perhaps not absolutely correct in a quantitative sense with respect to pressure and temperature, but it illustrates that the diagram for the system H_2O is not complete as shown in Figure 2.2. The boundary line between Ice_I and liquid is the connecting link between the two diagrams. It should also be noted that several polymorphic forms of "Ice" exist, each having its own field of stability. Whereas only one triple point is shown in Figure 2.2, the application of pressure on the system H_2O generates several additional triple points, some of

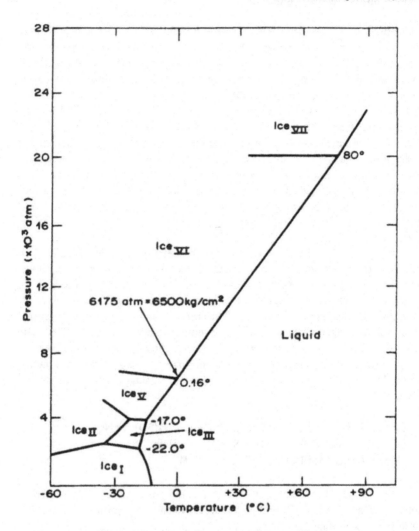

Figure 2.3. The Bridgman Ice Diagram (1, 2)

which involve equilibrium between three SOLID phases. Other triple points involve two solid phases and liquid H_2O. The definition of polymorph becomes obvious when using this diagram. A polymorphous compound or material is a single chemical substance which may exist in two or more crystalline structures. Polymorphism is very common in ceramic and metal systems and its significance is extremely important because, in general, when the crystal structure of a single substance changes, so do all the attendant chemical and physical properties. In a given situation, the properties may be either "worsened" or "improved", depending on what is being sought in the material.

It is very satisfying to have available the complete map of the stability relations for any one-component system, as for example for H_2O as shown in Figures 2.2 and 2.3.

It is very important to realize that, based on the P–T relationships as shown in two dimensions, any number of three dimensional phase diagram-property diagrams can be generated by plotting the property in the third dimension. For example, some of the most fundamental phase diagram-property diagrams would involve the following:

1. P–T–V (Volume)
2. P–T–G (Gibbs Free Energy)
3. P–T–d (Density)
4. P–T–S (Entropy)
5. P–T–η (Viscosity)
6. P–T–R.I. (Refractive Index)
7. P–T–K (Dielectric Constant)
8. P–T–α (Thermal Expansion)

Such complete spatial models have been worked out for only a very few systems. Ricci (3) shows some of these diagrams for H_2O in Chapter 2, pages 30–32. Even though the diagrams are only schematic, they are extremely helpful in visualizing in three dimensions the P–T–V, P–T–G, and P–T–η, relationships for H_2O. For ceramic materials, many chemical, thermal, mechanical, electrical, optical, nuclear and other properties beyond those listed above could result in an enormous number of phase v.s. property diagrams or property handbooks such as the Handbook of Chemistry and Physics (CRC Press, Inc.), Powder Diffraction Files (International Centre for Diffraction Data), and the Cindas publications on material properties (Center for Information and Numerical Data Analysis and Synthesis, Purdue University).

Much of the research done in materials science (ceramic, metals, polymers) involves composition vs property measurements over a very limited temperature and pressure range (usually room temperature and one atmosphere). In the future measurements will be made over wider temperature and pressure intervals, as technological problems require knowledge of properties at ultra-low and ultra-high temperature and pressure conditions.

VI. EXAMPLES OF REAL SYSTEMS

A. General

If one examines the collections of Phase Diagrams for Ceramists (1964, 1969, 1975 or 1981), it will be noted that the number of one-component oxide P–T diagrams is very limited. The reason that such a limited number exists is, of course, due mainly to the fact that the vapor pressures of ceramic oxides and compounds are relatively low, even at temperatures above $1000°C$. Furthermore, even if one wished to measure the vapor pressure of the more refractory oxides between $1000°C$ and $2000°C$, severe problems with containers would be encountered immediately. At $1500°C$ most container materials would be somewhat plastic, somewhat permeable and somewhat prone to become part of the system whose vapor pressure one wished to measure. The vapor pressure of several refractory oxides does not become appreciable until temperatures of $1800°C$ are reached. Examples include the divalent metal oxides such as BeO, MgO, CaO, the trivalent oxides such as Al_2O_3, Y_2O_3, and some rare earth oxides, and finally, SiO_2, TiO_2, ZrO_2, and HfO_2 among the tetravalent metal oxides.

On the other hand some oxides do have appreciable vapor pressures between $700-1500°C$, but there has been no special incentive to accurately measure these. Examples of such "ceramic" oxides include the alkali oxides, Tl_2O, CuO, ZnO, CdO, PbO, HgO, B_2O_3, Ga_2O_3, As_2O_3, Bi_2O_3, Cr_2O_3, SeO_2, TeO_2, P_2O_5, V_2O_5, MoO_3, WO_3, and some platinum group oxides.

The consequence of the fact that vapor pressures are not known or have not been measured is that the pressure axis in P–T diagrams for many simple ceramic oxides is only hypothetical, while the temperature axis is real. P–T diagrams are constructed on a "probable" basis, using whatever meager data are at hand, especially with respect to the pressure variable. However, even the postulated, "probable", or "not-impossible" diagram is extremely useful because it represents the best version of what the stability relationships ought to be in the light of all available data. Most often, the ceramist is content to diagram the solid state relationships (polymorphs) in a one-component system giving secondary importance to the solid-liquid relationships. If solid-vapor or liquid-vapor relationships are shown, they are most often hypothetical, as mentioned above.

To illustrate these statements, it is very instructive to consider the "simple" oxide systems SiO_2, TiO_2, GeO_2, and ZrO_2 and two systems of interest to materials scientists and engineers, Fe and C. The reader should be forewarned that none of the diagrams shown for any of these systems is the complete and final version of the stability relationships. Due to various kinds of experimental difficulties and the great persistance of metastable phases in some of the systems, the diagrams are incomplete, tentative and partially or totally schematic.

B. Silica

It is unfortunate that the ceramic scientist or engineer has to begin by trying to understand the "simple" system, SiO_2. It is the basis for all silicate chemistry

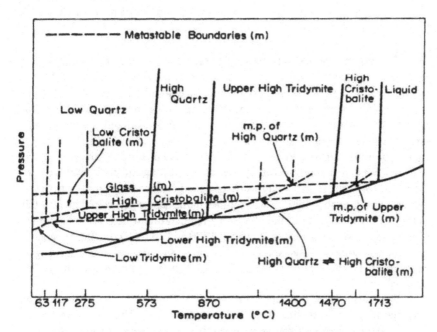

Figure 2.4. Stability Relations of the Silica Minerals; Fenner (8)

and obviously of the utmost importance, but it is actually one of the most complicated and difficult to understand in spite of the fact that complete books (4, 5, 6) have been written on the subject. Furthermore, as mentioned previously in Chapter 1, the system SiO_2 is basically an end member in the two-component system $Si-O_2$, (or $Si-SiO_2$), for which a diagram is available (7).

It is appropriate to first show the classical P–T diagram for the common silica phases, as drawn by Fenner (8) in 1913 (Figure 2.4). The temperature axis is real, but the pressure axis is only schematic, pressures being assumed to be near one atmosphere in an open system. Many of the events shown in the diagram are well-known and well established, such as:

1. The rapid, reversible $\alpha \rightleftarrows \beta$ quartz inversion at 573°C
2. The melting point of β-cristobalite at 1713°C
3. The undercooling of liquid SiO_2, resulting in silica glass at room temperature
4. The undercooling of (supposedly metastable) β-cristobalite below 1470°C, and its rapid, reversible inversion to α-cristobalite at temperatures around 275°C
5. The undercooling of (supposedly metastable) β'' tridymite below 870°C

and its rapid, reversible inversion to β' and α tridymite at 117°C and 63°C, respectively

The dashed lines indicate metastable events and therefore, the diagram suggests that superheated β-quartz can transform to undercooled β-cristobalite around 1200°C, and that melting to undercooled liquid silica would take place around 1450°C. One must realize that the temperatures mentioned above (1200 and 1450°C) are not to be regarded in the same sense as a fixed equilibrium inversion temperature such as the 573°C $\alpha \rightleftarrows \beta$ inversion of pure quartz or the melting of pure β-cristobalite at 1713°C. Metastable inversions, conversions, or melting will be mainly influenced by the rate of heating and cooling, as well as particle size effects, surface conditions, and the presence of small amounts of gaseous, liquid, or solid impurities. Another metastable event would involve the melting of high tridymite to liquid at some temperature which would be in the stability region of β-cristobalite (between 1470 and 1713°C). During the commercial preparation of fused silica (glass) from various natural quartz starting materials, it is likely that metastable melting of quartz to liquid silica does take place below 1713°C.

The diagram does not supply information on the rate at which events happen, especially the crystalline inversions and conversions. The $\alpha \rightleftarrows \beta$ inversions in quartz and cristobalite are known as a "displacive" or "snap" type, since only changes in silicon-oxygen bond angles or bond lengths are involved during the transformations and no bonds need be broken and reformed during the transitional period. In contrast, the conversion between β-quartz and β-cristobalite is described as "reconstructive" since the bonds in hexagonal β-quartz have to be broken and reformed into the cubic symmetry of β-cristobalite. This type of inversion is sluggish, requires higher energies for the transformations, and is often not reversible or only very slowly reversible at one atmosphere. It is not appropriate here to discuss the rate of transformation of the various forms of silica or the properties which are associated with each crystalline, glassy, or amorphous form. References 4 and 5 contain abundant information on the kinetics of the changes and the chemical and physical properties of the natural and synthetic phases of SiO_2. A summary of a few of the properties of some of the better known forms of SiO_2 is given in Table 2.1.

Since the work of Fenner, many papers have appeared on the silica phases and two new high pressure polymorphs have been discovered, coesite (9) and stishovite, (10) leading to a new version of the stability relations as shown in Figure 2.5. This figure is only schematic and is a result of the composite data from Yoder (11), Boyd and England (12), Tuttle and England (13), and Ostrovsky (14).

By far the most controversial portion of the relationships shown in Figure 2.5 involves the tridymite phase. Some authors feel it is a part of the diagram at one atmosphere pressure and others feel it does not have the region of stable existence shown between 870 and 1470°C at one atmosphere. One can only summarize the most important points as follows:

1. The relationships portrayed are for PURE SiO_2 under various conditions of P and T; the presence of H_2O or species of H_2O are forbidden
2. The presence of alkalies such as Li^+, Na^+, K^+, or any other impurity ion is forbidden
3. Silica liquid can be easily undercooled (a classical example) and it always crystallizes between 900–1713°C at one atmosphere to β-cristobalite (never tridymite or quartz) which later inverts to α-cristobalite during cooling
4. Ultra pure silica-tridymites (no water) have never been investigated under dry P–T conditions
5. The diagram infers that only moderate pressure might be necessary to produce PURE tridymite phases
6. At one atmosphere, it has been always necessary to add alkalies, sodium tungstate, or other impurities to bring about the formation of tridymite
7. Under hydrothermal conditions, tridymite has been produced by using ultra-pure silicon metal as a starting material (15). (H_2O or species of H_2O present.)
8. Natural tridymites always contain appreciable quantities of foreign ions such as Na^+, Ca^{2+}, and Al^{3+} (16, 17)
9. Recently, tridymite and cristobalite of 99%+ purity have been found on the lunar surface (18) and a quotation from the paper is instructive

At first glance it is remarkable that the lunar tridymite and cristobalite, in rocks which crystallized more than three billion years ago, have not inverted to quartz, as has occurred in most ancient terrestial rocks. Tridymite and cristobalite have also survived in meteorites, for which ages of about 4.6 billion years have been established. The survival of these high temperature polymorphs in meteorites and lunar rocks can probably be ascribed to the extremely anhydrous nature of these materials; the catalytic action of water in promoting crystallization and equilibrium is well documented.

In addition to the important point about the extremely anhydrous nature of lunar tridymite and cristobalite, it is also to be noted that the pressure seen by lunar rocks over geological time spans has probably been much lower than that experienced by these minerals on the earth.

Other noteworthy points about the system silica from a ceramic standpoint are as follows:

Table 2.1 Some Properties of Well Known Phases of Silica

Form	Structure	Density	R.I.	Preparation	Remarks
α-Quartz	Hex.	2.6510	ϵ, 1.553 ω, 1.544	Never without fluxes, pressure or both	Most abundant mineral source of SiO_2, purity 99.7%+
β-Quartz	Hex.	2.53 (600°C) 2.54 (1100°C)	ϵ, 1.540 (580°C) ω, 1.533 (580°C)	Stabilized to room temperature by solid solution of small ions Li^+, Be^{++}, Mg^{++}, Zn^{++}, Al^{3+}, etc.	
Tridymite S Low, Sl, α Middle, S2, B' High, S3, B"	Orth. Hex? Hex	2.262, 0°C 2.22, 200°C (calculated)	1.478–1.481	From quartz, silica glass, silicic acid with alkalies, $Na_2WO_4 \cdot H_2O$ and pressure.	Always impure, as found in nature Na^+, Ca^{3+}, Al^{3+}, others.
α-Cristobalite	Tetr	2.32	ϵ, 1.487 ω, 1.484	From inversion of β-crist.	Inversion temper. sensitive to previous thermal history of β form.
β-Cristobalite	Cubic	2.20 (500°C) (calculated)		By heating pure quartz to 1600°C for 10 hours. By heating silicic acid to 1300°C for 10 hours. By crystallization of silica	Thermal expansion and other properties similar to silica glass.

Silica Glass	Isotropic	2.20	1.46	glass at 1500°C for 10 hours.	Can be compacted at high temperature to produce glasses with unusual properties
Coesite	Monoclinic	2.93–3.01	α, 1.593; β, ?; γ, 1.597	From fusion of quartz, silica gel, silicic acid. 500–800°C 35,000 atm. 15 hours with flux sodium metasilicate, diammonium phosphate, boric acid, $NH_4 VO_3$	Relatively insoluble in HF
Stishovite	Tetragonal Rutile structure	4.35	ϵ, 1.826; ω, 1.799	1200–1400°C > 160 kbs	Only form of SiO_2 in which Si^{4+} is six coordinated with O.

NOTE: Coesite was synthesized in 1953 by Coes and Stishovite in 1961 by Stishov and Popova. Coesite was discovered in nature in 1960 at Meteor Crater, Arizona and in 1962, Stishovite was discovered in the coesite which had been extracted from the impacted sandstone. Usually, a natural mineral is known and studied long before its synthesis is attempted in the laboratory. In the case of coesite, stishovite and mullite, the pure, synthetic mineral was first made in the laboratory and then searched for and found in the earth. These three cases are unusual and their occurrence in nature is rare.

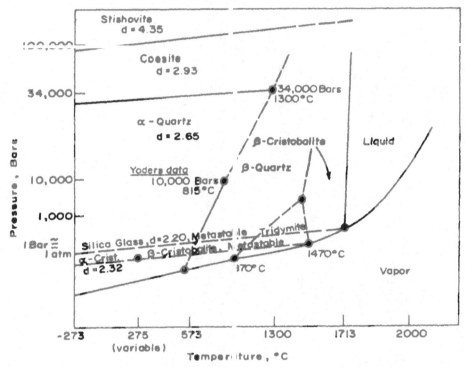

Figure 2.5. Schematic Composite p–T Diagram for the **System** SiO_2 (11–14)

1. It has often been observed by silicate phase equilibrium experimentalists that β-cristobalite can be brought to room temperature if it is surrounded by a glassy film and quenched rapidly. Destruction of the glass film by grinding allows the restraint to be removed and the cristobalite then reverts to the α form

2. The synthesis of α-quartz is usually done under hydrothermal conditions; that is, in the presence of H_2O and moderate pressures. Furthermore the sodium ion is usually present. The Bell Laboratories, Inc. commercial process for growing large α-quartz crystals involves seeding, hydrothermal pressure of about 15,000 psi., the presence of Na^+, and times measured in hours, days, or weeks. It is important to note that pure α-quartz has probably never been synthesized without the impurities mentioned above; H_2O, Na^+ or other alkalies, or a combination of water and impurities and the application of pressure. Naturally occurring α-quartz is very pure, usually containing more than 99.7% SiO_2

3. It was once thought that quartz would not take other materials into

solid solution, but it is now known that β-quartz especially will accept combinations of the smaller ions such as Li^+, Be^{2+}, Mg^{2+}, Zn^{2+}, Al^{3+} and that metastable β-quartz of various compositions can be obtained at room temperature by a "stabilization" process. The β-quartz polymorphs of GeO_2 and SiO_2 form a continuous series of solid solutions (19)

4. Commercial processes involving the silica phases include:

   a. The Fused Silica Glass Industry: Various grades of silica glass are made ranging from ultra pure to 99% SiO_2, starting with natural quartz or chemically pure sources of silica such as gels, silicic acid or $SiCl_4$. Fusion temperatures range from $1750°$-$2000°C$, using batch or continuous processes. The exceptionally high viscosity of molten silica makes the removal of entrapped gases one of the most difficult problems in the manufacturing process

   b. The Growth of α-Quartz Crystals: Even the best obtainable natural Brazilian quartz has been found to be less than satisfactory for the manufacture of piezoelectric oscillator plates and a process for the synthetic growth of α-quartz has now been substituted and standardized

   c. Silica Refractories; At one time, silica brick were in great demand in the steel industry as an open hearth roof. It has been replaced for this application by the basic brick, resulting in a tremendous decline in production of silica brick, but many special uses remain. Natural quartz raw materials to which 1-2 wt. % of CaO has been added are fabricated and fired to temperatures which result in a final product which contains residual quartz, tridymite, cristobalite, and glassy phase. Details may be found in textbooks on refractories, such as Chesters (20)

   d. Whiteware Products: Natural quartz is a standard constituent of many whiteware bodies, in amounts ranging from 5-35 wt. %. During firing quartz either remains as quartz (which subsequently goes through the $\beta \rightleftarrows \alpha$ inversion on cooling), converts to β-cristobalite (which usually goes through the $\beta \rightleftarrows \alpha$ inversion on cooling), or partially or totally dissolves in glass phases which develop at the highest temperatures of the firing cycle

   e. Glass-Ceramics: The composition of glass-ceramic bodies can be adjusted so that the major crystalline phase after final heat treatment is metastable β-quartz (21, 22, 23, 24, 25, 26) of complex composition as mentioned above. It is not pure SiO_2

   f. Semi-Refractory Clays and Structural Clay Products: Quartz is often a major impurity in semi-refractory clays and buff and red burning structural brick clays. During heat treatment, the quartz either re-

tains its identity as quartz, converts to cristobalite or partially dissolves in the developing glass phase. These events are influenced by
the kind and amount of oxide impurities present in the clay.

Except for its presence in silica brick as one of the major
phases, no major commercial applications of tridymite have been
developed (based on the properties of relatively pure tridymite).
Stishovite and coesite would be interesting abrasives, were it not for
the high cost of producing these phases under high pressure. For a
discussion of the many other natural and synthetic phases of silica,
references 4 and 5 are highly recommended.

Finally, it is appropriate to mention the many compounds which are iso-
structural with one or more of the silica phases, as originally detailed by Buerger
(27, 28, 29, 30) and demonstrated and expanded by many others in subsequent
papers.

C. Titania

i. Phase Relations
Four TiO_2 structures have been recognized; anatase, rutile, TiO_2 II, and brookite.
The phase relationships have been determined by Dachille, Simons, and Roy
(31) as shown in Figure 2.6. The TiO_2 II phase has the α-PbO_2 structure. In the
same paper, the authors show a second diagram involving the relationships between brookite, TiO_2 II, and rutile. There is some question as to whether or not
brookite is a PURE TiO_2 phase. Furthermore, it is thought to be metastable
with respect to the other TiO_2 phases; therefore, Figure 2.6 is sufficient to show
the probable equilibrium relationships in the system.

Jamison and Olinger (32) claim that thermochemical data show that
anatase is everywhere metastable with respect to rutile and TiO_2 II, which would
eliminate two boundary lines in Figure 2.6 and once again demonstrate that
equilibrium is often very difficult or impossible to achieve experimentally (in
relatively short times). The claim for equilibrium should be tested by approaching from as many directions as possible, but in many cases the equilibrium can
seemingly be approached from only one direction.

Many other phases have the rutile structure or a distorted or derivative
rutile structure (SnO_2, TaO_2, NbO_2, MoO_2, WO_2, OsO_2, IrO_2, RuO_2, VO_2,
CrO_2, MnO_2, GeO_2, MgF_2, ZnF_2, MnF_2, CoF_2, $AlAsO_4$, $CrVO_4$, Cr_2WO_6,
Cr_2TeO_6, $InSbO_4$, and $ScSbO_4$), implying the need for a study of the stability
relationships under various P–T conditions.

ii. Non-stoichiometric Phases
It has been known for many years that many oxide phases are non-stoichiometric; that is, instead of common oxides such as RO, R_2O_3, RO_2, R_2O_5, and

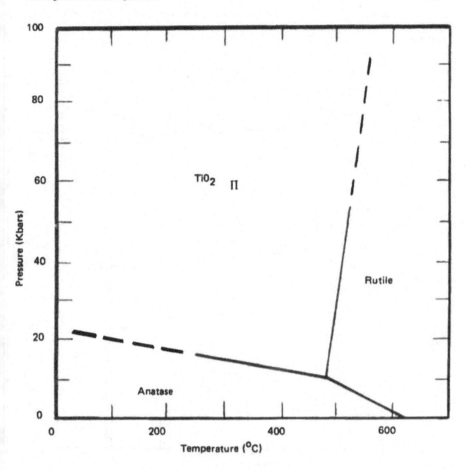

Figure 2.6. The p–T Diagram for TiO_2 (31)

RO_3, it is possible to have an oxide of iron such as wüstite, $FeO_{0.98}$, an oxide of titanium such as $TiO_{1.98}$, an oxide of vanadium such as V_2O_{5-x}, or an oxide of tungsten such as WO_{3-x}, where X is some small number, for example in the range between 0.001 and 0.1.

The system $Ti–O_2$ (Figure 2.7) (33) is an excellent one to illustrate this principle. Rutile is well known to exist as an oxygen-deficient phase to an oxygen content of at least 1.983. However, in the range between $TiO_{1.983}$ and TiO_2, the rutile structure is preserved and the interval is spoken of as the "homogeneity range" of the rutile phase or as oxygen-deficient rutile solid solutions. Many compounds with two or more cations can be produced as non-

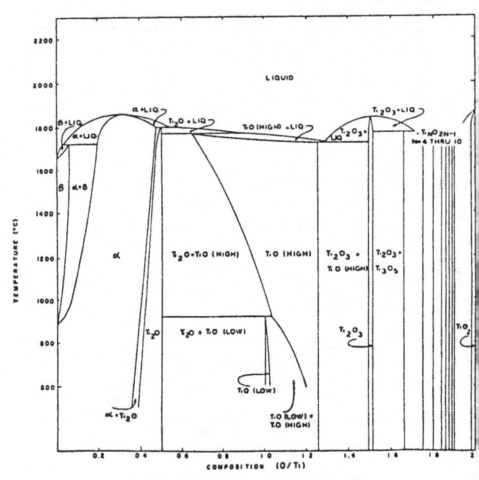

Figure 2.7. Titanium-Oxygen Phase Diagram (33)

stoichiometric or oxygen-deficient phases, as well as "simple" oxides. The electrical properties of these oxides or compounds are most sensitive to the non-stoichiometry and the effect of small changes in composition on the electrical (or magnetic) properties is the subject of many research papers. Certain other properties are not much affected by changes in stoichiometry, for example, the thermal expansion coefficient.

It is interesting to note that the Ti_3O_5 phase ($Ti_2O_3 \cdot TiO_2$) exists as a pseudobrookite ($Fe_2O_3 \cdot TiO_2$, $Al_2O_3 \cdot TiO_2$) structure, the Ti_2O_3 phase as a corundum (Al_2O_3) structure, and the TiO phase as an NaCl structure, similar to MgO, MnO, FeO, and CoO.

The less common oxides of certain elements will frequently have solid solution ranges or "homogeneity regions" as well as the common oxides. For example TiO, Ti_2O_3, Pr_7O_{12}, Pr_9O_{16}, Pr_5O_9, and Pr_6O_{11}.

D. Germania

The phase relationships in the system GeO_2 have been postulated by Sarver and Hummel (34), based on the discovery of a (metastable) $\alpha \rightleftarrows \beta$ inversion in the quartz phase. Using data of previous investigators, the diagram shown in Figure 2.8 was constructed. The pressure axis is totally hypothetical and implies the urgent need for a "dry" experimental investigation which would correctly identify the extent of the fields of stability of the phases shown and possibly even uncover new phases of GeO_2 such as coesite or stishovite.

Hill and Chang (35) have investigated the system under hydrothermal conditions (P_{H_2O} up to 1800 psi, temperature 950–1150°C) and found that the quartz phase has a wedge-shaped field of stability which terminates around 1100 psi and 1083°C, permitting the rutile form to exist in equilibrium with liquid above 1100 psi. The slope of the rutile $GeO_2 \rightleftarrows$ liquid curve is negative above the triple point, inferring that liquid GeO_2 has a greater density than rutile-type GeO_2 at elevated pressures. It is appropriate to consider the radius ratio for simple $AX_2(GeO_2)$ structures. The anion to cation radius ratio (R_A/R_x) necessary for tetrahedral coordination lies between 0.225–0.414, and for octahedral coordination between 0.414 and 0.732. Assuming a radius of 0.53 for Ge^{4+}, $R_A/R_x = \dfrac{0.53}{1.32} \cong 0.40$, both four or six coordinated structures might be expected. Experimentally GeO_2 has been found to exist as the hexagonal, four-coordinated quartz structure (d = 4.280) and the more dense tetragonal six-coordinated rutile structure (d = 6.277).

E. Zirconia

The phase diagram for the system ZrO_2 has been proposed by Ruh and Rockett (36) as shown in Figure 2.9. Some extremely important work by Smith and Kline (37) and Masdiyasni, Lynch, and Smith (38, 39) made this postulated diagram possible. Smith and Kline, using high temperature X-ray diffraction, showed that PURE ZrO_2 undergoes a rapid, reversible tetragonal to cubic inversion at 2300°C. Masdiyasni, et al. showed that PURE cubic ZrO_2 could be brought to room temperature in a metastable state if prepared from the alkoxide. When heated to about 300°C, it transforms to a metastable tetragonal form which in turn will transform to the stable monoclinic phase at 400°C. The tetragonal form may be brought to room temperature in a metastable state using the alkoxide or other starting materials, and it will transform to the stable monoclinic form if reheated between 400–700°C. Wittels and Sherrill (40) had previ-

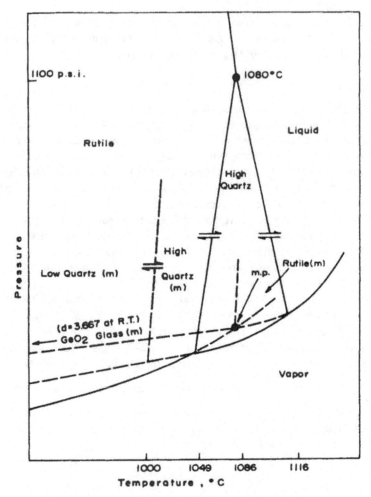

Figure 2.8. Schematic p–T Diagram for GeO_2 (34, 35)

ously produced cubic ZrO_2 at room temperature by fast neutron bombardment of monoclinic ZrO_2, but this work did not contribute to an understanding of the stability relations at that time. However, ideas were generated on the conversion mechanism.

The kinetics of the important monoclinic $\rightleftarrows$ tetragonal inversion have recently been reviewed (41) following earlier research by Maiti, Gokhale and Subbarao (42). On heating, crystallites ranging in size from 325 to 1235 Å inverted from monoclinic to tetragonal around 1170°C. It was found that the reverse

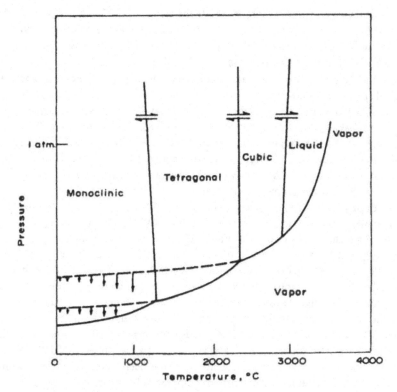

Figure 2.9. Proposed Diagram for the System ZrO_2 (36)

tetragonal-monoclinic transformation temperature was strongly dependent on crystallite size, shifting upward as the crystallite size increased. (851°C for 325 Å size, 975°C for 1235 Å size). For sizes below about 1000 Å, the kinetics are isothermal, above 1000 Å, athermal. The strain induced by the monoclinic → tetragonal upcycle strongly affects the transformation behavior on the down-cycle.

Mitsuhashi and Fujiki (43), using single crystals of ZrO_2, found the transformation range for ≈ 60 crystals was between 1160–1190°C on heating, but initial transformation on cooling ranged between 1070 to 1100 °C. Although the crystals remained transparent, during heating they were no doubt largely converted to polycrystalline material. R. C. Garvie (44) has written an excellent summary on the ZrO_2 phases and their properties. Volume 3 of Advances in Ceramics; Science and Technology of Zirconia (45) was a result of the First International Conference on this subject and it contains the most recent information on many aspects of ZrO_2.

The radius ratio for ZrO_2 is 0.57 if the radius of Zr^{4+} is taken as 0.76 Å, well within the octahedral range of 0.414–0.732, which would predict a rutile structure for the zirconia phases. Actually, all three forms are based on an eight coordinated fluorite-type structure.

In 1965, non-stoichiometric zirconia was reported by Livage and Mazieres (46, 47). A black material, said to be tetragonal $ZrO_{1.96}$, was obtained by heating hydrous zirconia in vacuum (10^{-3} torr) at 900°C. Andreeva, Gropyanov, and Kozlovskii (48) claimed that ZrO_2 fired at 2923°C for 1 hour lost 0.07% O, corresponding to an oxygen vacancy concentration of 4.3×10^{-18} per gram mole and a formula of $ZrO_{1.995}$. A density change from 5.58 to 5.77 was taken as an indication that the monoclinic ZrO_2 was tending toward tetragonal symmetry in the off-stoichiometric material.

Camiglia, Brown and Schroeder (49) showed that zirconia was probably oxygen-deficient in the range between 1000–1900°C and <1 atm P_{O_2}. The lattice dimensions of monoclinic and tetragonal zirconia decreased with increasing oxygen deficiency, but the lattice thermal expansion and rate of transformation were not appreciably affected. Nicholson (50) has emphasized the experimental difficulties attending the determination of the tetragonal-cubic inversion temperature by high temperature vacuum X-ray techniques and questions whether or not such techniques can give a reliable estimate of the transformation temperature of pure stoichiometric ZrO_2. Reworking of existing data for a transformation temperature of 2400°C and an oxygen partial pressure of 10^{-6} torr, gives a stoichiometry of $ZrO_{1.970}$. Although it has been stated (49) that the rate of the tetragonal-cubic transition is not appreciably affected by oxygen deficiency, it is logical to expect that the temperature of the transformation would be affected.

F. Iron

The estimated P–T diagram for Fe at ordinary pressures is shown in Figure 2.10 (51). Rhines, Chapter 2, Unary Systems, states "Since it is customary to handle metals in the open, where the gas phase can exist, the representation of the corresponding phase equilibria without reference to the gas phase constitutes an incomplete expression of the state of equilibrium. For most purposes this omission is inconsequential".

The high pressure P–T diagram for Fe is shown in Figure 2.11 (52). In addition to the three triple points shown in Figure 2.10, the high pressure diagram provides two more involving (1) L, δ, and γ, and (2) γ, α, and ϵ. Gordon (53) states, "The $\gamma \rightarrow \alpha$ transformation on cooling is without doubt the most important allotropic transformation known, since it is responsible for the ability of Fe-C alloys (steels) to attain their unusually high strengths, without which our modern civilization could not have been developed".

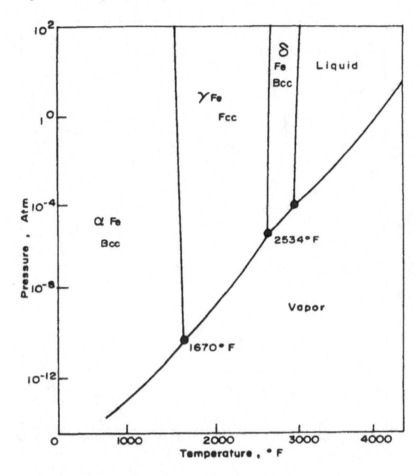

Figure 2.10. Stability Diagram for Iron at Ordinary Pressures (51)

Two comments are in order after this quotation. First, from a purely academic standpoint, the allotropy of iron as it occurs and as it is manipulated in the TWO-COMPONENT system Fe-C is what has so far been of such great importance commercially in our civilization. In future years the allotropy of iron as PURE iron in the ONE-component system Fe may become very important commercially, but as yet, this is not the case.

Second, and once again from mainly an academic or scientific viewpoint, the allotropy of iron is most important to the metallurgist, but a ceramist would consider the polymorphism of silica more important and an organic chemist would place the transitions in carbon or various polymers at the top of the list.

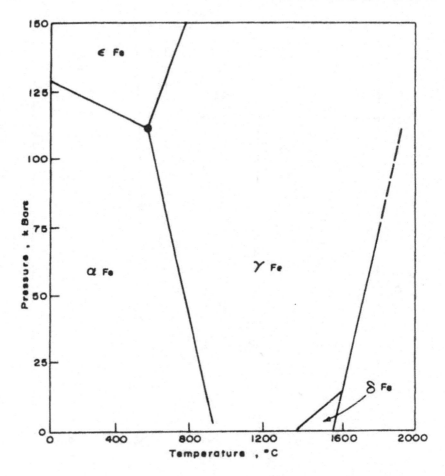

Figure 2.11. Stability Diagram for Iron at High Pressures (52)

G. Carbon

The stability diagram for the system C is shown in Figure 2.12 (54, 55, 56). High pressure favors the formation of diamond (mol. vol. = 3.42 cm^3/mol) over the less dense graphite phase (mol. vol. = 5.33 cm^3/mol). The region for practical diamond synthesis (57) is outlined in Fig. 2.12. A very high pressure form of carbon with metallic properties can be produced.

It is interesting to note the reversal in curvature of the solid-liquid boundary, indicating that graphite and diamond are less dense than the liquid at very high pressures. Natural and synthetic diamonds both exist in a metastable condi-

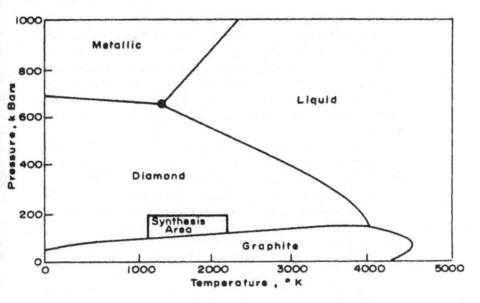

Figure 2.12. Stability Diagram for Carbon (54, 55, 56)

tion at room temperature and one atmosphere, also at temperatures greater than room temperature and pressures greater than one atmosphere (up to $10-10^2$ kilobars).

It is interesting to note that "vitreous carbon" is a commercial product and that its structure (58), like that of vitreous silica, is still in doubt. Furthermore, several new forms of carbon such as chaoite, lonsdaleite, carbon VI, and possibly other as yet unidentified forms are not shown in Figure 2.12, indicating the tentative nature of this one-component system.

There has been a great deal of interest in the system because of the quest for synthetic diamond, which began around 1800. The first successful production of industrial diamonds was announced by the General Electric Company in 1955 (59). The majority of industrial diamonds are small, usually less than a half millimeter in diameter, but useful in a large number of grinding, polishing and sawing operations. The ceramic industries use about one-third of all the industrial diamonds produced. Many of the fusion cast refractories are trimmed to close tolerance by diamond-impregnated cutting wheels.

VII. SUMMARY

The previous sections have illustrated the following important considerations:

1. The understanding of chemical or ceramic one-component systems is based on isothermal and isobaric analyses or consideration of the simultaneous effect of temperature and pressure. Ideally, one-component systems, depending upon the chemical nature, would be examined from $0°K$ and 0 bars to temperatures of $3000°K$ and extremely high pressures in order to produce a large map of the phase relations. However, many practical considerations limit the temperature and pressure region which can be studied.

2. Much more temperature v.s. pressure research is needed for "simple" one-component oxide, sulfide and fluoride diagrams. If the most common valence state for each solid element in the Periodic Table (from lithium to fermium) is considered, this means at least 90 oxide, 90 sulfide and 90 fluoride systems have to be thoroughly examined for p–T relationships, including common oxides such as SiO_2, TiO_2, GeO_2, ZrO_2, and several others. If less common valence states of each element are considered, far more than 90 oxide, 90 sulfide and 90 fluoride systems would have to be examined.

3. The effect of pressure on melting and polymorphic transitions in one-component (one common valence state) ceramic systems has only rarely been considered and much research needs to be done to produce one-component diagrams which can be considered as nearly complete p–T maps of oxide (etc.) systems. Eventually, the sublimation (solid-vapor) relationships should be included, at least from an academic standpoint, on one-component systems.

4. The temperature and magnitude of the volume changes which occur during solid state polymorphic transitions and the aggregate and axial thermal expansion of each polymorph is extremely important to the use of ceramic materials.

   The volume changes attending the polymorphic (or allotropic) transitions in the six materials mentioned in Section VI above range from very great to only incidental significance during their processing into shapes which are useful in commerce.

   The volume changes attending the $\alpha \rightleftarrows \beta$ quartz, $\alpha \rightleftarrows \beta$ cristobalite, and monoclinic $\rightleftarrows$ tetragonal zirconia inversions are so great that it is impossible to fabricate shapes which will survive a normal firing operation. For many years, ZrO_2 and β-quartz were not considered to be useful materials, due to the disruptive volume changes attending the rapid, reversible displacive inversions. By solid solution stabilization it is now possible to bring cubic zirconia and hexagonal β-quartz to room temperature. However, by adding a second, third, or fourth component to SiO_2 or ZrO_2, one no longer is dealing with the PURE oxide and two, three, and four component systems are involved.

A detailed discussion of the volume changes attending the inversions of SiO_2, TiO_2, GeO_2, ZrO_2, Fe, and C and the thermal expansion of all phases which appear in these systems is not appropriate at this time, but it is well known that these properties are of the utmost importance in the use of these materials as fabricated ceramic shapes.

This discussion of volume changes and thermal expansion is only one example of the need to determine certain chemical and physical properties of oxide polymorphs and, in many cases, the properties of related liquid phases.

5. The need to be alert to minor or "massive" nonstoichiometry was only briefly discussed with respect to the system $Ti-O_2$. However, such an example gives rise to possible return of "simple" oxide, sulfide, fluoride, etc. systems to binary metal-oxide, metal-sulfide, metal-fluoride, etc. systems as far as basic phase equilibrium is concerned. It also emphasizes the importance of "defect" structures, crystal structure analysis and the mechanism and kinetics of reactions as well as the equilibrium relations.

PROBLEMS

1. Define and illustrate the following terms:

1. Components	6. Homogeneous Systems
2. Critical Point	7. Invariant Point
3. Degrees of Freedom	8. Le Chateliers Principle
4. Devitrification	9. Phase
5. Equilibrium	10. Phase Rule

2. Sketch the stable and metastable p–T relationships in the system SiO_2 or the system ZrO_2. Be sure to label fields of stability, show boundary lines, show and label invariant points and indicate metastable relationships with dash lines. Discuss the preparation of the phases, some of the more important chemical and physical properties of the phases and the qualitative rates of inversion or conversion of the crystalline polymorphs. Discuss any unusual features of the p–T diagram or unusual characteristics or properties of the common phases of SiO_2 or ZrO_2.

3. What outstanding facts do you associate with the following phases:

1. Coesite	3. Cubic zirconia
2. Stishovite	4. Fused silica glass

4. Listed below are a number of possible states for the system SiO_2, under one

atmosphere pressure. Classify each according to the definition of 1) stable, 2) metastable, 3) unstable. A metastable phase is sometimes very unstable and may tend to quickly revert to a stable form; in some cases a metastable phase may persist indefinitely and appear to be stable under conditions where it is really not the thermodynamically stable phase. If the condition described below can be placed into more than one category, place it into all that you think are appropriate.

1. Molten silica at 1800°C
2. Hi-Tridymite at 1800°C
3. Quartz at 1800°C
4. Fused silica glass at room temperature
5. Fused silica glass at 1200°C
6. α-Quartz at room temperature
7. β-Quartz at 1500°C
8. High tridymite at 500°C
9. α-Cristobalite at room temperature
10. β-Cristobalite at 1500°C

5. The following phases are known to exist in a system:

Cordierite ($2MgO \cdot 2Al_2O_3 \cdot 5SiO_2$)
Mullite ($3Al_2O_3 \cdot 2SiO_2$)
Forsterite ($2MgO \cdot SiO_2$)
Protoenstatite ($MgO \cdot SiO_2$)
Periclase (MgO)

1. What are the components of the system?
2. Could all of the above-listed phases coexist at equilibrium? Explain your answer.
6. Sketch the p–T diagram for a one-component oxide system with the following characteristics:

1. The oxide exists as a rutile-type structure up to 1049°C, at which temperature it inverts to a β-quartz type structure.
2. Metastable melting takes place at 1086°C.
3. The β-quartz form will melt at 1116°C.
4. On cooling, the β-quartz form inverts rapidly at 1000°C to an α-quartz form which persists at room temperature. By differential thermal analysis methods, this inversion was found to be rapid and reversible during heating and cooling.

5. The densities which have been measured at room temperature for the different forms of the oxide are as follows: α-quartz, 4.28 gm/cc, rutile, 6.28 gm/cc, and liquid (glass), 3.67 gm/cc.

With this information, draw the boundary lines in the system, including solid-vapor, liquid-vapor, solid-liquid, and solid-solid equilibria. Show stable equilibrium relationships in solid lines and metastable relationships in dashed lines. Be sure and label all fields of stability and indicate clearly the positive or negative slope of the solid-solid and solid-liquid boundary lines and what phases are in equilibrium across boundary lines.

If the ionic radius of the cation in this oxide is about 0.53 Å, what is the probable valence of the cation and what is the probable composition of the oxide.

7. Le Chatelier's Principle states that, if the equilibrium in a system is disturbed by a change in the external conditions, the system reacts to oppose the effect of the disturbance. Stated in another way, it says that, if a constraint is imposed on a system in equilibrium, the equilibrium will shift in such a way as to relieve the effect of the constraint.

Draw enough of the system sulfur or the system SiO_2 to illustrate the qualitative use of Le Chatelier's Principle.

8. A system has the following characteristics: (one component)

   1. 3 solid phases, 1, 2 and 3
   2. 1 liquid phase
   3. 1 vapor phase
   4. Solid 1 has a higher density than Solid 2.
   5. Solid 2 has a higher density than Solid 3.
   6. Solids 2 and 3 are in equilibrium with liquid at 200°C and 5 atm.
   7. Under atmosphere pressure there is a solid phase transition from S_1 to S_2 at 50°C.
   8. S_2 sublimes at 100°C.
   9. At 2 atmospheres and 150°C, S_3, liquid and vapor exist in equilibrium.
   10. There are 4 stable triple points.
   11. The liquid can be supercooled and is involved in a metastable triple point at 1.5 atm. and 75°C.

Construct the equilibrium diagram showing metastability in dashed lines.

9. The heat of inversion of alpha to beta quartz (low to high) is about 4.3 cal/gm. Using the Clausius-Clapeyron equation, show how you would calculate the volume change (and its direction) during the inversion.

Yoders data

Pressure (bars)	Inversion Temp. C
0-1	572
1,000	599
5,000	704
10,000	815

1 bar = 0.987 atm = approx. 1 atm = 10^6 dynes/cm^2
ΔV = ?
1 atmosphere = 1033.3 gm/cm^2 = approx. 10^3 gm/cm^2
1 calorie = 42,670 gm-cm = approx. 42 X 10^3 gm-cm

Use only the approximate values and be satisfied with a rough estimate of the volume change. How does this figure square up with what you know about the thermal expansion behavior of quartz?

10. Using graph paper provided, draw the following hypothetical one-component system (p–T diagram):

1. Five solids S_1, S_2, S_3, S_4, and S_5, liquid and vapor.
2. S_1 has a greater density than any of the other solids.
3. The critical temperature and pressure are 300°C and 6 atms. (Denote by Tc)
4. There are nine stable triple points.
5. S_2 never exists in equilibrium with liquid or vapor.
6. S_5, S_1 and liquid are in equilibrium at 150°C and 5 atmospheres.
7. The melting points of S_4 and S_5 are not affected by pressure change.
8. S_2 is more dense than S_3.
9. S_3 sublimes at 100°C and 3/4 atm.
10. S_5 is more dense than S_4.
11. S_5 never exists in equilibrium with S_2 or S_3.
12. S_1, S_4 and S_2 are in equilibrium at 87.5° and 2.5 atmospheres.

11. Use the Clausius-Clapeyron equation to calculate the effect of pressure on the temperature change during the transition of rhombic ⇄ monoclinic sulfur.

ΔV = 0.01395 cc/gm
Q = 3.12 cal/gm
Δp = 1 atm = 1033.3 gm/cm^2
T = 95.5°C, 368.5°K.

Compare the calculated ΔT to that produced by the one atmosphere influence on ΔT during the $ICE_1 \rightleftarrows$ liquid transition.

12. Using graph paper provided, draw the following hypothetical one-component system (p–T diagram)

1. Three solids S_1, S_2, S_3, L and V
2. L has greater density than any of the solids
3. 5 stable triple points
4. At 5 atm and $0°C$, S_1, S_2 and L are in equilibrium
5. S_2 is more dense than S_1
6. S_2 is more dense than S_3
7. At 1 atm, S_1 transforms to S_2 at $100°C$, S_2 transforms to S_3 at $150°C$ and S_3 sublimes at $200°C$
8. The critical temp. and pressure are: $300°C$ and 5 atm.
9. Include in the diagram a possible metastable triple point and explain what phases exist in "metastable equilibrium" at that point.

Assume reasonable relationships when data are lacking. The lack of data need not prevent the presentation of a reasonable diagram.

A convenient list of one-component systems in "Phase Diagrams for Ceramists" which can be used for class discussions:

1964 Edition

System	Page	Figure
SiO_2	84	153, 154, 155, 156, 157
GeO_2	83	151

1969 Edition

ZnO	76	2269
SiO_2	76	2270, 2271, 2272

1975 Edition

TiO_2	76	4258
ZrO_2	76	4259
HfO_2	75	4256
Nb_2O_5	78	4260
TℓF	353	4800
CdS	419	4954
$CaCO_3$	293–294	4658, 4659

1981 Edition

SiO_2	84	5113, 5114
$MgCO_3$	283	5511
$CaCO_3$	283	5509, 5510
$RbCO_3$	282	5507
Tl_2CO_3	282	5508

REFERENCES

1. George C. Kennedy and William T. Holser, Pressure-Volume-Temperature and Phase Relations of Water and Carbon Dioxide in Sidney P. Clark (Ed.) Handbook of Physical Constants, G.S.A. Memoir 97, pp. 371-383, 1966.
2. P. W. Bridgman, Water in the Liquid and Five Solid Forms, Under Pressure, Proc. Am. Acad. Arts Sci., Vol. 47, pp. 441-558, 1912.
3. John E. Ricci, The Phase Rule and Heterogeneous Equilibrium, D. Van Nostrand Company, Inc., New York, 1951.
4. Dana's, The System of Mineralogy, Volume III, Silica Minerals, John Wiley and Sons, 1962.
5. The Phases of Silica, Robert B. Sosman, Rutgers University Press, New Brunswick, New Jersey, 1965.
6. R. K. Iler, The Chemistry of Silica, The Center for Professional Advancement, P.O. Box 994, East Brunswick, New Jersey 08816.
7. R. E. Johnson and Arnulf Muan, Phase Diagrams for the Systems Si-O and Cr-O, Journal Amer. Cer. Soc., 51 (8) pp. 430-433, 1968.
8. C. N. Fenner, J. Wash Acad, Sci., 1912, 2, 471-480 (preliminary report). Amer. J. Sci., 1913, 36, 331-384. Z. anorg. Chem., 1914, 85, 133-197.
9. L. Coes Jr., Science 118, 131-132, 1953.
10. S. M. Stishov and S. V. Popova, A New Dense Modification of Silica Geokhimiya, 10, 837-839, 1961.
11. H. S. Yoder, Jr., High-Low Quartz Inversion up to 10,000 Bars, Trans. Amer. Geophysical Union, 31, pp. 827-835, December, 1950.
12. F. R. Boyd and J. L. England, The Quartz–Coesite Transition, J. Geophys Res. 65, 749-756, 1960.
13. O. F. Tuttle and J. L. England, Preliminary Report on the System SiO_2-H_2O, Bull Geol. Soc. Amer. 66, 149, 1955.
14. I. A. Ostrovsky, PT-diagram of the System SiO_2-H_2O, Geol. Journ. Vol. 5, Pt 1, pp. 127-134, 1966.
15. V. G. Hill and R. Roy, Silica Structure Studies: VI, On Tridymites, Trans, Brit. Ceram. Soc. 57 (8), pp. 496-510, 1958.
16. Brian Mason, Tridymite and Christensenite, Amer. Min (9), pp. 112, 1953.
17. Rock Forming Minerals, Vol. 4, Framework Silicates, by Deer, Howie, and Zussman, p. 179, Silica Minerals.

18. Brian Mason, Lunar Tridymite and Cristobalite, Amer. Min. 57, Nos. 9 and 10, P. 1530, Sept.-Oct., 1972.
19. W. S. Miller, F. Dachille, E. C. Shafer and Rustum Roy, The System GeO_2-SiO_2 Amer. Min. 48 (9-10) 1024-32, 1963.
20. J. H. Chesters, Steelplant Refractories, Chapter II, Second Edition, January, 1957, Percy Lund, Humphries and Company, Ltd., London.
21. T. I. Prokopowicz and F. A. Hummel, Reactions in the System Li_2O-MgO-Al_2O_3-SiO_2 : II, Phase Equilibria in the High-Silica Region, J. Amer. Cer. Soc. 39 (8) 266-278, 1956.
22. W. Schreyer and J. F. Schairer, Metastable Solid Solutions with Quartz-Type Structures on the Join SiO_2-$Mg Al_2 O_4$, Z. Krist, 116, 60-82, 1961.
23. G. H. Beall, B. R. Karstetter and H. L. Ritter, Crystallization and Chemical Strengthening of Stuffed β-Quartz Glass-Ceramics, J. Amer. Cer. Soc. 50 (4) 181-189, April 1967.
24. P. W. McMillan, Glass-Ceramics, 229 pages Academic Press, Inc., New York, 1964.
25. T. J. Veasey, Recent Developments in the Production of Glass Ceramics, Minerals Science and Engineering 5 (2) 92-107 April, 1973.
26. George H. Beall, Structure, Properties, and Applications of Glass-Ceramics; Advances in Nucleation and Crystallization in Glasses, American Ceramic Society Special Publication No. 5, 1972. Also in Corning Research, 1973, p. 99.
27. M. J. Buerger, The Silica Framework Crystals and Their Stability Fields, Zeit. Krist, (90), P. 186, 1935.
28. M. J. Buerger, Derivative Crystal Structures, Journal Chemical Physics (15), pp. 1-11, 1947.
29. M. J. Buerger, Crystals Based on the Silica Structures, Amer. Min., (33), Nos. 11 and 12, pp. 751-52, 1948.
30. M. J. Buerger, Stuffed Derivatives of the Silica Structures, Am. Min., 39, Nos. 7 and 8, pp. 600-15, 1954.
31. Frank Dachille, P. Y. Simons, and Rustum Roy, Pressure-Temperature Studies of Anatase, Brookite, Rutile, and TiO_2 II, Am. Min. (53), p. 1929-39, Nov.-Dec., 1968.
32. John C. Jamison and Bart Olinger, Pressure-Temperature Studies of Anatase, Brookite, Rutile, and TiO_2 II: A Discussion, Amer. Min. 54, Nos. 9-10, p. 1477, Sept.-Oct., 1969.
33. Phillip G. Wahlbeck and Paul W. Gilles, Reinvestigation of the Phase Diagram for the System Titanium-Oxygen, J. Amer. Cer. Soc. 49 (4) 180-184, April 1966.
34. James F. Sarver and F. A. Hummel, Alpha to Beta Transition in Germania Quartz and A Pressure Temperature Diagram for GeO_2, Journ. Amer. Ceram. Soc. 43 (6), P. 336, June, 1960.

35. V. G. Hill and Luke L. Y. Chang, Hydrothermal Investigation of GeO_2, Amer. Min. 53, Nos. 9 and 10, p. 1744, Sept.–Oct., 1968.
36. Robert Ruh and Thomas J. Rockett, Proposed Diagram for the System ZrO_2, J. Amer. Ceram. Soc. 53 (6), P. 360, June, 1970.
37. D. K. Smith and C. F. Cline, Verification of Existence of Cubic Zirconia at High Temperature, J. Amer. Ceram. Soc. 45, (5), p. 249–50, 1962.
38. K. S. Mazdiyasni, C. T. Lynch, and J. S. Smith, Preparation of Ultra-High-Purity Submicron Refractory Oxides, J. Amer. Cer. Soc. 48 (7), p. 372, July, 1965.
39. K. S. Mazdiyasni, C. T. Lynch, and J. S. Smith, Metastable Transitions of Zirconium Oxide Obtained from Decomposition of Alkoxides, J. Amer. Cer. Soc. 49 (5), P. 286–87, 1966.
40. M. C. Wittels and F. A. Sherill, Irradiation-induced Phase transformation in Zirconia, J. Appl. Phys. 27, 643–4, 1956.
41. E. C. Subbarao, H. S. Maiti and K. K. Srivastava, Martensitic Transformation of Zirconia, Phys. Status Solidi A, 21(1) 9–40, 1974.
42. H. S. Maiti, K. V. G. K. Gokhate, and E. C. Subbarao, Kinetics and Burst Phenomena in ZrO_2 Transition, J. Amer. Ceram. Soc. 55 (6), p. 317–322, June, 1972.
43. T. Mitsubashi and Y. Fujiki, Phase Transformation of Monoclinic ZrO_2 Single Crystals, J. Amer. Cer. Soc. 56 (9) 493, Sept. 1973.
44. High Temperature Oxides, Part II, Oxides of Rare Earths, Titanium, Zirconium, Hafnium, Niobium, and Tantalum, Chapter IV, by R. C. Garvie.
45. Advances in Ceramics; Science and Technology of Zirconia, Vol. 3, American Ceramic Society, Columbus, Ohio, 43 214, 1981.
46. J. Livage and C. Mazieres, Amorphous Zirconia and Nonstoichiometric Zirconia, Compt. Rend 260(19) (Groupe 8) 5047–8, 1965.
47. Jacques Livage and Charles Mazieres, Nonstoichiometric Zirconia and Stabilization of Tetragonal Zirconia, Compt. Rend 261 (21) (Groupe 8) 4433–5, 1965.
48. N. A. Andreeva, V. M. Gropyanov, and L. V. Kozlooskii, Change in the Structure of Zirconium Dioxide at High Temperatures in a Vacuum, Izv Akad Nauk SSSR, Neorg. Mater 5(7) 1302–3, 1969.
49. S. C. Carmiglia, S. D. Brown and T. F. Schroeder, Phase Equilibria and Physical Properties of Oxygen-Deficient Zirconia and Thoria, J. Amer. Cer. Soc. 54(1) 13–17, Jan. 1971.
50. Patrick S. Nicholson, Influence of Reduction on Estimation of the ZrO_2 Tetragonal-Cubic Transformation Temperature, J. Amer. Cer. Soc. 54 (1) 52–53, Jan. 1971.
51. Frederick N. Rhines, Phase Diagrams in Metallurgy, McGraw-Hill Co., Inc., 1956.
52. T. Takahashi and W. A. Bassett, The Composition of the Earths Interior,

Scientific American, 212 (6), p. 100, 1965. All rights reserved by Scientific American, Inc.

53. Paul Gordon, Principles of Phase Diagrams in Materials Systems, McGraw-Hill Book Company, Inc., 1968.

54. H. M. Strong, Amer. Scientist 48:58, 1960.

55. F. P. Bundy, Science, Melting Point of Graphite at High Pressure: Heat of Fusion 137, 1055-1057, 1962.

56. C. G. Suits, Amer. Scientist, 52:395, 1964.

57. F. P. Bundy, H. T. Hall, H. M. Strong and R. H. Wentdorf, Jr., Man-made Diamonds, Nature Vo. 176, 51-55, 1955.

58. A Greenville Whittaker and B. Trooper, Single-Crystal Diffraction Patterns from Vitreous Carbon, J. Amer. Cer. Soc. 57 (10) 443-446, October, 1974.

59. Ralph J. Holmes, Synthetic and Other Man-Made Gems, Foote Prints 32, (1), p. 3, 1960.

3

THE TWO COMPONENT SYSTEM

I. INTRODUCTION

A. The p–T–x Diagram

A system consisting of two components is called a BINARY SYSTEM or TWO-COMPONENT SYSTEM. The general two component system is controlled by three variables (pressure, temperature and composition) and it requires a three-dimensional model (a p–T–x diagram) such as that shown schematically in Figure 3.1. The end members, or components, of the system are A and B and are represented as a p–T diagram of the kind discussed in Chapter 2. The Phase Rule for the general two component system is:

$$F = C - P + 2$$
$$= 2 - P + 2 = 4\text{-}P$$

This means that F may have the following values:

Four coexisting phases, $F = 4 - 4 = 0$, invariant
Three coexisting phases, $F = 4 - 3 = 1$, univariant
Two coexisting phases, $F = 4 - 2 = 2$, bivariant
One phase, $F = 4 - 1 = 3$, trivariant

At ordinary (not extremely high) pressures, the phases present would be a single vapor phase, one or two liquid phases and one or two solid phases. In other words, an invariant system (P = 4, F = 0) could commonly consist of

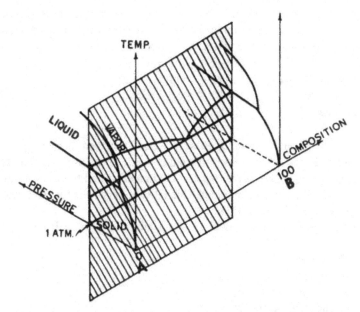

Figure 3.1. Intersection of Plane of Constant Pressure with Three Dimensional (p–T–x) Binary System

vapor, liquid and two solids, vapor, two liquids and a solid, two liquids and two solids, etc. If immiscibility of liquids and polymorphism of solids is considered, these events would give rise to several combinations of four phase, three phase and two phase equilibria. If only a gas, a liquid or a solid phase was present ($P = 1$, $F = 3$), pressure, temperature and composition (% A, % B) would have to be specified to completely define the system.

B. The "Condensed" System

In many ceramic systems the vapor pressures of the liquid and solid phases are negligible or so slight between room temperature and 1500–2000°C that the pressure variable is ignored and the solid-liquid phase relations are determined at atmospheric pressure (p = constant). The system is then called a "condensed" system and the Phase Rule reverts to $F = C - P + 1 = 3 - P$. In Figure 3.1 a plane has been passed through the diagram at a position corresponding to a constant pressure of one atmosphere. The intersection of this plane with the fusion curves of the end members and with those of all the intermediate mixtures of A and B results in the phase diagram for a two-component system at constant pressure (T-x).

One must realize that elimination of the pressure variable leads to a tremendous simplification of the viewpoint and representation of a binary system. At constant pressure, the Phase Rule and degrees of freedom are as follows:

Three coexisting phases, $F = 3 - 3 = 0$, invariant
Two coexisting phases, $F = 3 - 2 = 1$, univariant
One phase, $F = 3 - 1 = 2$, bivariant

The elimination of the vapor phase (p = const), tremendously simplifies the number and types of phases which would coexist in equilibrium. Temperature-composition diagrams can be represented in a plane and the need for three-dimensional visualization is eliminated. One must be very alert for cases where such simplification is *NOT* justified.

Two-component, "condensed" (T-x) systems may be classified as follows:

1. The simple eutectic
2. Intermediate compounds
   a. Congruent melting
   b. Incongruent melting
   c. Dissociation
3. Solid solution
   a. Complete
   b. Partial

Three additional considerations in these binary systems are:

1. Immiscibility in the liquid region
2. Unmixing of solid solutions or "exsolution"
3. Polymorphism of solids

If the end-members, intermediate compounds or solid solutions exist in two or more polymorphic forms, this phenomenon is superimposed on the three basic types of diagrams listed above and they become, in some cases, much more complex.

II. THE BINARY EUTECTIC SYSTEMS

A. Definitions and Application of the Phase Rule

A phase diagram for a condensed binary system of the simple eutectic-type is shown in Figure 3.2. This is a "composition vs. temperature" diagram with composition on the abscissa and temperature on the ordinate. In this hypothetical

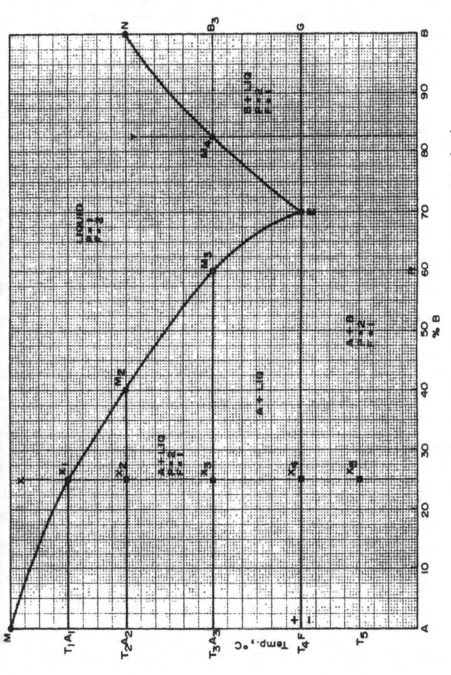

Figure 3.2. The Simple Binary Eutectic System (p = const, usually 1 atmosphere)

system, it is assumed that both end-members melt congruently. A congruently melting compound is one which exists in equilibrium with a liquid of the same composition at a fixed temperature at a fixed (usually atmospheric) pressure. Compositions between the end members may be represented on a mole percent or weight percent basis. In research, representation on a molar basis is usually preferred, where as in the pilot plant or production, representation on a weight percent basis is often advantageous or necessary.

Additions of B to A cause the freezing point to be lowered from M toward E, giving rise to the LIQUIDUS curve, ME. The liquidus temperature is the temperature above which no crystal can exist; or below which an infinitesimal amount of crystals coexist with essentially 100% liquid. It is the locus of temperatures at which crystals first begin to appear on cooling the melt under equilibrium conditions. Below line FG no liquid exists, and compositions are entirely crystalline. This line is called the *Solidus*. The intersection of liquidus curves ME and NE with the solidus line FG results in the binary EUTECTIC point E, where $F = 3 - 3 = 0$, the invariant point. A, B and liquid of composition E are in equilibrium at temperature T_4. The diagram is thus separated into three types of areas, 1) a liquid region, 2) two regions where liquid and a solid phase coexist, and 3) a region where crystals of A and B coexist.

a. In the single phase liquid region, the number of variables which must be specified to define the system are:

$$F = C - P + 1$$
$$F = 2 - 1 + 1 = 2$$

Both the temperature and the composition must be specified in order to uniquely describe the liquid phase.

b. If both liquid and solid phases coexist, such as A + liquid or B + liquid, the number of variables which must be specified in order to define the system is $F = 2 - 2 + 1 = 1$. Either the composition or the temperature is all that is necessary to define the system in terms of the variables which the system possesses. If the composition of the liquid phase is given (the liquid phase in this system is the only phase of variable composition), the temperature is automatically fixed because there is only one temperature at which that particular liquid composition is in equilibrium with a solid phase. An isothermal line (A_3-M_3 or M_4-B_3) drawn through the two-phase regions shown in Fig. 3.2 connects the compositions of the solid and liquid phases which are in equilibrium at that temperature. This line is called a *tie line* or *conjugation line* and connects the two *conjugate* phases which are in equilibrium. In Fig. 3.2 the lines $A_1 X_1$, $A_2 M_2$, $A_3 M_3$, and $M_4 B_3$ are examples of tie lines.

If the temperature is the variable specified, the liquid composition in equilibrium with crystals of A or with crystals of B is automatically fixed. The two-phase region may be thought of as containing an infinite number of tie lines the extremities of which indicate the compositions of the two phases which are in equilibrium at that particular temperature.

If two solid phases coexist, once again $F = 1$ and either temperature or composition must be specified to completely define the system.

c. Finally, as mentioned above, the eutectic point E is an invariant point where $F = 0$ and A, B and liquid of composition E, coexist at constant temperature ($P = 3$).

B. The Lever Rule

The simple binary eutectic system is ideal for the introduction of quantitative calculations by the Lever Rule. When a particular composition such as X (75.00% A, 25.00% B) in Figure 3.2 separates into only two phases at temperature T_3 (solid A_3 and liquid M_3), the original composition X_3, A_3 and M_3 are colinear. Furthermore, the Lever Rule states that the amounts of the two coexisting phases A_3 and M_3 are INVERSELY proportional to their distances from X_3. Thus in Figure 3.2, the amount of solid $A_3 = \dfrac{X_3 M_3}{A_3 M_3}$ and the amount of liquid $= \dfrac{X_3 A_3}{A_3 M_3}$. In this particular diagram the composition of the liquid phase M_3 is 40% A, 60% B. Therefore the percent of solid A_3 is $\dfrac{35}{60} \times 100 = 58.33\%$ and the percent of liquid M_3 is $\dfrac{25}{60} \times 100 = 41.67\%$.

The proportions of the A and B end members in the two coexisting phases A_3 and M_3 should always equal the proportions of A and B in the original composition X, which is 75% A and 25% B. Therefore,

$$
\begin{aligned}
\cdot \% \text{ A in } A_3 &= 100\% \\
\% \text{ A in } M_3 &= 40\% \\
\% \text{ B in } M_3 &= 60\% \\
100 \times 58.33 &= 58.33\% \\
40 \times 41.67 &= \underline{16.67} \\
&\quad\; 75.00\% \text{ A} \\
60 \times 41.67 &= 25.00\% \text{ B}
\end{aligned}
$$

The relationship may be visualized as a lever in which the fulcrum is X_3 and the lever arms are $A_3 X_3$ and $X_3 M_3$ for the relative quantities of liquid and solid, respectively.

C. The Isoplethal (Constant Composition) Analysis

In the one-component system p–T diagram, the analyses are either isothermal or isobaric (constant T or constant p). In a "condensed" two-component system where pressure is considered constant, the analyses are either isothermal or isoplethal (a vertical line of constant composition, % A, % B). The tracing of a cooling melt of composition X (an isoplethal analysis) will demonstrate previous definitions and prove how valuable the quantitative calculations can be for all simple eutectic systems and those containing intermediate compounds. The use of the Lever Rule in systems containing liquid immiscibility and complete or partial solid solution will be described later.

In Figure 3.2, composition X is a homogeneous liquid at all temperatures above the liquidus curve. At point X_1, an infinitesmal amount of crystals of A appear and the phases involved are a liquid containing 75% A and 25% B and crystals of 100% A (at temperature T_1).

Cooling from T_1 to T_2 under equilibrium conditions results in the movement of the liquid composition from X_1 to M_2 and the development of an amout of crystals of A determined by the Lever Rule. The amount of liquid of composition 60% A, 40% B is $\dfrac{A_2 X_2}{A_2 M_2}$ and the amount of solid A is $\dfrac{X_2 M_2}{A_2 M_2}$. Therefore, % liquid $= \dfrac{25}{40} \times 100 = 62.5\%$, and % A $= \dfrac{15}{40} \times 100 = 37.5\%$. Once again, the proportions of A and B in the two coexisting phases A + liquid should be 75% A and 25% B:

$$
\begin{aligned}
\% \text{ A in } A_2 &= 100\% \\
\% \text{ A in } M_2 &= 60\% \\
\% \text{ B in } M_2 &= 40\% \\
100 \times 37.5\% &= 37.5 \\
60 \times 62.5\% &= \underline{37.5} \\
&\quad\; 75.0\% \text{ A} \\
40 \times 62.5\% &= 25.0\% \text{ B}
\end{aligned}
$$

The calculation at temperature T_3 has already been used to demonstrate the Lever Rule. When the system is cooled to T_4, the eutectic temperature, crystals of A, the eutectic liquid and an infinitesmal amount of B would be in equilibrium. If the temperature was an infinitesmal amount ABOVE T_4, the Lever Rule could be used to calculate the relative amounts of A and eutectic liquid in equilibrium:

$$\% \text{ Liquid } (30\% \text{ A, } 70\% \text{ B}) = \frac{25}{70} \times 100 = 35.7$$

$$\% \text{ A } (100\% \text{ A}) = \frac{45}{70} \times 100 = 64.3$$

If the temperature was an infinitesmal amount BELOW T_4, % A = 75, % B = 25, identical with the starting liquid composition (X). If temperature continues to decrease, the mechanical mixture of 75% A, and 25% B will continue to coexist.

The above analyses show that it is possible to make quantitative analysis JUST ABOVE and JUST BELOW the eutectic temperature. As one passes through the eutectic temperature, it is generally not possible to make a quantitative analysis of the amounts of eutectic liquid, crystalline A and crystalline B, because the amount of heat withdrawn and the heat capacities of the phases may not be known.

It should be understood that the isoplethal analyses can be made starting with solids at low temperatures and raising the temperature until the composition becomes completely liquid at temperatures above the liquidus curve. More often, the analyses are made starting in the liquid region and lowering the temperature until the system consists of solids.

The liquidus curves (ME and NE) in Figure 3.2 are solubility curves similar to solubility curves for salts in aqueous solution. At constant pressure, the solubility of salts vary with temperature and are USUALLY, but not always, more soluble in hot water than cold water. At a constant temperature a solubility or saturation limit is reached.

Similarly, in high viscosity borate, silicate and phosphate liquids, crystalline phases have definite equilibrium solubility or saturation limits which vary with temperature. In Figure 3.2, the liquidus ME represents the solubility limit for A and the liquidus NE represents the solubility limit for B, at various temperatures. At the eutectic temperature T_4, the eutectic composition represents the mutual saturation of the liquid by A and B crystals.

D. Polymorphism of Solids

A or B or both solids may exist in two or more polymorphic forms. The important point is the temperature at which inversions or transformations occur. If, for example, an inversion in A takes place at T_2, a horizontal line ranging from end member A to M_2 would become a part of the phase diagram. At temperatures above T_2, the high temperature (α) form of A would coexist with liquid and below T_2, the lower temperature (β) form of A would coexist with liquid.

If the inversion in A occurred below the eutectic temperature T_4, a horizontal line would run between the A and B end members and B would coexist with higher temperature (α) A above the inversion temperature. Below the inversion temperature solid B would coexist with lower temperature (β) A. If no solid solution exists in a binary system, the temperature of the polymorphic inversion in an end member or intermediate compound is not affected by the presence of the other end member or another compound.

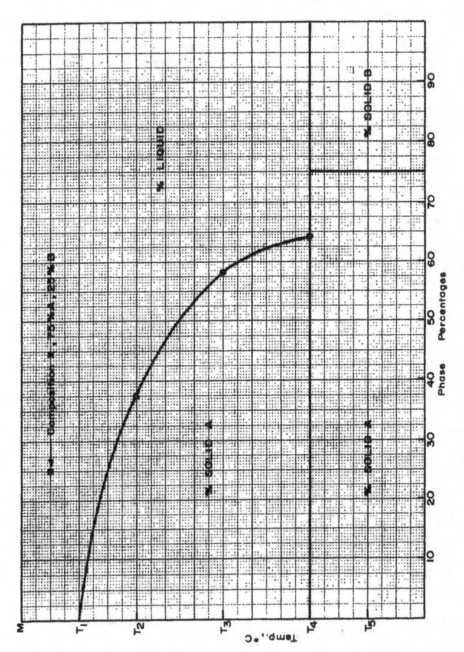

Figure 3.3.   The Phase Analysis Diagram

E. Phase Analysis Diagrams

The isoplethal analysis can lead to a phase analysis diagram where the percentage of existing phases is plotted against temperature, as in Figure 3.3. Using composition X from Figure 3.2, the amounts of A in equilibrium with liquid at temperatures T_2, T_3 and T_4 are 37.5%, 58.33% and 64.3%, respectively. After disappearance of the liquid below the eutectic, the amounts of solid A and solid B are 75% and 25%, respectively. Thus, Figure 3.3 gives the relative percentages of phases at all temperatures between those points where the composition is 100% liquid to those temperatures below the eutectic where the composition is 100% solid (% A + % B). Such a diagram may be constructed for any composition between 100% A and 100% B, using the type of calculations given in the previous sections 2 and 3. Simple isoplethal analyses and phase analysis diagrams can be made for the eutectic liquid composition (30% A, 70% B) and liquid composition Y (17.5% A, 82.5% B) in Figure 3.2 to increase ones understanding of the Lever Rule and the construction of the phase analysis diagram.

F. Microstructure Relations

When crystals are produced by cooling liquid compositions under equilibrium or non-equilibrium conditions to the point (temperature) where the liquid disappears, important factors are the rate of growth, the size and the morphology or form of the crystals. In general, the combination of liquid and one or more crystalline phases are two or more crystalline phases in binary systems is part of the very large subject of "microstructures" which is beyond the scope of this text. The book by Fulrath and Pask (1) is an excellent example of the importance of this topic to many areas of glass and ceramics.

The microstructure of liquid and crystalline phases or combinations of crystalline phases is extremely important during the experimental construction of a binary phase diagram, especially at certain critical temperatures or compositions.

III. INTERMEDIATE COMPOUNDS

In binary systems, compounds are composed of various ratios of the two end members or the basic components of the system. The thermal behavior of intermediate compounds is of three basic types, congruent melting, incongruent melting or dissociation.

A. Congruent Melting

In discussing the simple eutectic system in section B1, it was assumed that the end members melted congruently; that is, the crystalline phases A or B coexisted with a liquid of composition A or B at constant temperature (the melting temperature) at constant (usually atmospheric) pressure. Figure 3.4 shows an intermediate compound, A_2B_3 which also melts congruently. That is, at temperature T_4, compound A_2B_3 coexists at one atmosphere with a liquid (40% A, 60% B) of identical composition. Eutectic E_1 (P = 3, F = 0) involves coexistence of A, A_2B_3 and liquid (60% A, 40% B) at temperature T_8 and eutectic E_2 involves coexistence of B, A_2B_3 and liquid (30% A, 70% B) at temperature T_6. Obviously, the congruently melting compound A_2B_3 divides the system A-B into two smaller simple eutectic binary systems, A-A_2B_3 and B-A_2B_3. Isoplethal analyses of liquid compositions X (80% A, 20% B), Y (50% A, 50% B), and Z (10% A, 90% B) and construction of the corresponding phase analysis diagrams would strengthen the understanding of the examples in previous sections B3 and B5, using the simple binary eutectic system.

For example, for composition X at T_1 : an infinitesmal amount of A is in equilibrium with liquid of composition 80% A, 20% B,

at T_2 : 20% A
    80% liquid (75% A, 25% B)
at T_4 : 33.33% A
    66.67% liquid (70% A, 30% B)
at T_9 : 66.67% A
    33.33% A_2B_3

By making calculations for composition X at temperatures T_1 to T_9, sufficient data for the construction of a "phase analysis" diagram would be accumulated. (Temperature v.s. Phase Percentages). For every calculation, it should be remembered that the percentage of A and B in the coexisting phases should equal 80% A and 20% B.

For composition Y, analyses can be made between T_5 and T_9 and for composition Z, between T_3 and T_9.

B. Incongruent Melting

The definition of incongruent melting: At a specified pressure, it is the temperature at which one solid phase transforms to another solid phase and a liquid phase both of different chemical compositions than the original composition.

The fundamental relationships are shown in Figure 3.5. Point P is called a peritectic point and represents an invariant situation (P = 3, F = 0) similar to a eutectic relationship where three phases coexist at constant temperature and pressure. In this particular diagram, the compound A_3B_7 (30% A, 70% B) decomposes at temperature T_4 to solid B and a liquid of composition 50% A, 50% B. As temperatures increase, solid B coexists with liquid until complete melting

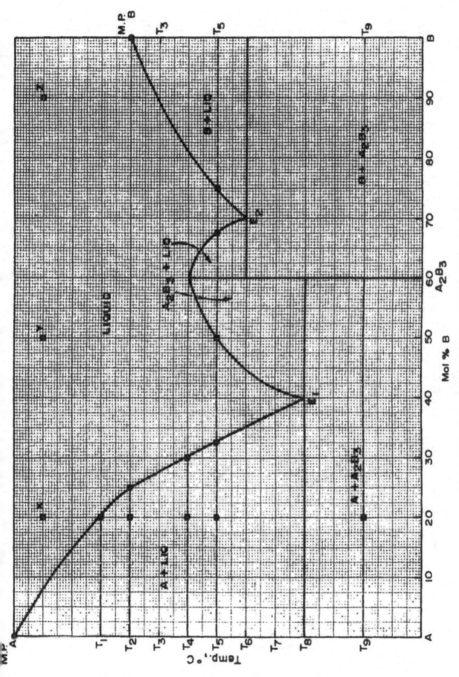

Figure 3.4. Congruently Melting Intermediate Compound

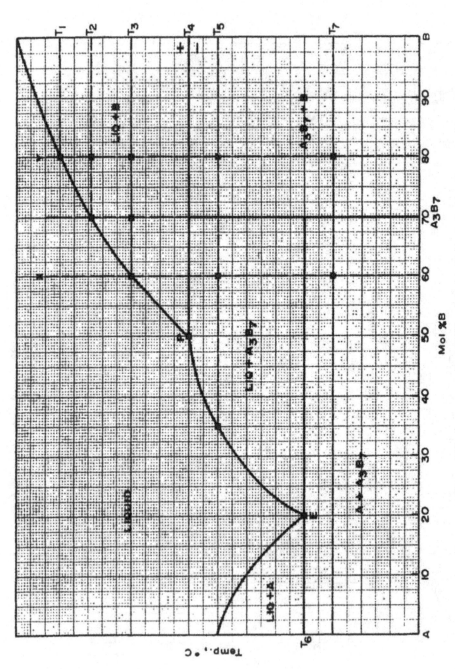

Figure 3.5. Incongruently Melting Intermediate Compound

occurs at T_2. An isoplethal analysis of liquid composition 30% A, 70% B involves liquid, liquid + B (from temperature T_2 to T_4) and finally, after the peritectic reaction at T_4 (liquid P + B $\rightleftarrows$ A_3B_7), the compound A_3B_7. For example, at T_3, 25% B is in equilibrium with 75% of liquid (40% A, 60% B). At temperature T_4^+, 40% B is in equilibrium with 60% peritectic liquid (50% A, 50% b) and at T_4^- liquid and B have reacted to form 100% A_3B_7.

Isoplethal analyses of liquid compositions X and Y contribute to a further understanding of the incongruent melting type of diagram. For example, liquid X when cooled is in equilibrium with an infinitesmal amount of solid B at T_3. From T_3 to T_4, liquids of various compositions from X to P will be in equilibrium with B. After the peritectic reaction, at T_5, 4 parts of liquid (65% A, 35% B) will coexist with 10 parts of A_3B_7. After the eutectic reaction at T_6, at T_7, a 100% solid mixture of A (14.3%) and B (85.7%) will coexist.

A liquid of composition Y when cooled will coexist with an infinitesmal amount of B at T_1. From T_1 to T_4, liquids of various compositions from Y to P will coexist with B. After the peritectic reaction, at T_5, 33.33% B will coexist with 66.67% A_3B_7, likewise at T_7.

From the standpoint of "refractoriness", all compositions to the right of compound A_3B_7, from A_3B_7 to B, are much superior to those to the left of A_3B_7, since no liquid forms until the peritectic temperature is reached. In compositions to the left of A_3B_7, liquid will form (under equilibrium conditions) at the eutectic temperature T_6.

C. Dissociation

Intermediate compounds may decompose on heating or cooling, forming two other crystalline phases. Either end member component phases or other intermediate phases may be involved in the dissociation of a particular intermediate compound. Three types of dissociations are possible:

1. Lower temperature limit of stability
2. Upper temperature limit of stability
3. Upper and lower temperature limit of stability

Figure 3.6 demonstrates type (a). The compound A_7B_3 dissociates to A + B at temperature T_4. It is assumed (for simplicity) in this particular diagram that A_7B_3 melts congruently at T_1. It could also melt incongruently. The important invariant reaction at T_4 is A + B $\rightleftarrows$ A_7B_3. (P = 3, F = 0). This point (A_7B_3, T_4) is an invariant point similar to E_1 and E_2 where three phases coexist at constant temperature. If crystalline A_7B_3 is cooled, it is assumed that sufficient time will be taken to allow the compound to decompose to 70% A and 30% B at T_4. If a mixture of 70% A and 30% B is heated to T_4, it is assumed that sufficient time

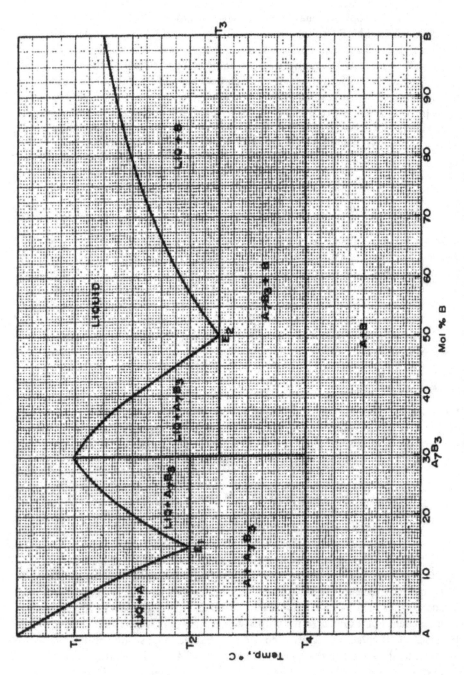

Figure 3.6.   Lower Temperature Limit of Stability of Intermediate Compound

will be taken under equilibrium conditions to permit the formation of 100% A_7B_3. Many liquid compositions between A and B could be chosen for quantitative isoplethal analyses and construction of "phase analysis" diagrams.

Figure 3.7 demonstrates type (b). A_3B_2 has an upper temperature limit of existence at T_2 and dissociates to αA + B. The diagram also demonstrates the polymorphism of end member A. At T_3, there is an $\alpha \rightleftharpoons \beta$ inversion in A which creates a horizontal boundary line between A and A_3B_2 because the inversion temperature of A is not affected by the presence of A_3B_2.

If an isoplethal analysis of a liquid composition 60% A, 40% B is made, crystals αA will coexist with liquids ranging from X to E at temperatures from T_0 to T_1. At T_1, following the eutectic reaction, 60% αA will coexist with 40% B. At T_2, under equilibrium conditions, αA will react with B to form A_3B_2.

Finally, as illustrated in Figure 3.8, an intermediate compound may have both lower and upper limits of stability, a much rarer case than Figures 3.6 and 3.7. An isoplethal analysis of a liquid containing 40% A, 60% B would involve coexistence of liquids (between X and E) with solid B from T_0 to T_1, coexistence of A and B from T_1 to T_2, existence of A_2B_3 from T_2 to T_3, dissociation of A_3B_2 to 40% A and 60% B at T_3, and finally coexistence of 40% A and 60% B below T_3.

D. Polymorphism, Inversions, Intermediate Compound Formation

It is important to note several features connected with polymorphic changes in crystals at this point. The Greek letters $\alpha, \beta, \gamma, \Delta, \kappa, \theta$, etc, are usually used to designate various polymorphic forms of crystalline compounds (or solid solutions). There has been no fixed convention about the use of these symbols for low, intermediate or high temperature forms of a particular crystal. That is, the letter α is often used to designate the highest temperature form of a crystal (α Ca_2SiO_4), but at other times it may signify the lowest temperature form (α - quartz, α - spodumene). One must simply know by experience whether the Greek letter used means the low, intermediate or high temperature form.

Secondly, it is important to realize that, in binary systems without solid solution, an inversion in an end member or intermediate compound takes place at the same temperature as it would if it was not in the presence of a second phase. For example, the $\alpha \rightleftharpoons \beta$ quartz inversion takes place at 573°C. In all binary systems involving the quartz phase, where the second component does not enter into solid solution in quartz, the inversion will take place at precisely 573°.

Finally, the phase diagram never offers any information on the kinetics of polymorphic changes. There are examples of displacive inversions in binary systems where the rate of inversion is virtually instantaneous regardless of whether the temperature is increasing or decreasing. Other inversions may be very sluggish

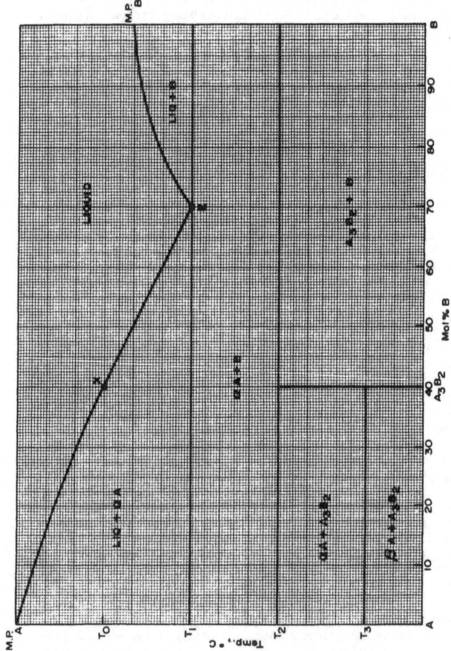

Figure 3.7. Upper Temperature Limit of Stability of Intermediate Compound

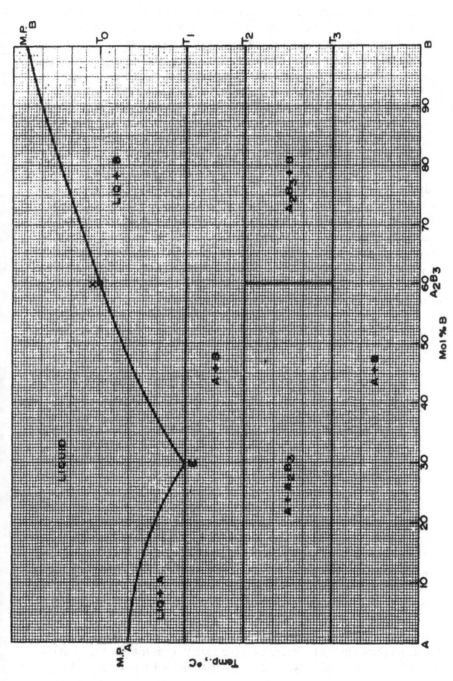

Figure 3.8. Intermediate Compound with Upper and Lower Temperature Limits of Stability

as the temperature is raised, but virtually instantaneous with decrease in temperature. Others may be instantaneous on heating or more or less sluggish on cooling. There are cases where the transformation is sluggish in both directions.

In older texts the words "enantiotropic" meant "reversible" inversion and "monotropic" meant "non-reversible" inversion or one which went in only one direction with temperature. With the advent of high pressure technology as a common technique in the investigation of phase equilibria, the word "monotropic" is of doubtful significance. Many inversions which were once thought to go in the "up-direction" only have now been shown to be reversible with the simultaneous application of temperature and pressure. This is especially true of many low temperature, high density forms of natural minerals which were once thought to undergo "monotropic" inversions. The laboratory synthesis of α-eucryptite, α-spodumene and petalite is a good example of the use of pressure to reverse the inversions which were once thought to be "monotropic". Many other examples of mineral structures and chemical compounds could be cited to illustrate the usefulness of high and ultra-high pressure techniques in reversing inversions which were once thought to be non-reversible. As previously mentioned, the kinetics and mechanisms of polymorphic changes is a separate large field of material science which is beyond the scope of this text.

Some very fundamental questions may be raised about the various kinds of "intermediate compound" type diagrams which exist between two metals, two oxides, two fluorides, two sulfides, etc. For example,

1. Is there any way to account for the number and stoichiometry of the compounds which exist between two end members?
2. Can the lowering of the liquidus temperature of an end member or intermediate compound by another end member or intermediate compound be explained quantitatively? (Freezing point depression)
3. What causes a compound to melt congruently, incongruently or to dissociate at a particular temperature?

The last two questions are beyond the scope of this text, but the first question may be dealt with in the following way with respect to oxide systems. In 1942, Dietzel (2) published a paper which ranked the elements according to their field strengths as ceramic oxides. A summary of the concept is shown in Table 3.1. The field strength for silicon is the valence of Si^{4+} divided by the square of the radius, $4/(0.39)^2 = 26$. The electrical potential for Si^{4+} is the valence divided by the radius $4/0.39 = 10$. The field strength at "clearance of anion" is simply the valence divided by the sum of the radii of Si^{4+} and 0^-, corrected for the actual coordination of silicon (4) = $4/(1.60)^2 = 1.57$. Using the last column shown in the table, Dietzels rule simply states that for binary oxide systems, if the difference between the numbers in this column is greater than

Element	Valency z	Radius of Ions KZ = 6 r	Ion Clearance a for KZ = 6	Actual Coordination Number KZ	a for Actual KZ	Field Strength z/r²	Elec. Potential z/r	Field Strength at Clearance of Anion z/a²
K	1	1.33	2.65	8	2.77	0.56	0.75	0.13
Na	1	0.98	2.30	6	2.30	1.04	1.02	0.19
Li	1	0.78	2.10	6	2.10	1.65	1.28	0.23
Ba	2	1.43	2.75	8	2.86	0.98	1.40	0.245
Pb	2	1.32	2.64	8	2.74	1.15	1.51	0.27
Sr	2	1.27	2.59	8	2.69	1.24	1.57	0.28
Ca	2	1.06	2.38	8	2.48	1.78	1.88	0.33
Mn	2	0.91	2.23	6	2.23	2.42	2.2	0.40
Fe	2	0.83	2.15	6	2.15	2.90	2.4	0.43
Zn	2	0.83	2.15	4	2.03	2.90	2.4	0.49
Mg	2	0.78	2.10	6	2.10	3.3	2.6	0.45
				4	1.96			0.53
Zr	4	0.87	2.19	8	2.28	5.3	4.6	0.77
Be	2	0.34	1.66	4	1.53	17.4	5.9	0.86
Fe	3	0.67	1.99	6	1.99	6.7	4.5	0.76
				4	1.88			0.85
Al	3	0.57	1.89	6	1.89	9.2	5.3	0.84
				4	1.77			0.96
Ti	4	0.64	1.96	6	1.96	9.8	6.3	1.04
B	3	0.20	1.52	4	1.50	75.	15.	1.34
				3	1.36			1.63
Si	4	0.39	1.71	4	1.60	26.	10.	1.57
P	5	0.34	1.66	4	1.55	44.	15.	2.1

(r + 1.32)

approximately 0.5, is it likely that compounds between the two oxides will form. For example, for SiO_2 and BaO, the difference is $1.57 - 0.245 \simeq 1.33 \gg$ 0.5 and compounds should form, according to Dietzel. Actually, six compounds have been found between BaO and SiO_2. It should be pointed out that exceptions to this rule can be found in binary oxide systems and that it does not predict the number or stoichiometry of the compounds, only that compounds are LIKELY to exist if Δ is greater than 0.5.

IV. ISOTHERMAL ANALYSIS; ISOTHERMAL EVAPORATION

Much emphasis has been placed on the isoplethal analysis up to this point; that is, the equilibrium cooling of a liquid of a certain bulk composition or the equilibrium heating of a solid phase composition (a stoichiometric compound or mixture of solids) to the point where complete liquifaction occurs.

In the beginning of the discussion in Section 2C, it was mentioned that an isothermal (constant temperature) analysis is the second possibility in binary systems (corresponding to isothermal and isobaric analyses in one-component systems).

For example, in Figures 3.2, 3.4, 3.5, 3.6, 3.7 and 3.8, isothermal analyses can be made at any constant temperature whatever in the diagrams. However, it is useful and appropriate to use Figure 3.4 to illustrate a specific isothermal analysis. Starting at temperature T_5 and end member A, various amounts of A and a liquid of composition 67.5% A and 32.5% B will coexist until sufficient B has been added to produce 100% liquid. At that point, as B is added to A, only liquid will exist until composition 50 A, 50 B is reached. Between 50 and 60% B, various ratios of A_2B_3 and liquid of composition 50 A, 50 B will coexist. At 60% B, the compound A_2B_3 exists as a single phase at T_5. Between A_2B_3 and a liquid of composition 32.5% A, 67.5% B, various ratios of these two phases will coexist. Single phase liquid exists between 67.5% B and 75% B, and finally at 75% B, various ratios of B and liquid (25% A, 75% B) will coexist until the composition becomes 100% B.

Similar analyses can be made for all of the specific temperatures listed in the six diagrams mentioned above.

Another special kind of isothermal event may take place in a binary system, but it occurs with less frequency and it is of less importance than the isoplethal analysis or the "standard" isothermal analyses which was just discussed. Instead of holding the bulk composition constant while temperature varies, the temperature is held constant while one component evaporates. Isothermal evaporation is best illustrated with the aqueous system $Fe_2Cl_6 - H_2O$ shown in Figure 3.9.

Starting with dilute solution x, as H_2O is evaporated at $30°$, the solution would crystallize to 100% $Fe_2Cl_6 \cdot 12H_2O$, become completely liquid, crystallize

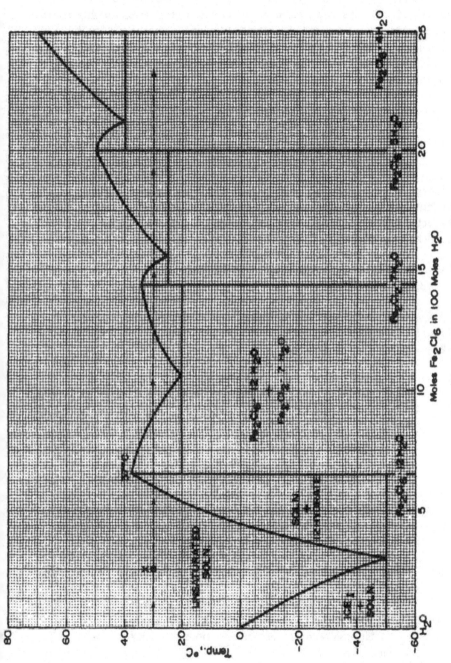

Figure 3.9. The Aqueous System Fe_2Cl_6–H_2O

to $Fe_2Cl_6 \cdot 7H_2O$, become completely liquid, and finally solidify to 100% $Fe_2Cl_6 \cdot 5H_2O$. This binary diagram not only illustrates the relationship between aqueous binary systems and high temperature binary ceramic systems, but also provides an example of how water soluble salts lower the melting point of ICE_1 and make winter driving of cars and trucks a somewhat safer event when salt is distributed over icy roads.

This kind of process could occur in oxide systems with one volatile component such as PbO, CdO, CuO, HgO, B_2O_3, P_2O_5, etc.

During a study of the relations among double oxides of trivalent elements, Keith and Roy (3) found that B_2O_3 evaporated at 1000°C from $CrBO_3$ after 7 days to give hexagonal grains of Cr_2O_3 about 0.1 mm in diameter. After a few minutes at 1700°C, $InBO_3$ lost B_2O_3 to yield well formed yellowish brown cubes of In_2O_3. Isothermal evaporation could thus be adapted to the growth of relatively large crystals of refractory oxides or compounds, using a carefully selected volatile oxide for the particular temperature level and kind of process which was involved.

V. METASTABLE RELATIONSHIPS IN SIMPLE BINARY SYSTEMS

A few metastable, non-equilibrium situations which may occur in simple binary systems are illustrated in Figure 3.10 by the use of dashed lines. When dashed lines are seen in a phase diagram, they usually mean

1. That the author is trying to show a metastable condition or,
2. that the author is uncertain about his data and therefore is reluctant to show the equilibrium boundary lines as solid lines. The dashed lines in this case are equilibrium boundary lines, but tentative or uncertain due to insufficient data.

Non-equilibrium events usually arise due to heating or cooling a system more rapidly than required by true thermodynamic equilibrium. For example, in Figure 3.10, the compound α AB_4, which ordinarily melts incongruently at T_3, if heated very rapidly (superheated) may melt congruently at T_1, leading to a metastable eutectic relationship between α AB_4 and B at T_2.

Under equilibrium conditions, the compounds β AB_4 transforms to α AB_4 at T_8. If heating is very fast, the metastable β to α inversion may not take place until some higher temperature, say T_7.

Finally, on the left hand side of the diagram, a metastable eutectic relationship is shown between end member A and AB at temperature T_6. The true equilibrium relationships involve eutectic reactions between A and A_7B_3 at T_3 and A_7B_3 and AB at T_4.

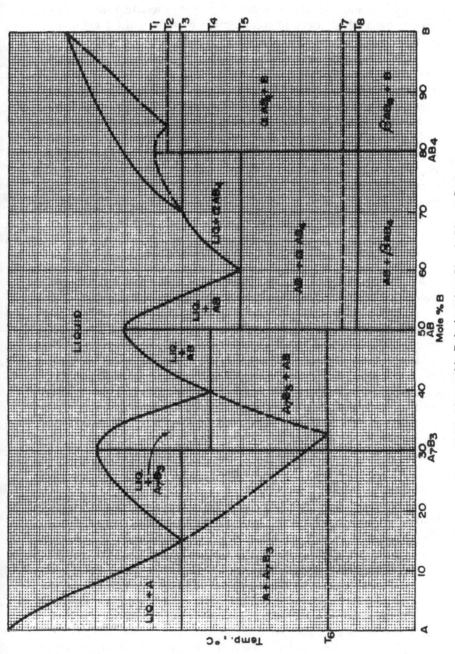

Figure 3.10. Metastable Relations in a Simple Binary System

Many other metastable conditions are possible, especially when three co-existing phases are involved during eutectictic and dissociation relationships. However, metastability can also occur during inversions, melting and other situations where only one or two phases are present.

Tables 1 and 2 in Appendix A provide a convenient list of real systems from "Phase Diagrams for Ceramists" (1964, 1969, 1975, 1981 Editions) which can be used for class discussions on simple binary eutectic systems and binary systems with intermediate compounds which melt congruently, melt incongruently or dissociate. It should be noted that several systems in Table II have a combination of two or three types of intermediate compounds. For example the system $Li_2O-B_2O_3$ (p. 91, Figure 180, 1964 edition) has congruently and incongruently melting compounds and three dissociating compounds with lower temperature limits of stability. Many of the diagrams have illustrations of phase transitions in end members or intermediate compounds and a few have some indications of metastable relationships.

VI. COMPLETE OR CONTINUOUS SOLID SOLUTION

A solid solution is defined as a single crystalline phase which may vary in composition within finite limits without the appearance of an additional phase. This definition is satisfactory, but many illustrations have to be made to understand the complex nature of solid solutions. Some phase equilibrium specialists feel that there could never be a simple binary eutectic system as shown in Figure 3.2 because B may be soluble in A to the extent of say 10^{-2} to 10^{-8} molar percent, or vice versa. The same comment may be applied to all subsequent diagrams which contain intermediate compounds, as discussed in section III. However, from a practical standpoint, solid solution of one phase in another usually means 10^{-1} to 100%. This statement is often based on the experimental ability to detect solid solubility to the extent of 0.1 to 1%, or even to as much as 3–5%. It should be emphasized that the figures mentioned above are highly dependent upon the chemical nature of the binary system and the ability to experimentally detect minor amounts of solid solubility.

With respect to the "ordinary" types of solid solutions which are involved in oxide, silicate, fluoride, sulfide, etc. diagrams the solubility may be of two basic types:

1. Substitional solid solution
   The solute atoms (cations or anions) fill lattice positions normally occupied by the solvent atoms. The solute atoms "substitute" for solvent atoms.
2. Interstitial solid solution
   The solute atoms fit into voids or interstices which are present in the solvent crystal structure.

In many oxide, silicate, fluoride, sulfide or other crystal structures involved in ceramic systems, a combination of substitutional and interstitial solid solution takes place, but not as frequently as types (a) and (b).

Solid solubility is obviously related to atomic or ionic sizes of elements, cations and anions and many years ago Hume-Rothery developed a "15% rule" which indicated that solid solution is unlikely if atomic size of elements (metals) differ by more than 15%. The rule also roughly applies to ionic crystal systems, but there are many exceptions. In general, high temperature favors solid solubility or increased compatibility and there are many examples of systems which are completely soluble at high temperatures, but which "unmix" or "exsolve" at lower temperatures. Pressure also affects the extent of solid solution. At this point, a recommendation should be made to the many textbooks, reference books, and papers on basic crystal chemistry.

Three types of diagrams involving complete solid solution are possible:

1. No temperature maximum or minimum in the liquidus curve
2. Maximum in liquidus
3. Minimum in liquidus

A. No Temperature Maximum or Minimum

i. Equilibrium Crystallization

Figure 3.11 illustrates the most simple type of solid solution diagram where end members A and B melt congruently, are completely intersoluble and where there is no maximum or minimum in the liquidus curve. The solidus curve between T_A and T_B creates an area where liquid and solid solutions, BOTH of variable composition, coexist and gives rise to the need to show how different the isoplethal analysis is, relative to all previous systems where no solid solution was present and all solid phases were of constant composition.

The equilibrium crystallization path of liquid X (70% A, 30% B) is as follows:

1. T_1, an infinitesmal amount of a solid solution containing 36.75% A and 63.25% B in equilibrium with original liquid.

2. T_2, 66.67% liquid (80% A, 20% B) in equilibrium with 33.33% solid solution (50% A, 50% B).

Note that the distribution of A and B is as follows:

% A in liquid = 0.6667 X 80 = 53.34
  % A in S.S. = 0.3333 X 50 = <u>16.66</u>

$$70.00\% \text{ A}$$

% B in liquid = 0.6667 X 20 = 13.33 +
  % B in S.S. = 0.3333 X 50 = <u>16.66</u> +

$$29.99 +\% \text{ B}$$

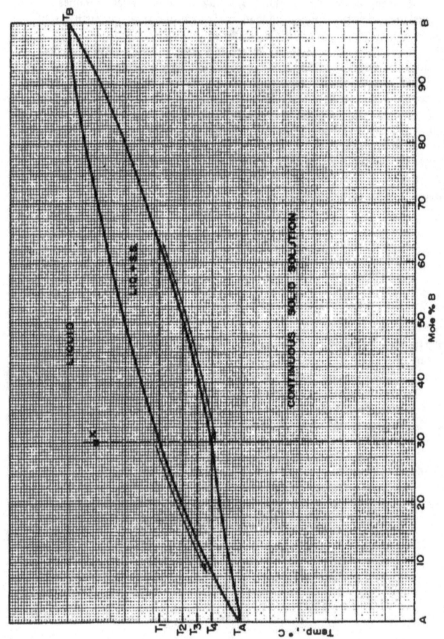

Figure 3.11. Continuous Solid Solution; No Maximum or Minimum in Liquids

3. T_3, 22/54 parts, 40.7% of liquid (86% A, 14% B) in equilibrium with 32/54 parts (59.3%) of S.S. (59% A, 41% B).

4. Finally, at T_4, as heat is withdrawn from the system, an infinitesmal amount of liquid of composition 91.5% A and 8.5% B will coexist with a solid solution containing (essentially) 70% A and 30% B. Below T_4, the single phase 70A, 30 B S.S. exists.

On heating a solid solution containing 70% A and 30% B under equilibrium conditions, it is only necessary to reverse the procedure until the composition is 100% liquid, containing 70% A and 30% B. Obviously the liquid composition will always follow the liquidus curve and the solid solution will always follow the solidus curve, under EQUILIBRIUM conditions.

It should be obvious that, in actual practice in metal, salt, oxide or silicate systems, equilibrium crystallization will rarely be achieved during cooling due to the generally slow diffusion in solids, relative to the rapid diffusion of ions or atoms in liquids. In other words perfect equilibrium crystallization demands that the solid solution change its composition in infinitesmal increments with each infinitesmal amount of temperature decrease. Rates of cooling in the laboratory or industrial processes are relatively rapid and perfect equilibrium crystallization has little or no chance to occur and as a result, successive layers of solid solutions are usually deposited on a nucleous, giving rise to "coring" , and an onion-skin effect.

The effect of coring can be overcome by long term "tempering", "normal-izing", or annealing at temperatures which are sufficiently high to permit rapid diffusion in the solid state.

ii. Fractional Crystallization

This process is similar to fractional distillation between liquid and vapor, as far as the Phase Rule is concerned.

Again referring to composition X in Figure 3.11, visualize a process where-by the composition is equilibrated at temperature T_2. Now, by some appropriate experimental technique, such as sieving out or separating the crystals by means of a platinum gauze, one completely and perfectly separates the crystals (50% A and 50% B) from the liquid which contains 80% A and 20% B. One now takes each of these "fractions" and equilibrates them at some convenient temperature in the two phase region (say at temperature T_1 for the 50-50 fraction and at temperature T_3 for the 80-20 fraction.

These two new fractions are now separated to create four fractions which are then further fractionated by holding at appropriate temperatures in the two phase region and separating to give eight fractions. After an infinite number of fractionations, the net result would be a complete separation of the original composition (70 A, 30 B) into the end members A and B. Such a tedious process was once the basis for the separation and purification of rare earth salts, now done by the rapid method of selective capture by ion exchange columns.

It is important to understand the fractionation process as applied to a simple binary system as a basis for understanding fractional crystallization processes in ternary systems.

B. Continuous Solid Solutions with a Maximum or Minimum in the Liquidus Curve

Two derivative cases of continuous solid solution are shown in Figures 3.12 and 3.13.

The maximum in the liquidus shown in Figure 3.12 corresponds to the "azeo-tropic" mixture between ethyl alcohol and water when liquid and vapor phases are involved.

The maximum in the liquidus essentially separates the system into two portions, but it must be emphasized that the solid solution region is continuous, as is the liquid region.

The Phase Rule may appear to be disobeyed at the temperature of the maximum of the liquidus, but an identity restriction is invoked at this point and if one applies the equation $F = 3 - P$, at constant pressure, a liquid and two solid solutions of identical composition are in equilibrium, making the point invariant. ($F = 0$) These maximum and minimum points are defined as "indifferent" points where two phases become identical and the system (apparently) loses one degree of freedom.

The same statements apply to Figure 3.13, where a minimum in the liquidus is shown. Solidus or liquidus boundary lines come into and leave the maximum or minimum point tangentially as shown in Figures 3.12 and 3.13.

Crystallization paths through the L + SS regions are similar to the one already discussed for the case where there is no maximum or minimum in the liquidus. If a liquid composition is located exactly at the maximum or minimum, it undergoes complete crystallization to a solid solution of the same composition, just like a congruently melting compound. Conversely, a solid solution with the maximum or minimum composition melts to a liquid of the same composition when heated.

It should be noted that perfect fractional crystallization processes in systems like these will yield an end-member and either the maximum or minimum composition, as the case may be, but never the two end members.

C. Unmixing of Solid Solutions, "Exsolution"

There are a number of cases where complete and continuous solid solution exists between two end-members at high temperatures, but UNMIXING occurs at some lower temperature when the system is cooled. Such a case is shown in Figure 3.14, where, at temperature T_1, the originally homogeneous solid solution undergoes decomposition or separation into two solid solutions, one rich in A and the

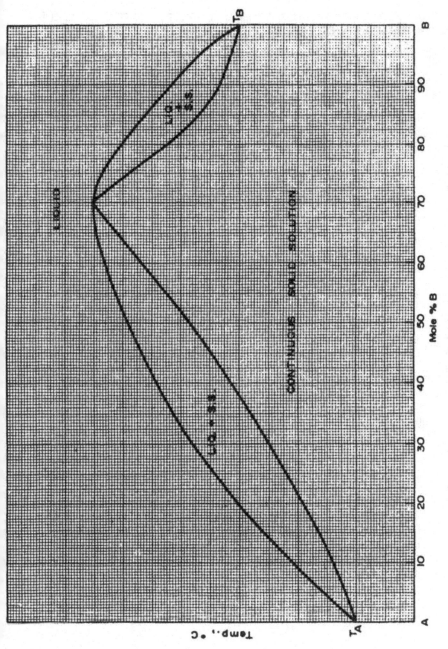

Figure 3.12. Continuous Solid Solution; Maximum in Liquids

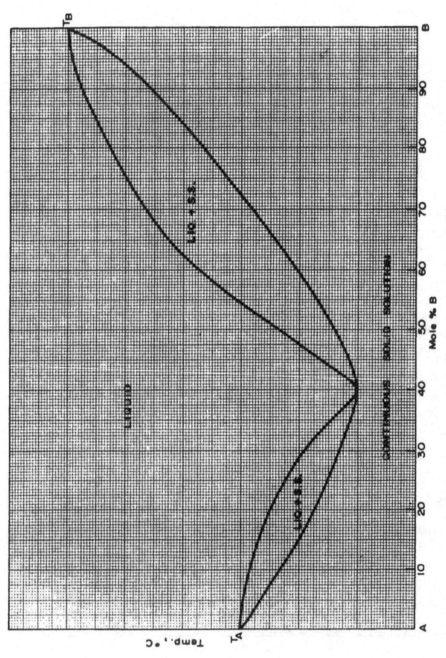

Figure 3.13. Continuous Solid Solution; Minimum in Liquids

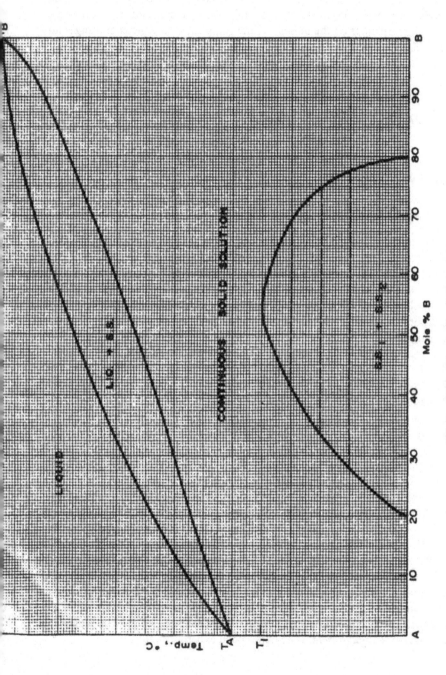

Figure 3.14. Unmixing of Solid Solutions

other rich in the B component. This is called an "immiscibility dome" and the tie lines drawn within the dome or boundary line, at constant temperature give, at their extremities on the boundary, the compositions of the coexisting solid solutions.

Some representative cases of continuous solid solution or unmixing in real systems are listed in Table 3 in Appendix A.

VII. PARTIAL SOLID SOLUTION

There are two basic cases of partial solid solubility, the eutectic and peritectic types as shown in Figures 3.15 and 3.16, respectively.

A. Eutectic Type

In Figure 3.15, a liquid of composition X (85 A, 15 B) will cool to the liquidus temperature (T_1) and precipitate an infinitesmal amount of SS_A (95 A, 5 B). It will, under equilibrium conditions, pass through the $L + SS_A$ region, crystallizing solid solutions which become richer in B as the liquid also becomes richer in B. Just below T_2, all the liquid disappears as heat is withdrawn and the final solid solution has the same composition (85A, 15 B) as the original liquid. Continued cooling merely brings the solid solution to room temperature, or below.

Composition Y (liquid, 50 A, 50 B) will cool until the liquidus temperature T_3 is reached, at which point it will precipitate an infinitesmal amount of solid solution containing 75A and 25 B. As the system cools under equilibrium conditions, both liquid and solid solution become richer in B. At temperature T_E, the last race of liquid of eutectic composition 40 A, 60 B disappears and the system becomes all solid, containing 25% of a solid solution of composition 20 A, 80 B and 75% of a solid solution of composition 60 A, 40 B. A simple multiplication will show that the amounts of A and B in these solid solutions are the same as in the original liquid (50 A, 50 B).

As heat is withdrawn from the system, the composition of the two co-existing solid solutions will change as indicated by the extremities of the tie lines at temperatures T_4, T_5, and T_6. As indicated by the solid boundary lines, the solubility of A in SS_B and of B in SS_A becomes less and less as the temperature is lowered. This is the most common case in metal, salt and oxide systems, but as indicated by dashed lines O-b and O-c, the solubility of B in SS_A may remain the same or even increase as the temperature decreases. There are no general principles as yet for the prediction of solubility behavior in oxide and silicate systems. The increase, decrease or constancy of solid solubility with decrease (or increase) in temperature must be determined experimentally.

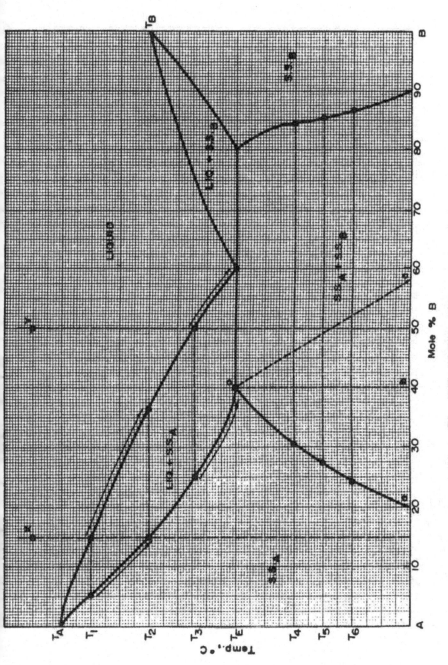

Figure 3.15. Eutectic Partial Solid Solution System

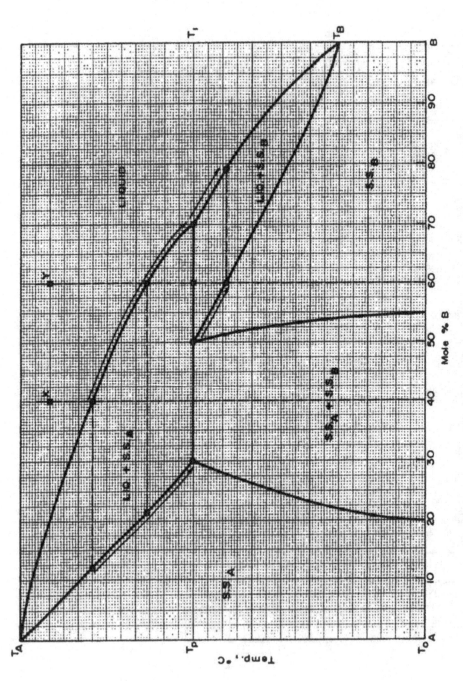

Figure 3.16. Peritectic Partial Solid Solution System

B. Peritectic Type

In Figure 3.16, a liquid of composition X (60 A, 40 B), will cool until it precipitates an infinitesmal amount of a solid solution containing (88A, 12 B) at the liquidus. As equilibrium cooling continues, the liquid and solid solution both become progressively richer in B as the peritectic temperature T_P is approached. At the peritectic temperature the liquid disappears and two solid solutions remain, 50% of a solid solution containing 50 A, 50 B and 50% of a second solid solution, containing 70 A, 30 B. As cooling continues, the two solid solutions gradually change composition as shown and at T_0, 3/7 parts of a solid solution containing 80 A, 20 B coexists with 4/7 parts of a solid solution containing 45 A and 55 B.

A melt of composition Y (40 A, 60 B) will cool to the liquidus temperature and precipitate an infinitesmal amount of a solid solution containing 78.5% A, 21.5% B. As cooling proceeds, the liquid and the solid solution will become richer in B and just above the peritectic temperature 25% of a solid solution containing 70 A and 30 B will coexist with a liquid containing 30 A and 70 B. As heat is withdrawn, SS_A disappears in favor of SS_B and just below T_p, 50% of liquid of composition 30 A and 70 B coexists with 50% of SS_B whose composition is 50 A, 50 B. Continued equilibrium cooling enriches the liquid and SS_B in B, and at T_1, the last trace of liquid disappears and solid solution SS_B (40 A, 60 B) remains. Further cooling merely withdraws heat from the solid solution.

It should be obvious that the major compositional changes which take place at the peritectic temperature for compositions X and Y would perhaps require times which are longer than those usually provided by laboratory or industrial cooling processes. If insufficient time is provided, there is ample opportunity for non-equilibrium situations to develop, even to the point where three phases would remain at room temperature (various assemblages of liquid, SS_A and SS_B, for example, where SS_A and SS_B would not necessarily have the composition they would ordinarily have at equilibrium).

VIII. PHASE TRANSITIONS IN THE END MEMBERS OR INTERMEDIATE SOLID SOLUTIONS; COMBINATIONS OF PARTIAL AND COMPLETE SOLID SOLUTION EXSOLUTION

So far, nothing has been said about the effect of polymorphism in end members or intermediate solid solutions on the configuration of the binary diagram, but this subject is of the utmost importance (in control of chemical and physical properties, for example). The occurrence of polymorphs causes the diagrams to become much more complex. Some of the most important cases which arise are shown in Figures 3.17, 3.18, 3.19, 3.20 and 3.21.

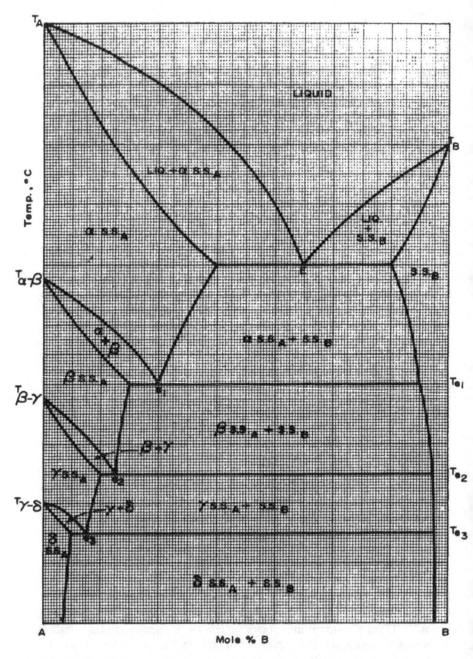

Figure 3.17. Polymorphism in Partial Solid Solution System (temperature decrease)

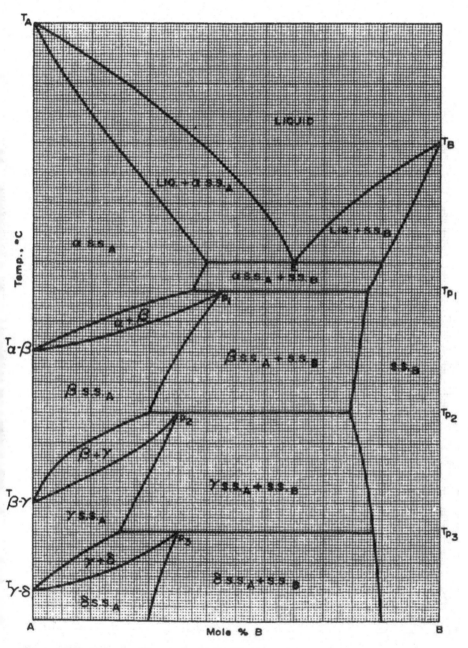

Figure 3.18. Polymorphism in Partial Solid Solution System (temperature increase)

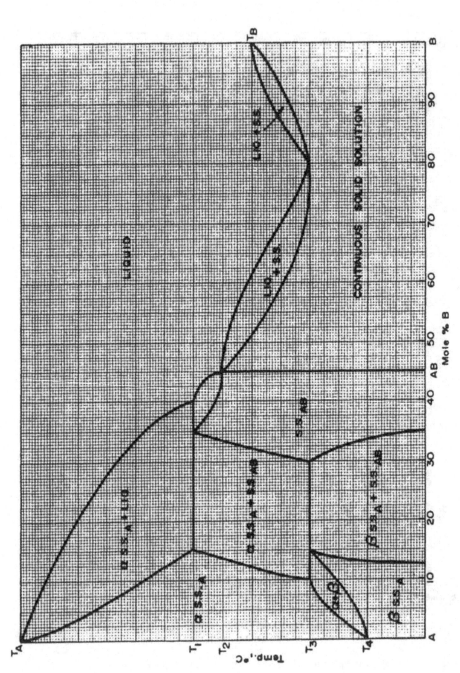

Figure 2.40. Combined Partial and Continuous Solid Solution System

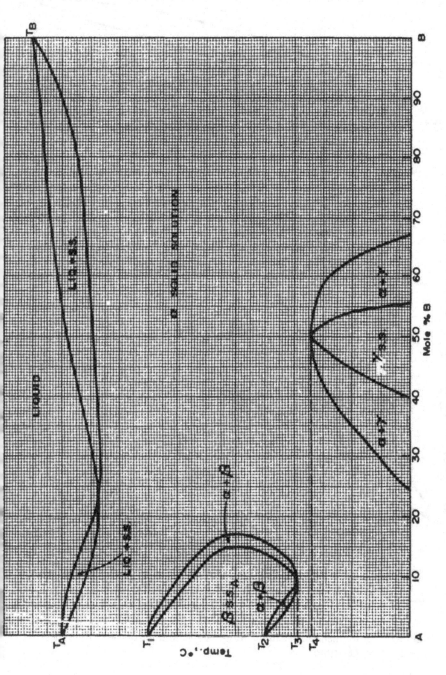

Figure 3.20. Combined Continuous Solid Solution and Exsolution

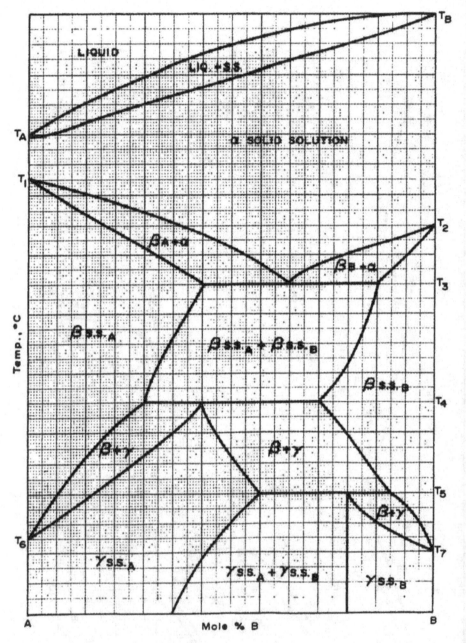

Figure 3.21. Combined Partial and Continuous Solid Solution with Eutectoid and Peritectoid Relations

In Figure 3.17, solid solution of B in the various polymorphs of A leads to a lowering of the transition temperature of the $\alpha \rightleftarrows \beta$, $\beta \rightleftarrows \gamma$ and $\gamma \rightleftarrows \delta$ inversions as shown. The inclusion of B in solid solution in the various forms of A often leads to a change in the rate of a particular inversion.

For example, in Figure 3.17, if the rate of the $\alpha \rightleftarrows \beta$ inversion is slowed down to such an extent that, on cooling, the α SS_A solid solution can be preserved at room temperature, the α SS_A solid solution phase is said to be "stabilized". Similar "stabilizations" of the β and γ solid solutions phases would be possible. Whereas the point E is referred to as a eutectic, points e_1, e_2 and e_3 are called eutectoids (three solid phases coexisting in equilibrium). It is now clear that the inversion of α and β solid solution phases would take place over a temperature interval from $T_{\alpha-\beta}$ to e_1, the β and γ solid solution phases between $T_{\beta-\gamma}$ and e_2 and the γ and δ solid solution phases betweem $T_{\gamma-\delta}$ to e_3. The rate of various transitions in solid solution phases may vary markedly in a given system or between systems and the phase diagram alone gives no data on rates. This kind of information must be obtained by reading the original paper which produced the phase diagram.

Figure 3.18 shows a solid solution system where transition temperatures are increased by inclusion of component B in solid solution in the various polymorphs of A. Large single phase regions of α SS_A, β SS_A, γ SS_A and δ SS_A exist separated by two phase regions containing two polymorphic forms of the solid solution. The points p_1, p_2, and p_3 are called peritectoids (three solid phases in equilibrium) and a characteristic peritectoid temperature is associated with each invariant point.

Figures 3.19, 3.20 and 3.21 show some of the possible variations in the solid solution diagram. Figure 3.19 shows a combination of a partial solid solution system between A and AB and a complete solid solution between AB and B with a minimum in the liquidus. The compound AB melts congruently, but takes A into solid solution (partially) and B into solution, completely.

Figure 3.20 shows a solid solution system with a minimum at the liquidus, a region of β SS_A involving transitions in pure A at T_1 and T_2 and a region of low temperature γ SS. The γ SS varies in the amounts of A and B contained in it with temperature and it is unstable above T_4. This type of diagram is encountered in several metal systems where A is metallic iron.

Figure 3.21 shows a combination of complete solid solution, eutectoid reaction (T_3) and peritectoid reaction (T_4) with transitions in both end members, at T_1 and T_6 for A and T_2 and T_7 for B.

IX. COMPLEX SOLID SOLUTION DIAGRAMS

Although Sections VI., VII., and VIII. have illustrated the basic concepts of continuous and partial solid solution, phase transitions and exsolution by the use of a reasonable number of relatively simple diagrams, many more types of binary solid solution diagrams are possible.

Table III in Appendix A provides a convenient list of real systems from "Phase Diagrams for Ceramists" which can be used for class discussions on various types of simple and complex solid solution diagrams. Figure X, page 13 in the 1964 edition shows a very useful complex solid solution diagram and the chapters on binary systems in the Rhines book "Phase Diagrams in Metallurgy" contain many special types of solid solution diagrams. Since metal systems are most often solid solution types, the collections of Hansen and Anderko (4), Elliott (5) and Shunk (6) are extremely useful to illustrate the basic concepts in binary solid solution diagrams. The collection of boride, carbide and nitride diagrams by E. Rudy (7) provides another useful source of real solid solution type diagrams.

X. LIQUID IMMISCIBILITY

A. Historical

The physical appearance or texture of two or more liquid separations is influenced by three major factors, the viscosity, density and surface tension of the separated phases. Dr. Joel H. Hildebrand* and Dr. S. L. Kittsley† have been engaged in a friendly context since 1934 to develop the greatest number of immiscible liquid layers in chemical systems, but their more serious objectives were to trace the history of immiscible liquid systems and illustrate how far solution theory lags experimentation. In 1962, Hildebrand (8) developed a 10-layer system consisting of mercury, gallium, a fluocarbon, phosphorus, silicone oil (Dow-Corning 710 fluid), mucilage (aqueous), polyvinyl alcohol (aqueous), methyl cellulose (aqueous), aniline and hexane. These liquids have low viscosities and differ sufficiently in density that they will separate into ten distinct layers after vigorous mixing. Centifuging will accelerate the separation.

The above history concerns the field of basic chemistry. In the case of oxides, silicates and the ceramic field, the classical work of J. W. Greig (9) (Geophysical Laboratory, Carnegie Institution of Washington) was the first to show (1927) that divalent oxides such as MgO, CaO, SrO, MnO, ZnO, FeO, NiO and CoO produce immiscible liquid melts in silica-rich mixtures. Binary silica-rich compositions were quenched from temperatures around 1700°C and examined at room temperature with the optical microscope. Electron microscopy was not yet available at this time, but the optical micrographs which were published

---

*University of California, Berkeley, former president of the American Chemical Society, winner of many Chemical Society awards.
†Marquette University, Milwaukee, Wisconsin.

proved for the first time that RO–SiO$_2$ liquids were definitely immiscible. Related ternary systems were also examined such as MgO–CaO–SiO$_2$. When electron microscopy became readily available for the examination of glass and ceramic materials, some of the earliest work on liquid immiscibility involved ternary borosilicate systems such as Na$_2$O–B$_2$O$_3$–SiO$_2$ (10, 11) (1954, 1959) and Li$_2$O–B$_2$O$_3$–SiO$_2$ (12) (1959).

Since 1959, an enormous amount of electron microscopy has been done on borate, silicate, oxide and fluoride stable and metastable liquid immiscibility as indicated by the publications of Werner Vogel (13) and Ernest M. Levin (14). The major reason for the far greater research and understanding of immiscibility since 1954 is the 10,000 to 100,000 magnification provided by electron microscopy. Classical optical microscopy was limited to magnifications of 500 to 2000 and in many cases, the liquid phase separations could not be detected, especially metastable immiscible types.

B. Basic Diagrams

Three fundamental cases of liquid immiscibility arise in binary systems as shown in Figures 3.22a, b and c.

Figure 3.22a shows the case of a pair of immiscible liquids with an upper temperature limit (T_c) of miscibility (or upper consolute point, according to chemical nomenclature), Figure 3.22b, the case of a pair of immiscible liquids with a lower temperature limit (T_c) of miscibility and Figure 3.22c, a classical case of two liquid immiscibility with both an upper and lower consolute point, where a complete loop encloses the two liquid region and where extremities of tie lines such as ℓ_1–ℓ_2, and ℓ_3–ℓ_4 indicate the composition of the two conjugate liquids. Thus, it is obvious 1) that the compositions of the two coexisting liquids vary with temperature and 2) that miscibility is achieved above and below certain critical temperatures as shown.

C. Stable Liquid Immiscibility

In ceramic systems, liquid immiscibility of the general type shown in Figure 3.22a is encountered in binary borate, silicate, and fluoride systems. (See Table 4, Appendix A). The general appearance of an immiscible liquid silicate system is shown in Figure 3.23. The point m is an invariant point which is called a monotectic (F = 3 - P = 3 - 3 = 0) and involves an equilibrium between two liquid phases and a crystalline phase (in contrast to a eutectic which has two crystalline phases in equilibrium with one liquid). The range of composition between n and m is often spoken of as the "extent of immiscibility" for the system (from 60% B to 95% B in this case), although it is obvious that the "extent" becomes less and less as the temperature of the system is increased above the monotectic temperature. For example at temperature T_1, 40% of a liquid of

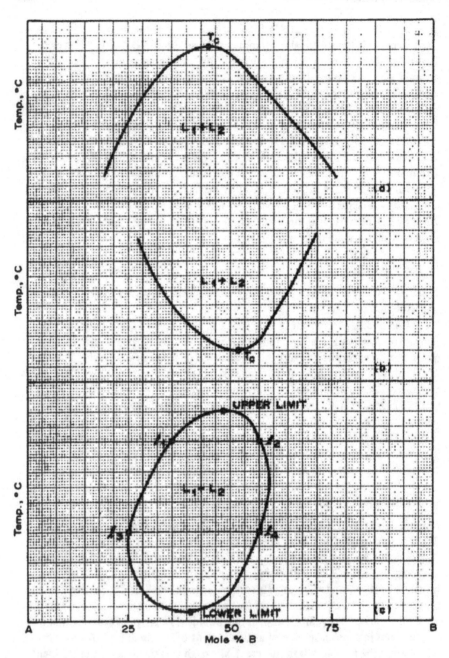

Figure 3.22. Basic Liquid Immiscibility Diagrams a) Upper consolute point, b) Lower consolute point, c) Upper and lower consolute point

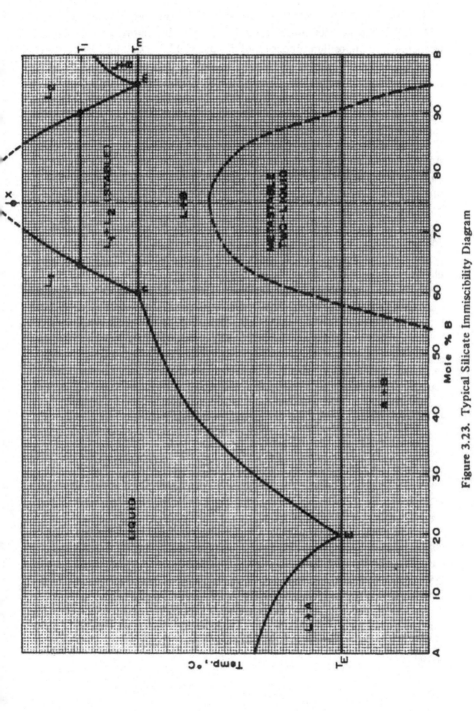

Figure 3.23. Typical Silicate Immiscibility Diagram

composition 10A, 90B is in equilibrium with 60% of a liquid of composition 35A, 65B, if the bulk composition is X (25A, 75B). Just above the monotectic temperature, composition X consists of 3/7 parts (42.9%) of a liquid of composition 5A, 95B and 4/7 parts (57.1%) of a liquid composition 40A, 60B. As heat is withdrawn at the monotectic, one liquid (L_2) disappears and just below the monotectic temperature 3/8 parts (37.5%) of crystalline B are in equilibrium with 5/8 parts (62.5%) of liquid of composition 40B, 60B (L_1). Further cooling and subsequent equilibration at lower temperatures will result in various ratios of crystalline B and L_1, with the composition of L_1 moving from n to E. At T_E, the usual eutectic reaction occurs, and the system becomes 100% solid, a mixture of A and B. If composition X is heated to a sufficiently high temperature it will move above the immiscibility "dome" and become a single phase homogeneous liquid.

It has been demonstrated in previous sections that a binary intermediate compound may melt congruently, melt incongruently or dissociate to two solid phases. It is also possible for an intermediate compound to melt to two liquid phases, as shown in Figure 3.24.

In two-component ceramic systems, immiscibility of borates and silicates has been studied extensively. Many borate systems (Tl_2O, MgO, CaO, ZnO, SrO, CdO, PbO, CoO, NiO, Sc_2O_3, R.E.$_2O_3$, Bi_2O_3, ThO_2 and Nb_2O_5) have been shown to have stable two-liquid regions near the B_2O_3-rich end. In the case of RO-B_2O_3 systems, the viscosity of the RO-rich liquid is low enough at the temperatures involved (900-1300°C) so that reasonably clean conventional separation can be made. For example, a composition in the two-liquid region of the system BaO-B_2O_3 can be used as a laboratory experiment for undergraduates to demonstrate many aspects of the two-liquid phenomenon. If the composition is melted in a platinum crucible, the two layered separation can be easily observed after cooling to room temperature. The high density, more fluid BaO-rich liquid will be found as a very clear glass at the bottom of the crucible and the low density more viscous B_2O_3-rich liquid will be found as a cloudy, vitreous upper layer due to the fact that it is difficult to free the B_2O_3-rich layer from H_2O and other gaseous inclusions.

The best RO-B_2O_3 systems for this demonstration are BaO-B_2O_3 and PbO-B_2O_3 due to the large difference in density of the immiscible liquids. The BaO-B_2O_3 is also extremely important because it is one of the few binary oxide systems in which the composition (15 wt % BaO) and temperature (1225°C) of the conjugate point in the immiscibility dome is known (15).

Finally, with respect to borate systems, Tl_2O-B_2O_3 is the only one which demonstrates the closed loop (Figure 3.22c type) with an upper consolute point at 720°C and a lower (metastable) consolute point at 505°C in the Tl_2O-rich region.

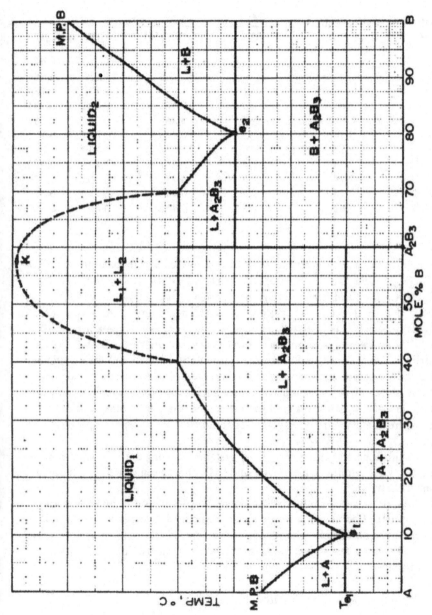

Figure 3.24. Melting of Intermediate Binary Compound to Two Liquids

Table 3.2 Relation of Liquid Immiscibility to Ionic Field Strength of B^{3+} and Si^{4+} and Divalent Cations

Cation ++	Cation Radius (Ahrens)	$\dfrac{Z^+}{CN}$	$\dfrac{Z}{(r_+ +1.40)^2}$	B^{3+}	Si^{4+}
				0.23	0.42
				1.129	1.208
				ΔIFS	ΔIFS
MgO	0.67	1/3	0.467	0.662	0.741
CaO	0.99	1/3, 1/4	0.350	0.779	0.858
ZnO	0.74	1/3	0.437	0.692	0.771
SrO	1.12	1/4	0.315	0.814	0.893
BaO	1.34	1/4	0.266	0.863	0.942
FeO	0.74	1/3	0.437	0.692	0.771
CoO	0.73	1/3	0.441	0.688	0.767
NiO	0.69	1/3	0.458	0.671	0.750

In binary silicate systems, stable liquid immiscibility is found in many RO-SiO_2 (MgO, CaO, ZnO, SrO, FeO, CoO, NiO, CuO, PbO), R_2O_3-SiO_2 (Fe$_2$O$_3$, Ga$_2$O$_3$, Cr$_2$O$_3$), and R.E.-SiO_2 systems, and in certain RO_2-SiO_2 (TiO$_2$, ZrO$_2$, ThO$_2$, UO$_2$) and R_2O_5-SiO_2 systems. Experimentally, the viscosity of silicate liquids (especially the SiO_2-rich liquid) is so high between temperatures of 700-1900°C that clean separations can never be achieved, unlike the BaO-B_2O_3 immiscible liquids. Density differences are also relatively small, but they would be sufficient to permit clean separation if it were not for the viscosity factor. The microstructure of separated binary silicate liquids therefore, manifests itself as a dispersion of irregularly shaped droplets of one phase in another when the liquids are quenched from temperatures above 1700°C, as for example in the MgO-SiO_2 and CaO-SiO_2 systems.

Unfortunately, only a few binary systems involving the two other most common glass-forming oxides, GeO_2 and P_2O_5, have been examined for liquid immiscibility (14).

As in the case of intermediate compound formation in binary systems, (Section IIID, Chapter 3), the influence of ionic field strength on the location and extent of liquid immiscibility has been examined (Levin (14), Table 1, page 162-163). In the case of RO-B_2O_3 and RO-SiO_2 systems, Table 3.2 shows the relationships.

D. Metastable Liquid Immiscibility

A typical metastable two liquid region is shown in Figure 3.23. To quote Levin (14), "Just as stable liquid immiscibility in oxide systems was once believed to

be rare and later was found extensively in borate and silicate systems, so metastable liquid immiscibility has been found to be prevalent in the same systems".
In the case of metastable immiscibility, the change and increase in detection was
mostly due to the ability to magnify at 5000-100,000 with the electron microscope and detect microliquation which could never be seen with the optical
microscope.

As far as real systems are concerned, it is now known that many binary
borate and silicate systems contain a region of metastable immiscibility. Li, Na,
K, Rb, and Cs borate systems, Li, Na, Cs silicate systems, and B_2O_3-SiO_2 and
Al_2O_3-SiO_2 are definitely known to contain very specific metastable two liquid
regions. (Levin (14), Table IV, page 186).

Some of the most significant early work on metastable immiscibility in
binary oxide systems has been done by Shaw and Uhlmann (16) (borates),
Charles (17) (silicates), Hammel (18) (Na_2O-SiO_2), Charles and Wagstaff (19)
(B_2O_3-SiO_2), Ganguli and Saha (20) (Al_2O_3-SiO_2) and MacDowell and Beal
(21) (Al_2O_3-SiO_2).

In many cases of metastable two liquid separation, which is developed by
holding for long times at temperatures between 500-1500°C, the droplets are
spherical or nearly spherical, due to surface tension relationships. In stable two
liquid separation, the phases are usually visible in the light microscope at magnifications of 100-1000X; metastable two liquid separation may require magnifications of 5000-100,000X in the electron microscope to be detected. For a collection of microstructures developed by two-liquid separation, the reader is referred
to Vogel (13) and many individual papers in the literature as quoted by Levin
(14). Table V in Appendix A gives a convenient list of references on binary and
ternary silicate liquid immiscibility. The list is incomplete since most of the
foreign literature has not been quoted (especially Russian, Japanese and
German) and some of the papers written in English have not been cited. However, the list can be helpful for teaching purposes to illustrate the immiscibility
phenomenon in real systems, especially in binary systems.

Finally, Figure 3.25 shows the behavior of liquid immiscibility with respect to density, viscosity, surface tension, stability and metastability, ranging
from low viscosity chemical fluids to high viscosity metastable silicate systems.

E. Spinodal Decomposition

When homogeneous liquids or solid solutions are cooled to the point where the
metastable liquids or the solid solutions unmix, Figure 3.26, the nonequilibrium
cooling gives rise to the "spinodal" curve. It is a kinetic phenomenon and
originally applied to metallic alloys, later to oxide glasses, and currently to oxide
solid solutions and metallic glasses.

The details of spinodal decomposition is beyond the scope of this text. If
more detailed class discussion is intended, the Jantzen and Herman (22) Chapter

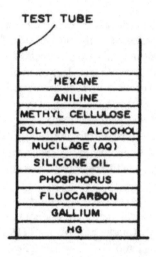

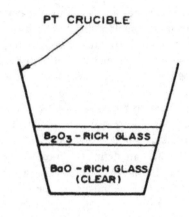

a. Low Viscosity Immiscible
   Chemical Compositions [8]

b. Low Viscosity BaO - B₂O₃
   Immiscible Liquids [15]

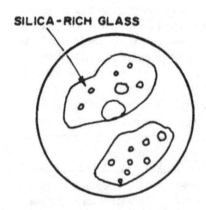

c. Photomicrograph (350 X)
   of Stable Liquid
   Immiscibility, MgO - SiO₂
   System [9] (High Viscosity)

d. Electron Micrograph [19] (30 Mol
   % SiO₂ - 70% B₂O₃) of
   Metastable Liquid Immiscibility
   (Heated 3 Weeks, 510°C)

Figure 3.25. General Appearance of Immiscible Liquid Systems, Ranging From
Those Visible to the Eye to Those Which Require 50-100,000 Magnification of
the Electron Microscope

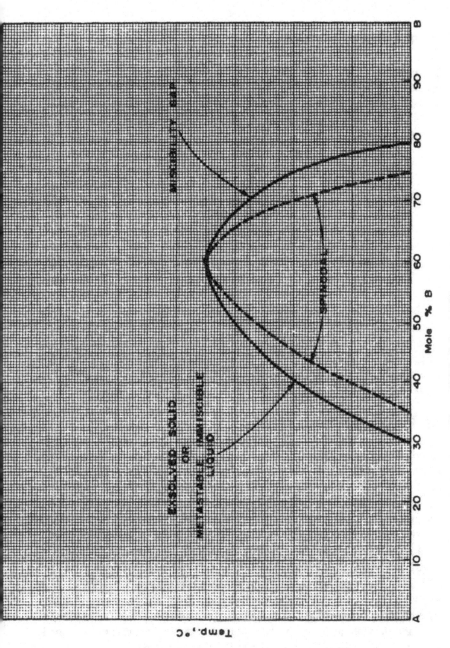

Figure 3.26. Spinodal Decomposition

3 in Vol. 6-V of Phase Diagrams: Materials Science and Technology should be used.

XI. REAL OXIDE SYSTEMS

The number of real systems which might be chosen to illustrate the nature and use of the binary system is very large, as indicated in Tables 1-4 in Appendix A. Among the more common borate, aluminate, silicate, phosphate, titanate, zirconate and niobate binary systems, the following would be recommended for class discussion:

$Li_2O-B_2O_3$	$MgO-SiO_2$	$ZrO-P_2O_5$
$Na_2O-B_2O_3$	$CaO-SiO_2$	$BaO-TiO_2$
$BaO-B_2O_3$	$Al_2O_3-SiO_2$	$Al_2O_3-TiO_2$
$MgO-Al_2O_3$	TiO_2-SiO_2	ZrO_2-TiO_2
$CaO-Al_2O_3$	ZrO_2-SiO_2	$MgO-ZrO_2$
Li_2O-SiO_2	$CaO-P_2O_5$	$CaO-ZrO_2$
Na_2O-SiO_2	$SrO-P_2O_5$	$Al_2O_3-ZrO_2$
K_2O-SiO_2	$Al_2O_3-P_2O_5$	$Ta_2O_5-Nb_2O_5$

In addition to a discussion of their relation to the basic types of equilibrium diagrams, these systems could illustrate water soluble glasses, commercial use of spinels, cement compositions, electrical ceramics, mullite and $Al_2O_3-SiO_2$ refractories, low expansion silica or silica-rich glasses, zircon refractories and gegemstones, luminescence, electronic ceramics, stabilization of zirconia and low linear expansion and axial thermal expansion of crystalline compounds.

XII. FLUORIDE MODEL SYSTEMS

Certain fluorides are regarded as "weakened" models of ceramic oxides. For example, BeF_2 (M.P. $\sim$ 550°C) is the weakened model for SiO_2, M.P. 1723°C, and NaF (M.P. $\sim$ 1000°C) is the weakened model for CaO (M.P. 2600°C). BeF_2 can be easily obtained as a glass and a quartz structure, corresponding to the silica forms. NaF and CaO have rocksalt structures. The binary system $CaO-SiO_2$ has four intermediate compounds, 1:1, 3:2, 2:1 and 3:1 (Figure 237, p. 104, 1964 collection of diagrams). The lowest melting compound is $CaSiO_3$, congruent at 1544°C. The system $NaF-BeF_2$ is the weakened model system for $CaO-SiO_2$, but it contains only two intermediate compounds $NaBeF_3$ (1:1) and Na_2BeF_4 (2:1), corresponding structurally to $CaSiO_3$ and Ca_2SiO_4. Thus, the system can be examined at more convenient temperatures than the system $CaO-SiO_2$ and data obtained on the two intermediate compounds can be related (with caution) to the refractory 1:1 and 2:1 silicate compounds.

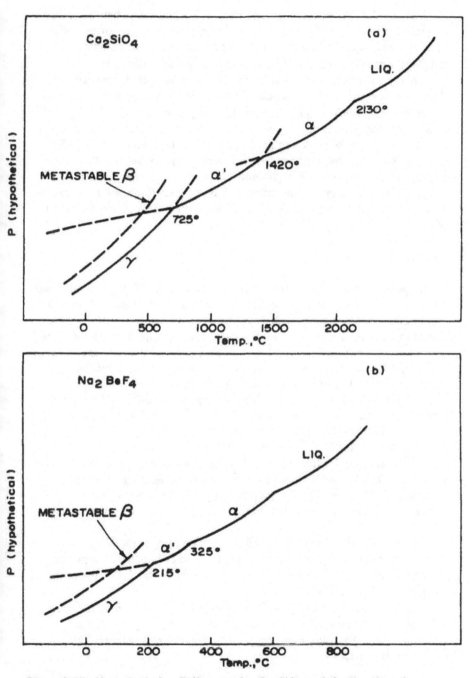

Figure 3.27. Hypothetical p–T diagrams for Ca$_2$SiO$_4$ and the "weakened model" Na$_2$BeF$_4$

Figure 3.27 shows the extremely close p–T relationships between calcium orthosilicate and Na_2BeF_4 polymorphs. The crystal structures of the polymorphs are identical and the inversion temperature relationships indicate how much more convenient it would be to study the fluoride structures.

Many such fluoride "model" systems can be visualized, especially with BeF_2 serving as the weakened model of SiO_2 or MgF_2 serving as the weakened model of TiO_2. For example, the systems MgF_2–BeF_2 and CaF_2–BeF_2 are reasonably close models of TiO_2–SiO_2 and ZrO_2–SiO_2, respectively. LiF–BeF_2 and LiF–MgF_2 are the weakened model systems for ZnO–SiO_2 and ZnO–TiO_2. The analogy cannot be carried too far as indicated by the fact that there is no correspondence between KF–MgF_2 and the important ceramic system BaO–TiO_2. The experimental determination and use of these fluoride systems as silicate model systems would be considerably greater if it were not for the toxicity of BeF_2.

PROBLEMS

Sketch phase diagrams for the following systems even though in some cases, data is not complete. When data are lacking, choose reasonable points and reasonable constructions which comply with Phase Rule requirements. Do not include features which are not required by the data. Lable the fields to demonstrate what phases are in equilibrium and draw cooling curves which correspond to the most important characteristics of each system. Place the compositions on a molar basis unless it is necessary to use a weight percent basis.

1. Na_3Bi melts at 790°, and NaBi decomposes at 450° into Na_3Bi and a liquid. The eutectic temperatures are 97° and 218°. Na melts at 98°, Bi at 273°.

2. Sodium (melting point 98°) and antimony (melting point 630°), form two compounds, Na_3Sb which melts at 823°, and NaSb which melts at 503°.

3. Magnesium (melting point 651°) and nickel (melting point 1452°) form a compound $MgNi_2$ which melts at 1180° and a compound Mg_2Ni which decomposes at 770° into a liquid containing 38 per cent Ni and the other compound. The eutectics are at 28 per cent Ni and 510°; and at 88 per cent Ni and 1080°.

4. Mercury (melting point –39°) and lead (melting point 327°) dissolve in all proportions in the liquid state, and they form no compounds. A liquid phase is in equilibrium at –40° with two crystalline phases, containing 35 per cent and 100 per cent Hg, respectively.

5. Iron (melting point 1535°) and Fe_3Sb_2 (melting point 1015°) form solid solutions in one another to a limited extent, and $FeSb_2$ decomposes at 728° into a liquid and the other compound. The eutectics are at 1000° and 628°; Sb melts at 630°.

6. Copper (melting point 1083°) forms solid solutions in CaCu$_4$ (melting point 936°), and CaCu$_4$ forms solid solutions in Ca (melting point 810°), but no other solid solutions form. The eutectics are at 6 per cent Ca and 910°, and at 38 per cent Ca and 560°.

7. Lead (melting point 327°) and palladium (melting point 1555°) form four compounds, PdPb$_2$ (melting point 454°), PbPd (decomposes at 495° into a liquid and Pd$_2$Pb), Pd$_2$Pb (decomposes at 830° into a liquid and Pd$_3$Pb) and Pd$_3$Pb (melting point 1240°). Solid solutions from 77 to 100 per cent Pd are formed, but there is only one liquid solution. The eutectic temperatures are 260°, 450°, and 1185°.

8. Cobalt melts at 1480°, CoSb melts at 1190°, CoSb$_2$ decomposes at 900° into CoSb and a liquid containing 91 per cent Sb and pure Sb melts at 630°. There are eutectics at 1090° and 40 per cent Sb, and at 620° and 99 per cent Sb, and a solid solution area exists up to 12 per cent Sb.

9. Aluminum (melting point 658°) and cobalt (melting point 1480°) form three compounds, of which AlCo melts at 1630°, Al$_5$Co$_2$ decomposes at 1175°, and Al$_4$Co decomposes at 943°. A complete series of solid solutions forms from AlCo to pure cobalt.

10. Bismuth (melting point 273°) and lead (melting point 327°) form no compounds, but solid solution containing 37 per cent and 97 per cent lead form. No liquid phase exists below 120°.

11. Nickel (melting point 1452°) and manganese (melting point 1260°) form a complete series of solid solutions. A liquid phase containing 55 per cent manganese solidifies at 1030°.

12. The compound CaAl$_3$ decomposes at 692° into two liquid phases, which become soluble in one another above 750°. No other compounds form, and there are no solid solutions; Ca melts at 810°, Al at 658°.

13. Label the phases present in each area of Diagram P.3.13 provided (use any suitable, reasonable type of nomenclature) and make a complete quantitative analysis of the phase or phases present at the 11 points indicated. If only one phase is present, give its composition; if two phases are present, give the amount and composition of each phase and demonstrate (for only three points) that the total composition in terms of A and B remain the same as that of the original liquid. Draw the lines at appropriate places in two phase regions to illustrate your analyses. If there are any incorrect features in the diagram, make a reconstruction.

14. Apply the Phase Rule to 1) a one phase region 2) a two phase equilibrium and 3) an invariant point in the above diagram.

15. Draw a solid solution type diagram for a binary system which contains ALL of the following features:

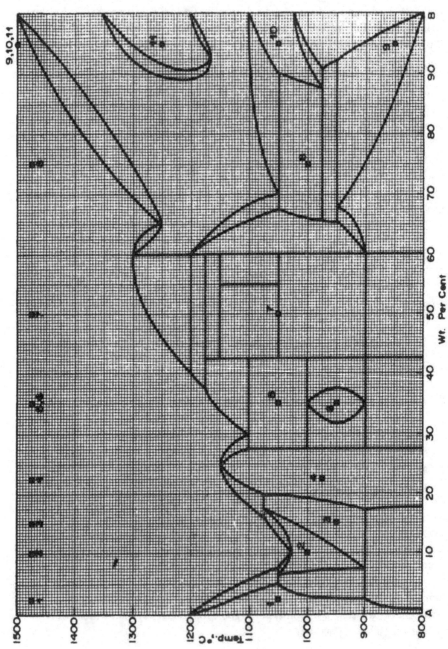

1. Three inversions in component A
2. Two polymorphic inversions in component B
3. Two intermediate compounds AB and BA
4. A peritectic reaction and a eutectic type reaction
5. Partial solid solubility of AB in component A, resulting in a decrease in inversion temperature of two of the A polymorphs and an increase in inversion temperature of the third.
6. Partial solid solubility of BA in component B, resulting in a decrease in inversion temperature of one of the B polymorphs and an increase inversion temperature for the second.
7. Complete solid solution with a maximum at the liquidus exists between AB and BA, except for a small solubility gap beginning 50° below the congruent melting temperature of AB and ending at 150°C below the congruent melting temperature of BA.

16. Draw the original Bowen and Greig (1924) diagram for the system Al_2O_3–SiO_2 and describe the major experimental difficulties encountered by various investigators as they attempted to revise the configuration. Draw the 1962 "best estimate" diagram of Aramaki and Roy and emphasize the major differences between the 1924 and 1962 diagrams. In addition to describing experimental difficulties, emphasize the concepts and changes in concepts which took place between 1924–1962.

Recent work of McDowell and Beal (1969) has proven the existence of metastable two liquid formation in the system. Show this on the Aramaki and Roy diagram using dashed lines.

17. Diagram P.3.17 provided shows a typical solid solution configuration. End members A and B exist in two polymorphic forms, high temperature α and low temperature β. Use appropriate Greek letter symbols to properly label the areas in the remainder of the diagram, clearly indicating what phase or phases exist in each area.

Make a complete quantitative analysis of the equilibrium for:

1. Composition 1 at 1800°C
2. Composition at 2 at 1200°C
3. Composition 3 at 1300°C
4. Composition 4 at 1300°C
5. Composition 5 at 700°C

Show in each case that the amounts of A and B in the final assemblage are the same as those of a homogeneous liquid of compositions 1, 2, 3, 4, and 5, respectively.

18. Draw ONE complex equilibrium diagram A–B (Diagram P.3.18) which contains every one of the following features:

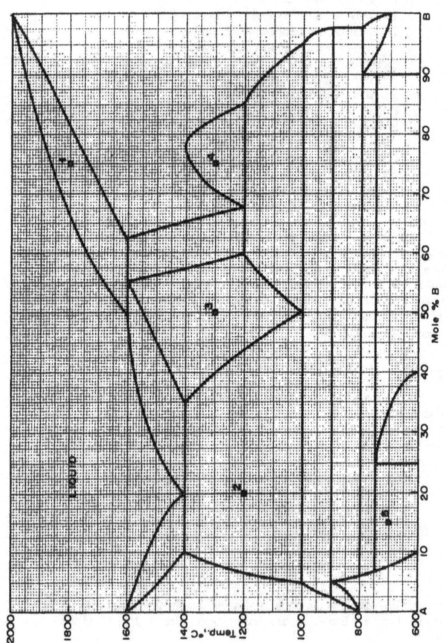

Diagram P.3. 17. Typical Solid Solution Configuration

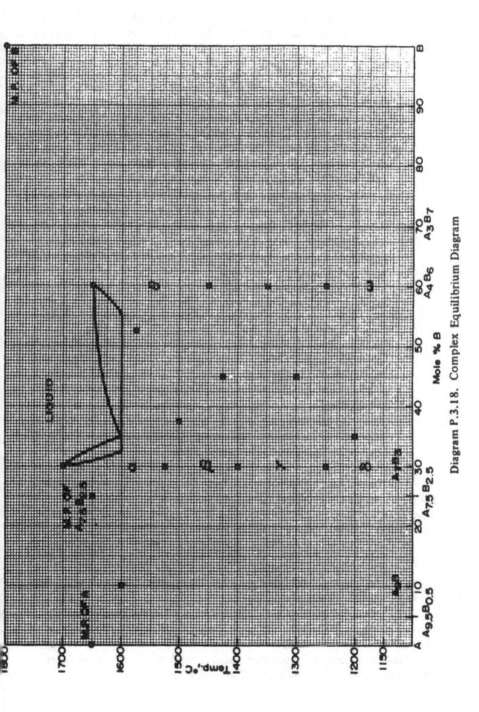

Diagram P.3.18. Complex Equilibrium Diagram

1. A congruently melting (1700°C) intermediate compound A_7B_3 exists in four polymorphic forms $(\alpha, \beta, \upsilon, \delta)$ with transition temperatures at 1250, 1400, and 1525°. All the transition temperatures are lowered by solid solubility of the adjacent compound A_4B_6. Eutectoid compositions and temperatures are at 37.5% B and 1500°, 45% B and 1300° C, and 35% B and 1200°C.

2. The adjacent congruently melting (1650°C) compound (A_4B_6) has two transitions at 1250° and 1450°, whose temperatures are increased by the solid solubility of A_7B_3. Invariant points are at 45 B and 1425° and 52.5 B and 1575°. Solubility limits between A_7B_3 and A_4B_6 at 1150° are at 32 B and 52 B. (Polymorphs are labeled θ, ϵ, ω).

3. A two liquid region exists on the B (M.P. 1800°) liquidus curve, giving an invariant reaction at 95B and 1650°. The two liquids become miscible at 1775° and their maximum extent is 15%.

4. A compound with a lower limit of stability at 1300° exists at A_3B_7 and melts incongruently at 1550° to a liquid which is 2.5% richer in A than the compound.

5. Complete solubility exists between $A_{7.5}B_{2.5}$ and A_9B, but there is a maximum in the liquidus at 1500° and unmixing occurs at 1400°, giving two conjugate solid solutions (87.5 A and 77.5 A) at 1200°.

6. $A_{9.5}B_{0.5}$ is only stable at 1400°. Dissociates to two other solid phases.

7. Show a metastable event in the region between A_7B_3 and A_4B_6 by the prolongation of phase boundaries as dashed lines. Explain the event.

8. Label each area in the diagram for "Phases present" using reasonable nomenclature.

9. Show two other metastable events or occurrences in any part of the diagram which you choose and explain your construction. Use dashed lines to indicate a metastable relationship.

19. The system Fe–O is shown as Figure 8, page 38 in Levin, Robbins and McMurdie. Phase Diagram for Ceramists, 1964. The caption for Figure 8 lists the components as $Fe–Fe_2O_3$. Neglecting the wüstite and magnetite solid solutions, and using the definition of the components of a system, show quantitatively that Fe and Fe_2O_3 are satisfactory as end-members of the system. (Assume that FeO and Fe_3O_4 are the only intermediate phases.) If one does *not* neglect the presence of wüstite and magnetite solid solutions, what would you prefer to use as the end-member components? Why?

20. Label the phases present in each area of the four binary phase diagrams provided in Diagram P.3.20 (A, B, C, and D), UNLESS you feel that corrections are necessary and that the diagram should be redrawn or corrected. If you redraw or correct any of the four diagrams, explain *why* it was necessary and label phases present in each area of the new diagram.

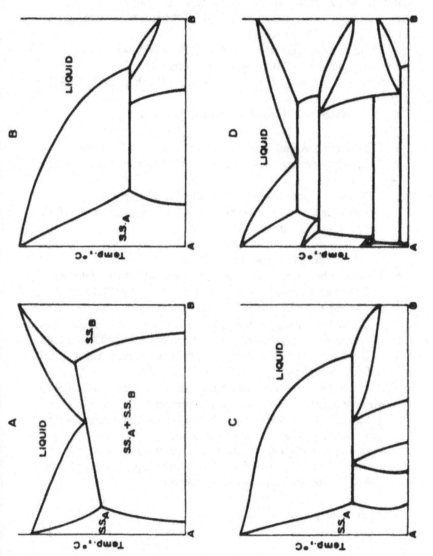

Diagram P.3.20. Binary Phase Diagram

21. Study and analyze the equilibria in the system $Ti-O_2$, including the structures of the intermediate compounds TiO (NaCl type), Ti_2O_3 (corundum type), Ti_3O_5 (pseudobrookite type) and TiO_2 (rutile). See 1964 and 1969 compilations, Figs. 22 and 2081 and 2082, Phase Diagrams for Ceramists.

22. Study in detail the features of any of the binary systems listed in Tables I, II, III, or IV of Appendix A.

23. Explain all the features of the system Pr-O as shown on page 8 as Fig. 2079 of the 1969 Phase Diagrams for Ceramists.

24. Study the system W-O as shown on page 10 as Fig. 2084 of the 1969 compilation.

25. In the system A-B (composition on molar basis):

1. End members A and B melt at $1400°$ and $1300°C$, respectively.
2. A_9B_1 melts incongruently at $1250°C$, the peritectic liquid contains 20% B.
3. A_7B_3 has an upper temperature limit of stability at $850°C$.
4. A_6B_4 has an inversion at $1000°$ and melts incongruently at $1100°C$. The peritectic liquid contains 45% B.
5. A_4B_6 has an inversion at $800°$ and melts congruently at $1100°$. It has a lower limit of stability at $750°$.
6. A eutectic between A_6B_4 and A_4B_6 exists at 50% A and $900°C$.
7. A_3B_7 has an inversion at $850°$ and melts incongruently to another compound and a peritectic liquid containing 35% A at $1000°C$.
8. A_2B_8 melts to two liquids at $1250°$. The immiscibility extends over a 20% compositional interval and the dome closes at $1400°$ and 19% A.
9. Of the three eutectics in the system, the one richest in B is at $1100°$ and the one richest in A is at $900°C$.
10. Draw the binary diagram using these data and label EACH AREA which you have created for "phases present". Use some kind of appropriate Greek letter nomenclature for low and high temperature polymorphs, such as α A_4B_6, β A_4B_6, etc.
11. Draw a "phase analysis" diagram for compositions containing 10, 20, 30, 40, 50, 60, 70, 80 or 90% A by using iso-plethal analyses at appropriate temperatures for each composition.

26. Define and illustrate the following terms:

1. Tie line	6. Peritectic point
2. Eutectoid	7. Peritectoid point
3. Incongruent melting point	8. Monotectic
4. Indifferent point	9. Exsolution
5. Isothermal analysis	10. Isoplethal analysis

27. The system CaO-SiO$_2$ is shown as Figure 237, page 104 in Levin, Robbins and McMurdie (1964 edition). Assume SiO$_2$ can be isothermally evaporated at 1600°C from a composition containing 20% CaO and 80% SiO$_2$ until there is no SiO$_2$ remaining. What is the phase composition of the starting mixture at 1600°C? Describe ALL of the major events which would occur during evaporation of the silica until the composition is 100% CaO. Be as quantitative in your descriptions as possible, making full use of the horizontal composition axis.

28. Draw ONE solid solution system A-B, based on the following data:

1. A melts at 700°C, transforms from γ to β at 550° and from β to α at 300°.
2. B melts at 1200°C, transforms from γ to β at 800° and from β to α at 650°.
3. There is complete solubility between the γ forms of A and B, with no max or min in the liquidus curve.
4. Peritectoid reaction occurs at 600° involving a γ solid solution containing 20% B and an intermediate δ solid solution containing 30% B. The βB solid solution contains 25% B.
5. At 700°, peritectoid reaction also occurs involving an intermediate δ solid solution containing 65% B and a βB solid solution containing 75% B. The γ solid solution contains 55% B at this temperature.
6. Eutectoid reaction occurs at 550° involving an αB solid solution containing 15% A, the intermediate δ solid solution containing 30% A, and βB solid solution.
7. At 300°C, the limit of solubility of B in βA is 15%, and of A in α B is 30%. At this temperature the δ solid solution extends from 26% B to 60% B.
8. At 200°, another eutectoid reaction occurs, involving an α A solid solution containing 10% B, a δ solid solution containing 25% B, and βB.

      Using the αA, βA, γ, αB, βB, and δ nomenclature, label each area of the diagram for phases present.

      At 400°, for a composition containing 77.5% A and 22.5% B, give the amount and composition of the equilibrium phases.

29. The number of diagrams in the 1964, 1969, 1975 and 1981 collections of "Phase Diagrams for Ceramists" are 2066, 2083, 850, and 591, respectively, a total of 5,590. Assume that ceramists or material scientists are interested in the most common valence state of 90 elements in the Periodic Table. Using the following equation (combination of n things, taken two at a time),

$$C_{(n,r)} = \frac{n!}{r!(n-r)!} = \frac{90!}{2!(90-2)!}$$

Calculate the total number of binary oxide, fluoride, sulfide, boride, carbide, nitride, etc. systems which would have to be produced. Ans.; 4,005, each.

If the less common valence states of the 90 elements are included (P^{3+}, Ti^{2+}, Ti^{3+}, V^{3+}, transition elements, rare earth elements, etc., etc.), the total number of binary systems would be enormous. Make assumptions and calculate the number of binary diagrams which still need to be determined.

REFERENCES

1. Richard M. Fulrath and Joseph A. Pask, Editors, Ceramic Microstructures, John Wiley and Sons, Inc., 1968.
2. A. Dietzel, Field Strengths of Cations and Their Relation to Devitrification Phenomena, To Formation of Compounds, and To Melting Points of Silicates, Zeit Fur Electrochemie 48, 9–23, 1942.
3. M. L. Keith and Rustum Roy, Structural Relations Among Double Oxides of Trivalent Elements, American Mineralogist 39, p 1–23, 1954.
4. Max Hansen and Kurt Anderko, Constitution of Binary Alloys, McGraw-Hill, 1948.
5. Rodney P. Elliott, Constitution of Binary Alloys, First Supplement, McGraw-Hill, 1965.
6. Francis A. Shunk, Constitution of Binary Alloys, Second Supplement, McGraw-Hill, 1969.
7. E. Rudy, Compendium of Phase Diagram Data, AFML-TR-65-2, Part V, Air Force Materials Laboratory, Metals and Ceramics Division, Wright-Patterson AFB, Ohio.
8. Joel H. Hildebrand and Robert L. Scott, Regular Solutions, Prentice-Hall, Inc., Englewood Cliffs, New Jersey, 1962, frontispiece.
9. J. W. Grieg, Immiscibility in Silicate Melts, Amer. Jour. Sci., Vol. XIII, pp. 1–44, January, 1927; Vol. XIII, p. 133–54, February, 1927.
10. A. F. Prebus and J. W. Michener, Electron Microscope Investigation of Glasses, Ind. Eng. Chem., 46(1) 147–53, 1954.
11. Mueno Watanabe, Haruo Noake, and Takeshi Aiba, Electron Micrographs of Some Borosilicate Glasses and Their Internal Structure, J. Amer. Cer. Soc., 42(12), 593–99, December, 1959.
12. B. S. R. Sastry and F. A. Hummel, Studies in Lithium Oxide Systems: III, Liquid Immiscibility in the System Li_2O–B_2O_3–SiO_2, J. Amer. Cer. Soc., 42(2), 81–88, 1959.
13. Werner Vogel, Structure and Crystallization of Glasses, 1965 German Edition, VEB Deutscher Verlag für Grundstoffindustrie, Leipzig; 1971 English Edition, Pergamon Press.
14. Ernest M. Levin, Liquid Immiscibility in Oxide Systems, p. 144–233 Phase

Diagrams, Vol. 3, The Use of Phase Diagrams in Electronic Materials and Glass Technology, Academic Press, 1970.

15. Ernest M. Levin and Given W. Cleek, Shape of Liquid Immiscibility Volume in the System Barium Oxide-Boric Oxide-Silica, J. Amer. Cer. Soc. 41(5), 175-179, May, 1958.

16. R. R. Shaw and D. R. Uhlmann, Subliquidus Immiscibility in Binary Alkali Borates, J. Amer. Cer. Soc. 51(7), 377-82, 1968.

17. R. J. Charles, Metastable Liquid Immiscibility in Alkali Metal Oxide-Silica Systems, J. Amer. Cer. Soc. 49(2), 55-62, 1966.

18. J. J. Hammel, Experimental Evidence of Spinodal Decomposition in Glasses of the Na_2O-SiO_2 System, Gordon and Breach, June, 1965.

19. R. J. Charles and F. E. Wagstaff, Metastable Immiscibility in the $B_2O_3-SiO_2$ System, J. Amer. Cer. Soc., 51(1), 16-20, 1968.

20. D. Ganguli and P. Saha, Macroliquation in the System $Al_2O_3-SiO_2$, Materials Res. Bull. 2, 25-36, 1967.

21. J. F. MacDowell and G. H. Beal, Immiscibility and Crystallization in $Al_2O_3-SiO_2$ Glasses, J. Amer. Cer. Soc., 52(1), 17, 1969.

22. C. M. F. Jantzen and H. Herman, Spinodal Decomposition—Phase Diagram Representation and Occurence, Chapter III, Vol. 6-V, Refractory Materials Phase Diagrams: Materials Science and Technology, Editor, Allen Alper, Academic Press, 1978.

4

EXPERIMENTAL METHODS
OF CONSTRUCTING PHASE DIAGRAMS

I. INTRODUCTION

After a ceramist or material scientist has learned to interpret one, two, three, and four component phase diagrams, and to make quantitative use of them, he may still fail to appreciate what has to be done to construct by experimental methods a diagram which will be essentially correct and withstand the test of a reasonable period of time. Every phase diagram which has been constructed and appears in a collection may be subject to minor changes from time to time and many of them will undergo major revisions as the years go by. Professionals who use the diagrams frequently should not:

1. Underestimate the time-consuming detailed analysis which is usually required to construct a diagram from experimental data and
2. consider even the most thoroughly investigated diagram as "perfect" and one which will remain constant for all time.

In some cases, a simple eutectic type binary system may be constructed and established within a two week period, but in general a binary diagram of even modest complexity will require six months to a year for completion by use of the more or less standard experimental techniques. It is not uncommon to take three years to work out the relationships in a three-component system and a four-component system could require ten years for completion if high standards of precision and accuracy were applied. It is often the case that only portions of specific systems are investigated for phase relations, as dictated by an interest in the chemical or physical properties of the phases in that limited

portion of the system. For example, in the alkali borate or silicate systems, it will be noticed that the relationships have been worked out on the high borate or silica side of the diagram, but not on the high alkali side, since there has been no incentive to use the alkali-rich compositions. This observation can be extended to ternary alkali borosilicate or aluminosilicate systems. The alkali-rich corner of these systems is invariably ignored. One must be highly motivated by either theoretical or (more often) practical considerations to consider the tedious work which is required to work out the relationships in even a portion of a system

If one could calculate the polymorphism, melting and dissociation of compounds, the location of fields of stability, positions of boundary lines, slopes of liquidus curves, limits of solid solubility, and the shapes of liquidus surfaces by the use of thermodynamic data, a great deal of labor could be saved and this method would of course be preferred. So far, the thermodynamic approach to the construction of equilibrium diagrams is limited to one and two component metal systems, or to one component oxide systems, and some relatively rare cases of two-component oxide systems.

In order to determine the equilibrium phase relations of a two, three or four component system it is necessary to establish the phases existing at various temperatures for selected compositions within the system. The positions of the invariant points, the liquidus lines or surfaces, and other points, lines, or surfaces at which phase reactions occur must be determined in such a manner that the data represent phase relations for *equilibrium* conditions.

II. MAJOR FACTORS IN EXPERIMENTAL METHODS

Several factors are especially important in the determination of phase-equilibrium relations.

A. Chemical Purity of Components

Starting materials should be of the highest purity. Foreign constituents may affect the system in such a manner that the system actually has, because of the foreign constituents, additional components and phases. Such additional components may cause large deviations from the phase-equilibrium relations that are intended to be determined. On the other hand, it is recognized that foreign constituents may be beneficial rather than detrimental when such materials act as "mineralizers", fluxes, or catalysts which increase reaction rates and prevent the formation of metastable phases. In earlier days (approximately from 1910–1945), problems of chemical purity of starting materials and components existed. At that time many oxides which were needed for phase equilibrium determinations either were not available or they contained 0.01 to 2% of impurities. Since 1945, the chemical industry has expanded tremendously, especially the "specialty" chemical industry, and practically every oxide, fluoride or sulfide needed for phase studies is now available in a high purity condition. Nowadays, it is common practice to use starting materials of purity 99.99% for phase equilibrium studies and in general, purities range from 99.9% to 99.9999%.

B. State of Subdivision or Physical Characteristics of Starting Materials

Equilibrium is more easily approached when the area of contact of raw materials is greater. Thus, fine-particled materials intimately mixed and in contact aid in the attainment of equilibrium. This principle was recognized by Körsöry in 1941 (1) and it is rediscovered and re-emphasized (2, 3, 4) periodically as a valuable means of promoting reactions. The choice of starting material with respect to state of subdivision may vary from ordinary 200 mesh ground powders, to precipitated chemicals, through a variety of gels, sols, and colloidal preparations. If necessary, production of solid or liquid phases through reaction of gases can be considered as a means of obtaining equilibrium.

When glass-forming systems are being investigated, the raw materials are melted, crushed, and remelted as many times as necessary to produce a very homogeneous starting material (glass) for "quenching" studies. However, phase separation in glasses is now known to be widespread, and starting glasses must be carefully examined to insure that they are single phase and homogeneous.

C. The Time Criterion

The definition of equilibrium in the 1964 edition of Phase Diagrams for Ceramists (p. 6) is quoted:

> From the theoretical, thermodynamic standpoint, the conditions for equilibrium can be exactly and precisely defined; because for any reversible process, no useful energy passes from or into the system.

From the practical, experimental standpoint, however, the actual attainment of an equilibrium state within a system may be very difficult to assess. Three criteria have been used variously either singly or together: (1) The time criterion, based on the constancy of phase properties with the passage of time; (2) the approach from two directions criterion, yielding under the same conditions phases of identical properties, e.g., from undersaturation and supersaturation, or from raising and lowering the temperature to the same value; and (3) the attainment by different procedures criterion, producing phases having the same properties when the same conditions, with respect to the variants, are reached.

None of these criteria are entirely adequate for excluding metastable relationships. In silica systems, in particular, metastable equilibrium is common and may persist for long periods of time and at high temperatures. In the final analysis, interpretation and judgment by the investigator are of prime importance."

The time at any temperature must be sufficiently long in order to permit completeness of reactions. To the inexperienced experimentalist, this is one of the most difficult pitfalls in phase equilibrium studies. The ceramist is familiar with the instantaneous (sometimes explosively violent) reaction of gases and the (usually) rapid diffusion and reaction of liquids. Many solid state reactions and phase transitions take place within seconds, minutes or hours, but there are some which require years and multiples of years and others which appear not to take place at all. The reactions in the binary systems $Li_2O-B_2O_3$, Li_2O-SiO_2, $Li_2O-P_2O_5$, Li_2O-WO_3, $CaO-WO_3$, and $CaO-P_2O_5$ take place within minutes or hours, but Schairer and Bowen (5) admit that equilibrium is NOT achieved in certain portions of the system $K_2O-Al_2O_3-SiO_2$ in three years. The attainment of equilibrium in many systems containing PbO is very slow. Thus, in some systems, the experimentalist must be prepared to wait years for equilibrium to be established in systems of high purity or he must find some way to accelerate the reactions, transitions, and crystallizations by chemical or physical methods which do not change the number of components or phases in the original system under consideration.

D. Constancy of Composition

After making the correct weight or molar calculations on starting materials which are required for one, two, three, four, etc. component systems, the compositions must remain constant as intended. Two major factors are involved, the first being the container in which the equilibrium reactions are made. Usually platinum or platinum alloy crucibles or containers are used for oxide, borate, silicate, phosphate and other types of ceramic systems. Platinum or its alloys are usually highly unreactive and inert to refractory oxides, borates and silicates, but when phosphate and other types of ceramic compositions are heated to temperatures above 1200°C, there may be mild to destructive reaction with the containers. When oxide containers such as fused silica, alumina, zirconia, mullite, zircon or thoria are used at high temperatures, it is absolutely essential to make chemical analyses of the compositions after the time-temperature equilibrium heat treatment.

The second factor involved during equilibrium heat treatments is volatilization of certain oxides, leading in some cases to compositions which are far different from the original calculations and the original starting material mixtures. In Chapter 1 (Section VI.A), a group of "ceramic" oxides was listed whose vapor pressures were very substantial between 700–1500°C. If phase equilibrium studies are done with oxides of this type, the compositions should be heat treated in enclosed platinum, platinum alloy or other containers which will not react with the oxide composition and prevent the volatile oxides from leaving the system.

Once again, if the compositions are not heat treated in closed containers which prevent loss of vapor, it is absolutely essential to make chemical analyses of the compositions after the equilibrium heat treatment.

If reaction temperatures between $1700°$-$2400°C$ are involved in systems containing BeO, MgO, CaO, Y_2O_3, La_2O_3, SiO_2, SnO_2 and Ta_2O_5, which are considered (for example) to be "very refractory" oxides, it is practically impossible to use an "unreactive" container and the loss of a certain amount of the oxide during heat treatments will again require chemical analyses.

Some of the very common types of systems where the vapor loss must be controlled or chemically analyzed are those which contain alkali oxides, CdO, PbO, B_2O_3, Ga_2O_3, Bi_2O_3, GeO_2, P_2O_5, MoO_3 and WO_3.

E. Chemical Composition and Chemical Analysis

In current phase equilibrium research, the composition and purity of starting materials are usually very well known. In some cases, spectrochemical and chemical analyses must be made. After equilibrium or non-equilibrium heat treatment of compositions in one-component and especially multicomponent systems are made, it is frequently necessary to chemically analyze the final product, especially if there has been small or large reaction with the container or vaporization of certain constituents at temperatures between 700-$2500°C$. In most cases, the spectrochemical and chemical analyses are relatively standard procedures, but occasionally, when less common elements are involved, the accuracy and precision of the analysis becomes a serious problem.

F. Phase Identification

i. Primary Methods
(a) Optical Microscopy: Historically, when the Geophysical Laboratory at the Carnegie Institution of Washington, D.C. first began to do silicate phase equilibria around 1910, the only apparatus used for phase identification was the optical microscope, particularly the polarizing microscope which had been used for decades prior to 1910 by mineralogists to identify natural minerals. The polarizing microscope was highly satisfactory to determine the refractive indices, birefringence, dispersion and other optical properties of both crystals and glass, providing both were of reasonable size to identify, using 100 to 1000X magnifications. Problems occured when it was necessary to determine solid solubility limits and liquid immiscibility in multicomponent systems. Current optical microscopy of ceramic phases and materials includes the use of transmission and

reflection techniques, using ordinary light, monochromatic light (such as sodium vapor lamp, $Na_D \lambda = 589 \, m\mu$), special light sources, and polarized light.

(b) X-ray Diffraction: Although Roentgen discovered x-rays in 1895, it was not until 1912 that M. von Laue discovered that crystals could act as diffraction gratings for x-rays. After these two marvelous discoveries, it was not until 1917 that Debye and Scherrer (Germany) and Hull (United States) found that characteristic x-ray diffraction patterns could be obtained from fine-grained crystalline aggregates. Around 1927, determination of crystal structures became a standard procedure as did identification of crystal compositions by powder diffraction. After 1927, identification of crystals by optical microscopy and x-ray diffraction became a standard procedure for identification of phases in all types of phase equilibrium studies, bringing together crystal structure analysts, geochemists, mineralogists, metallurgists and ceramists in a manner which had not previously existed. The number of textbooks available on optical microscopy and x-ray diffraction is so large, so well known and so easily obtainable that it is not necessary to provide lists at this time.

(c) Electron Microscopy: The general use of electron microscopy in various types of research began after World War II, and in the 1950's, the examination of glass and ceramics began. If one examines Table 5 in Appendix A, it is obvious that an enormous amount of electron microscopy was used to determine phase separation (liquid immiscibility) in glasses in the 1960's. At the same time, transmission electron microscopy was used in crystalline ceramics, glass-ceramics, ceramic composites, metal-ceramics, plastic-ceramics and any kind of system which involved ceramic materials. During the 1960's and 1970's, scanning electron microscopy began to be used as well as transmission microscopy and the electron microscopy techniques became so useful and powerful that, starting around 1975, the Journal of the American Ceramic Society has provided standard electron microscope photographs as part of the rear cover on each publication. Again, the number of textbooks available on electron microscopy is so well known that it is not necessary to provide a list.

ii. Secondary Methods
In addition to the very major optical and electron microscopic methods and x-ray diffraction techniques which are used for over 95% of direct phase identification in phase equilibrium studies, there are several secondary methods of phase identification which are often extremely helpful supplements to the three primary techniques. These methods are:

1. Thermal analysis, Hume-Rothery (6), Chapters 10, 11 and 14, Rhines (7), Chapter 21.
2. Differential thermal analysis, Mackenzie (8, 9), Smothers (10), Smothers and Yao (11) and Wendlandt (12).

3. Thermal gravimetric analysis.
4. Special electrical or electronic measurements.
5. Special optical property measurements, infrared, ultraviolet, etc.
6. Dilatometric measurements.
7. Mössbauer spectroscopy.

MacChesney and Rosenberg (13) have provided a chapter titled "The Methods of Phase Equilibria Determination and Their Associated Problems" which can be used as a supplement to the above 116 data.

III. GLASS-FORMING SYSTEMS V S. NON-GLASS FORMING SYSTEMS

A. General

The history of phase equilibrium studies of sluggish silicate systems by the Geophysical Laboratory of the Carnegie Institution of Washington, D.C. is extremely classical. Rapid chemical and metallurgical reactions had been studied in conventional ways for many years before the Geophysical Lab introduced the "quench" furnace, the "quench" technique and many other techniques for the phase equilibrium determination of sluggish silicate systems. The primary intention was to understand geological, geochemical, geophysical and mineralogical relations, and the origin and evolution of natural rocks and minerals. Classical papers by N. L. Bowen and his associates on the evolution of igneous rocks and other books and papers on rock and minerals emphasize the goals of the Geophysical Lab at that time. The purpose was only distantly related to commercial ceramic products in the period 1910-1920, but starting with the Al_2O_3-SiO_2 diagram by Bowen and Greig in 1924, the geochemical and mineralogical relationships to ceramic science and technology became parallel and close with respect to all silicate systems.

Since 1945, phase equilibria in ceramics has been involved with "rapid" systems as well as "sluggish" silicate systems, most of which are non-glass forming types. Titanate and zirconate systems are some of the best examples of relatively rapid, non-glass forming systems which require techniques other than "quenching" to establish equilibrium diagrams.

B. Glass-forming Systems

Many systems besides silicate systems are glass-forming types, for example, common types such as borate, germanate and phosphate, and relatively uncommon types such as tellurate, vanadate, aluminate, carbonate, As_2O_3, Sb_2O_3, Bi_2O_3, MoO_3, WO_3, fluoride and sulfide systems. At this point, reference to appropriate textbooks on glass composition and properties should be mentioned such as

Rawson (14), Scholes and Greene (15), Pye, Frechette and Kreidl (16), and Doremus (17).

If a system is glass-forming, this means that the liquids can be undercooled and brought to room temperature as glasses which can be crushed to a powder, placed in platinum envelopes, equilibrated to various temperatures and times, "quenched" to room temperature after equilibration and conveniently re-examined at room temperature for the crystalline and glass phases which had existed during the equilibrium (or non-equilibrium) heat treatment. This is a capsule description of the so-called "quenching" method for the determination of the phase equilibria in glass-forming systems. It has been described briefly by Levin, Robbins, and McMurdie (18) in the 1964 Compilation of Phase Diagrams for Ceramists (p. 29–31), which in turn contains the references to the original detailed descriptions of the method in several Geophysical Laboratory publications. It is essential to have platinum ware for original melting of the glasses, platinum foil for containing the crushed powders, suitable tube furnaces for equilibration, a means of precise temperature measurement, adequate quenching techniques, quenching liquids such as mercury, carbon tetrachloride, or water, and tools for phase identification. Another excellent description of the quenching technique which is used for ceramic systems was given by P. D. S. St. Pierre (19) during investigation of basic bone china systems. Photographs of the quenching furnace, temperature measurement, temperature controller and the entire system are provided.

An essential part of the process is to measure the refractive indices of the original glass melts in a binary, ternary, or quaternary system and make plots of refractive index versus composition. This procedure permits the determination of the composition of any glasses obtained during subsequent quenching trials by means of their refractive indices, even though the glass may be present with substantial amounts of one or more crystalline phases. This technique is usually much easier than mechanically separating the glass phase (which is often almost impossible) and analyzing it by standard chemical methods.

Temperature measurement at equilibrium is most often done by Pt-Pt, 10% Rh thermocouples, up to approximately 1650°C and with an accuracy of ± 0.5° to ± 2.0°C. At lower temperatures, up to 1200°C, Chromel-Alumel couples can be used with great accuracy, and above 1650°C, other platinum alloy combinations, iridium alloy combinations and W-Mo, W-Rh alloys can be used, but in the later cases, special atmospheres must be used to prevent oxidation. The most common temperature calibration standards are gold, 1062.6°C, diopside ($CaO \cdot MgO \cdot 2SiO_2$), 1391.5°C, $Li_2O \cdot SiO_2$, 1201°C, $\alpha CaO \cdot SiO_2$, 1544°C, and anorthite ($CaO \cdot Al_2O_3 \cdot 2SiO_2$), 1550°C. The synthetic oxide compositions must be of extremely high purity.

As temperatures are increased, container problems become very formidable and platinum and its alloys cannot be used above about 1700°C. Iridium is a

possible container material, but it is costly and not entirely unreactive above
1900°C. Above 2000°C, the problem of remaining "on composition" has to be
questioned because the vapor pressure of liquid and solid phases can no longer
be ignored and encapsulating techniques or other special methods must be
employed. In some systems, encapsulation must be used even at 1000°C. Borate,
phosphate, germanate, gallate, tellurate and molybdate systems develop appreciable vapor pressures from liquid and solid phases and one is forced into
platinum or platinum alloy encapsulation if reliable phase diagrams are expected.

The definition of a "glass-forming system" for phase equilibrium purposes
is highly dependent on the quantity of material used and the rate of cooling of
the melt. A glass technologist or engineer usually requires that he be able to obtain at least 100 grams of glass by relatively slow cooling, but in phase equilibrium studies, a milligram of glass obtained by "splat" cooling is often defined as
a suitable starting material. The rate of cooling of 0.1 to 0.0001 of a gram of
glass, glass plus one to 3 or 4 crystals, a crystal or a mixture of crystals during
quenching is dependent not only upon the heat capacity of the materials but
also upon the size of the sample, the size of the quench furnace, the quenching
technique and the cooling liquid or gas. Conventionally, the quenching rate of
0.01 gram of the above compositions from 1400–1700°C to room temperature
would be about 1500°C per second. The quench rate is actually extremely important because one is assuming that the identification of phases at room temperature is exactly equivalent to whatever phase or phases existed at high temperature. The technique of "splat-cooling" has been used by metallurgists for many
years and in 1973, Takamori and Roy (20) used splat-cooling and flame spraying
techniques for a closer look at the $Al_2O_3-SiO_2$ system. The cooling rates should
have been considerably greater than 1500°C/sec. in this case.

Now that metallurgists are seriously involved with metallic glasses, the
cooling rates are known to be around 10^4, 10^6 or even 10^{10}°C/second. A metallurgical quotation (21) is as follows: "The practice of metallurgy is centuries old,
while the art and science of rapid solidification technology (RST) is barely 20
years old. In spite of its recent emergence, RST is having a significant impact on
technological developments."

C. Non-glass Forming Systems

Many ceramic systems do not form glass, even in minute quantities or by extremely rapid cooling. Titanate, zirconate, and tungstate systems are good examples of systems which do not ordinarily form glass and for which the "quenching" method is not generally used unless it is necessary to "quench in" a high
temperature crystalline phase to room temperature.

Such systems have to be examined by the "dynamic" methods of thermal
analysis, differential thermal analysis and thermogravimetric analysis, or by the

"static" methods of high temperature microscopy, high temperature X-ray diffraction, or other high temperature methods whereby the system is examined for structure, composition, chemical, or physical properties AT some temperature between room temperature and 2500°C. Experimental difficulties generally become multiplied as the temperature is increased and fewer and fewer reliable methods are available. For example, during high temperature X-ray diffraction studies, very often the vigorous thermal motion of the atoms tend to interfere with the appearance of the normal diffracted intensities one would expect at room temperature and patterns become diffuse and more difficult to interpret.

Numerous books or papers (13) are readily available on the various high temperature techniques of DTA, TGA, light microscopy, electron and scanning electron microscopy, X-ray diffraction, electron diffraction, thermal expansion, and electron microprobe analysis, and no bibliography need be provided.

IV. SPECIAL TECHNIQUES

Three special techniques should be mentioned which will be of increasing importance to ceramic science and engineering as phase equilibrium research and diagrams continue to be produced.

A. High Pressure

Since the original pioneering work of P. W. Bridgman, various workers have extended the technique of synthesis and analysis of metals, oxides, salts, and polymers at high to ultra-high pressure. Some recent references to high pressure techniques, methods and materials are given in Table 6 of Appendix A.

B. Controlled Oxygen Pressure

Increasing emphasis will be placed on the investigation of phase relationships in ceramic systems using controlled oxygen pressure. It is of obvious importance for electronic and magnetic ceramics (titanates, ferrites, piezoelectrics) where properties are critically dependent on very small changes in stoichiometry and for ceramic pigments which are dependent on the control of the oxidation state of the transition metals and rare earths. Control of oxygen pressure has been used in phosphor research for many years.

The outstanding references on the technique and interpretation of controlled oxygen pressure are by Muan (22) and Muan and Osborn (23).

C. Hydrothermal Methods

Geologists, mineralogists, geochemists, and ceramists have all employed this technique to accelerate the rate of oxide reactions and produce an equilibrium state which otherwise may never be achieved with "dry" systems. Many ceramists and persons engaged in inorganic synthesis have long recognized that H_2O in the form of liquid or vapor can accelerate high temperature reactions at atmospheric pressure. The application of pressure in addition to H_2O and temperature will remain for the most part in the hands of research scientists or only be used in small scale production of relatively costly materials for many years in the future, but eventually the method will find a place in large scale production of ceramic products. The new textbook by Ehlers (24) deserves special mention at this point. Even the titles of the various chapters in this book should interest a ceramist and be suggestive of techniques which might well be used in future production of at least specialty products, if not high tonnage products.

In concluding the discussion of techniques for studying phase equilibria, it should be emphasized that normally no single technique may be completely satisfactory. Several methods are usually used to determine the phase relationships and the final evaluation of the data and construction of the diagram are based on information derived from all of these approaches. For example, X-ray diffraction analysis, while extremely convenient and yielding qualitative and sometimes extremely quantitative data on crystalline phases, may not be very useful in locating liquidus temperatures. The quenching method and subsequent use of the light microscope is the effective technique which easily and accurately determines the presence or absence of a small quantity of crystals and the (binary) liquidus curve or (ternary) liquidus surface of glass-forming systems.

V. DISCUSSION AND EVALUATION OF EXPERIMENTAL PROCEDURES ON REAL SYSTEMS

A. One Component Systems

i. Pressure Variable

Two items in Chapter 2 indicate that pressure research is needed for the development of a greater number of quantitative one component p–T diagrams: 1) the list of one component systems given on page 53 and 54 is extremely small, even though it includes practically all of the systems which are presented in the very large 1964, 1969, 1975 and 1981 compilations of "Phase Diagrams for Ceramists". 2) The discussion of the systems SiO_2, TiO_2, GeO_2, ZrO_2, Fe and C in Chapter 2 indicate that although much is known about the temperature variable and temperature-composition features of the systems, relatively little is known about the pressure variable. Each oxide, fluoride and sulfide system requires much more pressure (p–T) research in order to produce a really quantitative diagram. A statement by Donald H. Lindsley (25) about the system TiO_2 emphasizes the need for pressure research. When discussing TiO_2 he states, "Although many studies of the synthesis and inversion of these polymorphs (rutile, anatase, and brookite) have been made, no phase diagram is given here because most (probably all) published "phase boundaries" between rutile and anatase or brookite reflect *kinetics* rather than equilibrium."

ii. Reduction, Non-stoichiometry
In addition to the major need for pressure research, one has to be very alert for non-stoichiometry in one component systems. TiO_2 is once again an excellent example, as indicated by the $Ti-O_2$ diagram (Figure 2-7) shown in Chapter 2. Many other one-component oxides, fluorides and sulfides, which are conventional one component systems at room temperature may become two-component metal-oxygen, $M-F_2$, or $M-S_2$ systems at elevated temperatures. Even the most refractory, stable and inert ceramic oxide, ZrO_2, is now known to under go a slight reduction to ZrO_{2-x} at very high temperatures.

iii. Special Techniques
Now that ceramic science and technology has extended to fluoride, sulfide, nitride, oxyfluoride, oxysulfide, oxynitride (Sialons) and other special materials beyond the conventional oxide, borate and silicate systems, one must realize that special techniques are necessary to determine the fundamental phase diagrams which support the commercial use of such systems.

One of the most outstanding examples of need for special techniques is given in Chapters 4 and 5 of Sulfide Mineralogy, Mineralogical Society of America (26). The thirteen papers by Kullerud, especially two (27, 28) which deal with techniques, provide the kind of information which is needed to enter into sulfide synthesis and sulfide phase equilibria. Similar papers on fluoride systems have been provided by R. E. Thoma and his associates at the Oak Ridge National Laboratory.

B. Two Component Systems

i. Pressure Variable, Influence on Binary Phase Diagrams
In the beginning sections of Chapter 3, it was said that two component $p-T-x$ systems (which would require three-dimensional, pressure, temperature and composition diagrams to illustrate the equilibrium relationships) would be set aside in favor of the far more simple two dimensional T-x diagrams because most commercial ceramic products were produced at constant pressure, usually one atmosphere.

It should now be emphasized that pressure (as well as temperature) would have a significant influence on polymorphic transitions of crystals, melting behavior, dissociation behavior, invariant points, liquidus curves, densities of crystals and glasses, the extent of partial solid solution, and the extent of solid and liquid immiscibility in two component systems. Pressure would also have a substantial influence on the chemical and physical properties of both crystalline and glass phases.

Pressure-temperature diagrams are occasionally produced for binary intermediate compounds, but if one reviews the binary oxide diagrams in all four
issues of Phase Diagrams for Ceramists, essentially all of them are conventional
composition v.s. temperature diagrams (condensed systems). In the 1964 issue
(pages 96-97), the p–T diagram for kyanite-sillimanite-andalusite is shown, along
with a "condensed" Al_2O_3-SiO_2 diagram which was produced at 25.2 kbars
(29).

Diagrams which show the influence of pressure on incongruently melting
or dissociating (upper temperature limit) binary compounds are shown by Eitel
(30) in paragraph 30 (Figure 6) and by Ehlers (24) (Figure 118) on page 126 of
Chapter 6, "Systems Under Confining Pressure". The Eitel diagram shows that
an increase in pressure causes an incongruently melting binary compound to become a congruently melting compound. The Ehlers diagram shows that a
dissociating compound (upper temperature limit) becomes an incongruently
melting intermediate compound and finally a congruently melting melting compound as pressure increases. It also shows the change in the liquidus surface as
pressure increases. (Five diagrams are shown, $P_1 < P_2 < P_3 < P_4 < P_5$)

ii. The Influence of H_2O and CO_2 on Oxide Systems

When hydrates or carbonates are used as starting materials in synthesis of crystalline compounds or production of equilibrium diagrams, it is necessary to prove
that H_2O and CO_2 are not components of the system after equilibrium heat
treatments. Typical hydrates are alkali, alkaline earth and trivalent metal hydrates
and typical carbonates are alkali and alkaline earth types (Li, Na, K, Rb, Cs, Mg,
Ca, Sr, Ba). A specific example of both H_2O and CO_2 retention was created by a
recent paper (31) on "optical and magnetic properties of some transition metal
ions in barium phosphate glass". The use of $BaCO_3$, BaO, H_3PO_4 and P_2O_5 and
the subsequent heat treatment at 900°-1000°C was said to produce $Ba_3(PO_4)_2$
glass, even though it is well known that anhydrous "dry" $Ba_3(PO_4)_2$ is not glass
forming. An analysis (32) of the materials and heat treatments indicated that the
so-called "$Ba_3(PO_4)_2$ glass" must have contained H_2O and CO_2. A chemical
analysis would have indicated how much H_2O and CO_2 was contained in the
BaO-P_2O_5-H_2O-CO_2 composition which was used for several highly precise
optical and magnetic measurements. From a phase equilibrium or glass synthesis
standpoint, a study of glass formation in the ternary systems BaO-P_2O_5-H_2O
and BaO-P_2O_5-CO_2 and the quaternary system BaO-P_2O_5-H_2O-CO_2 would be
very informative.

It is well known that H_2O and CO_2 are difficult to eliminate from many
oxide, borate, silicate and phosphate systems, if the heat treatments are between
500°-1500°C. Therefore, as stated in Section IIE in this chapter, chemical and
compositional analyses after equilibrium or non-equilibrium heat treatments are
often vitally necessary before making chemical and physical property measurements.

C. Revision of Diagrams

i. General

In Section VA, it was emphasized that p-T diagrams of one component systems (for example SiO_2, TiO_2, GeO_2 and ZrO_2) have not yet been completed because of the need for more high pressure research.

In the case of conventional two component "condensed" systems, "constant pressure" diagrams need to be completed or revised due to experimental techniques other than pressure. Difficulties in determination of polymorphism, melting behavior, stoichiometry of intermediate compounds, dissociation behavior, invariant points (temperature and composition), the extent of partial solid solubility, the extent of unmixing of solid solutions, the extent of liquid immiscibility and the determination of liquidus curves are all factors which lead to eventual revision of binary phase diagrams. For example, one could always question the temperature-composition location of eutectic, peritectic, monotectic, eutectoid and peritectoid invariant points, or the indifferent points in continuous solid solution diagrams (maximum or minimum).

ii. Real Systems

(a) ZrO_2 binary systems: For many years, ZrO_2 was not thought to be a satisfactory commercial product due to the large volume change which occurred during the monoclinic-tetragonal inversion. In the period 1935-1955, it was found that additions of MgO, CaO, Y_2O_3 and other oxides produced a cubic form of a ZrO_2 solid solution which had a linear thermal expansion of 110 X 10^{-7} cm/cm/°C from room temperature to 1500°C. This led to commercial products of 100% cubic ZrO_2 or combinations of cubic and monoclinic ZrO_2 or cubic, tetragonal and monoclinic ZrO_2. Before 1962, binary phase diagrams involving ZrO_2 always revealed the influence of the second component on the monoclinic-tetragonal inversion of ZrO_2, and the stability fields of both polymorphs. However, the stability field of cubic ZrO_2 was never related to the ZrO_2 end member until Smith and Cline (33) proved by high temperature X-ray diffraction technique that a displacive, "snap" inversion between tetragonal and cubic ZrO_2 took place around 2300°C. After this outstanding experimental determination in 1962, all binary ZrO_2 phase diagrams have been revised as indicated by the Stubican and Hellman (34) paper in Science and Technology of Zirconia and the 1969, 1975 and 1981 editions of "Phase Diagrams for Ceramists". The tetragonal-cubic inversion of pure ZrO_2 is now shown, along with the stability field of the cubic ZrO_2 solid solution, starting at the 100% ZrO_2 end member. More information is now known about the influence of the second component on the stability fields of the tetragonal and monoclinic ZrO_2 phases, the intermediate compounds and the phase relationships in the system in the area of the second component.

In the case of La_2O_3-ZrO_2, three (possible) versions of the system are shown in the 1964 issue of "Phase Diagrams for Ceramists" (p. 136, figure 346).

In the 1969 issue (p. 103, figure 2374) and 1981 issue (p. 133, figure 5232), modifications of the 1964 diagrams are shown.

In the case of Nd_2O_3–ZrO_2 and Y_2O_3–ZrO_2, diagrams are shown in the 1964 issue (p. 105) and subsequent revisions are shown in the 1969, 1975, and 1981 issues.

In the case of UO_2–ZrO_2, the first group of diagrams is presented on pages 70 and 71 (figures 119-124) of the 1964 issue and subsequent modifications are shown in the 1969 and 1975 issues.

(b) Other Systems Which Have Undergone Revisions: Binary systems which have undergone revisions and appear in three of the four major collections of "Phase Diagrams for Ceramists" are:

Fe-S	Al_2O_3-Y_2O_3
MgO-H_2O	MgO-Ta_2O_5
CaO-CO_2	

Binary systems which have been revised and appear in all four collections are:

Fe-O	CaO-SiO_2
U-O	Al_2O_3-SiO_2

Ternary systems which appear in three collections are:

Fe-Mg-O	Na_2O-Fe_2O_3-SiO_2
Fe-Mn-O	MgO-CaO-Al_2O_3
Fe-Y-O	MgO-SiO_2-TiO_2
Fe-Si-O	CaO-BaO-SiO_2
Fe-Ti-O	MgO-MgF_2-SiO_2
LiF–NaF–AlF_3	CaO-CaF_2-SiO_2
Na_2O-B_2O_3-SiO_2	Na_2O-K_2O-P_2O_5

Four systems which appear in all four collections (with modifications and revisions) are:

MgO-CaO-SiO_2
CaO-SiO_2-P_2O_5
MgO-CaO-Al_2O_3-SiO_2
Na_2O-MgO-CaO-Al_2O_3-SiO_2

(c) The Al_2O_3–SiO_2 System: Like the one-component system SiO_2, the binary system Al_2O_3–SiO_2 is of tremendous historical and practical importance for ceramists and, from a teaching and learning standpoint, there is probably no other system which illustrates so well the difficulties involved in the experimental determination of phase equilibria in a binary system and the gradual evolution of a more correct picture of the phase relationships over a period of more than 60 years.

The natural minerals quartz (SiO_2), corundum ($A\ell_2O_3$), kaolinite ($A\ell_2O_3 \cdot 2SiO_2 \cdot 2H_2O$), "white", "buff" and "red-burning" clays, kyanite, andalusite, and sillimanite were the basis of historical ceramic products such as pottery, porcelain, bricks, tile, "whiteware" and refractories and, more recently, spark plugs, fusion cast refractories and fiber glass. In modern times, synthetic chemicals such as silicic acid, Cab-O-Sil, aluminum hydrates, aluminum nitrates, gels, etc. are used for commercial $A\ell_2O_3 - SiO_2$ ceramic products in addition to the above minerals. The history of the system $A\ell_2O_3 - SiO_2$, involving mullite and all of the above minerals, chemicals, and ceramic products would require a small textbook to properly record the information, similar to the R. B. Sosman book on "The Phases of Silica".

However, from the standpoint of the phase diagram of the system $A\ell_2O_3 - SiO_2$ and the understanding by the student of compositional control, heat treatment, container problems, quenching and phase analysis techniques, it is appropriate to very briefly comment on the history and current version of the phase diagram.

The first version of the diagram at one atmosphere was provided by Shepard, Rankin and Wright (35) in 1909 at a time when the intermediate compound was thought to be $A\ell_2O_3 \cdot SiO_2$ (kyanite, andalusite, sillimanite). In 1924, Bowen and Greig (36) showed that mullite (3 $A\ell_2O_3 \cdot 2$ SiO_2) was the intermediate compound at one atmosphere and their diagram indicated that mullite melted incongruently at approximately 1810°C. A short paper by R. B. Sosman (37) has an interesting history about the discovery of mullite as a natural mineral, its relationship to kyanite, andalusite and sillimanite and to the Bowen and Greig $A\ell_2O_3 - SiO_2$ research. For approximately 25 years, many people were involved with investigations of mullite, alumino silicate glasses and various aspects of the system, but it was not until 1962 that Aramaki and Roy (38) produced a new diagram which indicated that mullite melted congruently at approximately 1850°C and took $A\ell_2O_3$ into solid solution under stable and metastable conditions. (See 1964 collection of Phase Diagrams for Ceramists, page 123).

In 1963, Konopicky (39) noted that investigators who worked in air reported an incongruent melting point for mullite and those who worked in a special atmosphere, vacuum, or in sealed noble metal tubes reported a congruent melting point. He further stated that the system may not be a true binary system at high temperatures, due to reduction and volatility, and that it may be most properly represented as the ternary system $A\ell - Si - O_2$, permitting the composition of phases to be most easily and properly expressed.

Ganguli and Saha (40), MacDowell and Beall (41) and Takamori and Roy (42) provided data on the immiscibility (primarily metastability) of glasses in the system and in the last decade, Joseph A. Pask and his associates B. F. Davis, Ilhan A. Aksey, S. H. Risbud and V. F. Draper have provided several papers on stable and metastable relations in the system resulting in the most recent $A\ell_2O_3$-SiO_2 phase diagram (1981 Edition of "Phase Diagrams for Ceramists").

A few comments can be made on the 60 year (but mainly in the last 20 years) modification of the $A\ell_2O_3$-SiO_2 diagram.

(1) The pressure variable: If the system $A\ell_2O_3$-SiO_2 was examined by both temperature and pressure, a three-dimensional (p–T–x) solid model would represent the relationships between the silica phases (quartz, tridymite, cristobalite, coesite, stishovite and liquid (glass) and the alumina phases (primarily corundum and liquid). All silica phases, except stishovite, are four-coordinated oxygen structures and the primary $A\ell_2O_3$ phase (corundum) is a six-coordinated oxygen structure. The pressure variable would therefore produce interesting relationships between $A\ell_2O_3$ and SiO_2 phases, and expecially between corundum and stishovite (six-coordinated rutile-type structure).

(2) Oxygen pressure: The comment by Konopicky that the system $A\ell_2O_3$-SiO_2 might be more properly represented as the ternary system $A\ell$-Si-O_2 is significant. At temperatures around $1800°$-$2000°C$, certain experimental techniques and variables (atmospheres, containers, quenching, temperature and time) might result in compositional changes which bring the system off the binary join and into the $A\ell$-SiO-O_2 ternary.

In conclusion, in spite of all the excellent research which has been done on the system $A\ell_2O_3$-SiO_2, particularly in the last decade, even the one atmosphere diagram may need further revision on liquid and crystalline relationships. Based on structural considerations, it is likely that stable liquid immiscibility may exist on the $A\ell_2O_3$-rich side of the diagram and be related to the metastable immiscibility which is now diagrammed. It may also be necessary to revise the crystalline relationships in the mullite area. (stable and metastable)

In addition to the extreme care which must be considered on atmospheres, containers, quenching, temperatures and times, it would also be valuable in this case (at one atmosphere) to consider the use of a well chosen, satisfactory third component which could be used at a 1-5% level to help understand liquid and solid relationships in the mullite and alumina-rich area. Some useful third components could be Li_2O, B_2O_3 and P_2O_5, each of which could substantially lower equilibrium temperatures and times, provide easier atmosphere, container and quenching techniques, and enable equilibrium and non-equilibrium extrapolation toward the $A\ell_2O_3$-SiO_2 system (43).

VI. THERMODYNAMICS AND COMPUTER CALCULATION

As mentioned in the Preface, some of the most significant sources of basic and applied thermodynamics of phase diagrams are the texts of Gordon (44), Reisman (45) and Prince (46), especially on one and two component systems.

Two chapters in "Phase Diagrams, Materials Science and Technology", are also useful for understanding thermodynamic relationships in one and two component phase diagrams. The first is by Y. K. Rao on "Thermodynamics of Phase Diagrams", Chapter I in Volume 6-I, Theory, Principles and Techniques of Phase Diagrams (1970) and the second is by O. Taft Sorensen, on "Thermodynamics and Structure of Nonstoichiometric Binary Oxides", Chapter II in Volume 6-V.

Finally, computer calculation is described by Larry Kaufman and Harold Bernstein in "Computer Calculation of Phase Diagrams", Academic Press, 334 pp., 1970. The computing is primarily for metallic systems but there is special reference to refractory materials.

APPENDIX A

A Convenient List of Real Binary Diagrams which Illustrate Simple·Eutectic Diagrams, Congruent Melting, Incongruent Melting, Dissociation, Continuous Solid Solution, Partial Solid Solution, Polymorphism, Liquid Immiscibility and the Revision or Modification of Binary Diagrams. From "Phase Diagrams for Ceramists", American Ceramic Society (Tables 4.1-4.4). References on Liquid Immiscibility (Table 4.5).

Table 4.1 Simple Eutectic Systems

System	Page	Figure	System	Page	Figure
1964 Edition					
$Li_2O_3 \cdot 3TiO_2 - TiO_2$	92	185	$LiF - CeF_3$	421	1474
$Na_2O - Fe_2O_3$	93	189	$LiF - PuF_3$	421	1475
$MnO - P_2O_5$	114	278	$LiF - UF_3$	421	1476
$Al_2O_3 - P_2O_5$	123	318	$NaF - RbF$	422	1481
$LiF - KF$	415	1452	$NaF - PbF_2$	424	1488
$NaF - KF$	415	1453	$BeF_2 - MgF_2$	429	1514
$LiF - NaF$	419	1467	$MgF_2 - CaF_2$	430	1519
$LiF - CaF_2$	420	1470			

Table 4.1 (*Continued*)

System	Page	Figure	System	Page	Figure
		1969	Edition		
$BeO-WO_3$	83	2294	$LiF-NdF_3$	364	3371
$CaO-La_2O_3$	84	2299	$LiF-PrF_3$	365	3373
$LiF-KF$	360	3346	$LiF-SmF_3$	365	3375
$LiF-CeF_3$	364	3363	$NaF-PbF_2$	366	3381
$LiF-LaF_3$	364	3369	$NaF-BF_3$	367	3384
		1975	Edition		
$BeO-Sm_2O_3$	102	4306	$SnO-SiO_2$	126	4358
$CdMoO_4-MoO_3$	110	4322	$NaF-RbF$	356	4808
$NiO-Y_2O_3$	121	4346	MgF_2-CaF_2	361	4820
		1981	Edition		
$BeO-ThO_2$	94	5139			

Table 4.2 Systems with Intermediate Compounds

System	Page	Figure	System	Page	Figure
	a.	Congruent	Melting		
		1964	Edition		
K_2O-GeO_2	87	166	$NaF-BeF_2$	422	1482
K_2O-SiO_2	87	167	$NaF-FeF_2$	423	1485
Na_2O-GeO_2	93	190	$NaF-MgF_2$	423	1486
	93	191	$NaF-NiF_2$	423	1487
$BaO-Al_2O_3$	97	206	$NaF-ZnF_2$	424	1489
$MgO-P_2O_5$	113	272	$RbF-CaF_2$	427	1502
$PbO-P_2O_5$	117	288	BeF_2-PbF_2	429	1515
$ZnO-Nb_2O_5$	120	304			
		1969	Edition		
Cs_2O-SiO_2	77	2274	$Y_2O_3-SiO_2$	108	2391
Li_2O-GeO_2	79	2280	$LiF-CsF$	357	3332
$Li_2O-V_2O_5$	80	2282	$CsF-PbF_2$	357	3336
Na_2O-GeO_2	81	2285	$CsF-RF_3$	358–359	3337–3343
Rb_2O-SiO_2	82	2290	$KF-CaF_2$	360	3348
$BaO-WO_3$	83	2293	$KF-NbF_5$	363	3359
$CaO-WO_3$	86	2306	$LiF-ErF_3$	364	3365

Table 4.2 *(Continued)*

System	Page	Figure	System	Page	Figure
CdO–P$_2$O$_5$	86	2307	LiF–LuF$_3$	364	3370
SrO–B$_2$O$_3$	92	2333	LiF–TmF$_3$	365	3377
SrO–WO$_3$	94	2337	LiF–YbF$_3$	365	3378
Al$_2$O$_3$–Nb$_2$O$_3$	95	2342	RbF–YF$_3$	372	3408

1975 Edition

System	Page	Figure	System	Page	Figure
Cs$_2$O–GeO$_2$	81	4266	Cr$_2$O$_3$–La$_2$O$_3$	144	4397
Cu$_2$O–P$_2$O$_5$	81	4267	Cr$_2$O$_3$–Sm$_2$O$_3$	145	4399
Na$_2$O–TeO$_2$	91	4286	La$_2$O$_3$–Nb$_2$O$_5$	154	4421
Na$_2$O–MoO$_3$	93	4291	La$_2$O$_3$–Ta$_2$O$_5$	155	4422
Tℓ$_2$O–TℓBO$_2$	98	4299	Nb$_2$O$_5$–ThO$_2$	168	4450
BaO–Cr$_2$O$_3$	99	4300	KF–ScF$_3$	355	4805
PbO–WO$_3$	125	4356	LiF–CrF$_3$	356	4807
Cr$_2$O$_3$–Eu$_2$O$_3$	143	4395	MnF$_2$–BaF$_2$	360	4817
Cr$_2$O$_3$–Gd$_2$O$_3$	144	4396	ZnF$_2$–PbF$_2$	364	4827

1981 Edition

System	Page	Figure	System	Page	Figure
Ag$_2$O–V$_2$O$_5$	85	5115	SrO–TeO$_2$	112	5181
Na$_2$O–SiO$_2$	87	5122	SrO–MoO$_3$	113	5184
Rb$_2$O–TeO$_2$	90	5128	ZnO–Nb$_2$O$_5$	115	5187
BaO–WO$_3$	93	5137	Cr$_2$O$_3$–Gd$_2$O$_3$	121	5201
CaO–WO$_3$	97	5147	Cr$_2$O$_3$–La$_2$O$_3$	121	5202
MgO–TeO$_2$	103	5161	Cr$_2$O$_3$–Nb$_2$O$_3$	122	5203
PbO–GeO$_2$	106	5168, 5169	Cr$_2$O$_3$–Sm$_2$O$_3$	122	5204
PbO–SiO$_2$	107, 108, 109	5170, 5171, 5172, 5173	Cr$_2$O$_3$–Yb$_2$O$_3$	123	5207
PbO–CrO$_3$	109	5174			

b. Incongruent Melting

1964 Edition

System	Page	Figure	System	Page	Figure
K$_2$O–Nb$_2$O$_5$	88	171	SiO$_2$–P$_2$O$_5$	142	364
K$_2$O–P$_2$O$_5$	88	172	LiF–CaF$_2$	413	1445
K$_2$O–Ta$_2$O$_5$	88	173	CaF$_2$–ZrF$_4$	415	1450
K$_2$O–WO$_3$	90	177	KF–BeF$_2$	416	1456
Li$_2$O–B$_2$O$_3$	91	180	KF–NiF$_2$	417	1459
Na$_2$O–P$_2$O$_5$	95	197	KF–ZnF$_2$	417	1460
BeO–BaO	96	204	LiF–ThF$_4$	421	1478
CaO–SiO$_2$	104	237	LiF–UF$_4$	422	1479

Table 4.2 (*Continued*)

System	Page	Figure	System	Page	Figure
$NiO-V_2O_5$	114	279	MgF_2-BaF_2	428	1511
$Fe_2O_3-P_2O_5$	133	338	BeF_2-CaF_2	429	1513
SiO_2-ThO_2	141	359	Ag_2S-Tl_2S	519	1887
SiO_2-ZrO_2	141	361, 362			

1969 Edition

System	Page	Figure	System	Page	Figure
$Cs_2O-Nb_2O_5$	77	2275	$KF-LaF_3$	362	3355
$Na_2O-Al_2O_3$	81	2284, 2286	$KF-PrF_3$	362	3356
$CaO-Yb_2O_3$	84	2300	$LiF-DyF_3$	364	3364
$MgO-V_2O_5$	89	2319	$LiF-GdF_3$	364	3367
$SrO-P_2O_5$	93	2336	$NaF-LaF_3$	368	3390
$B_2O_3-Al_2O_3$	94	2339	$RbF-LaF_3$	371	3405
$Bi_2O_3-TiO_2$	99	2360	Na_2S-Cu_2S	526	3991

1975 Edition

System	Page	Figure	System	Page	Figure
$Cs_2O-B_2O_3$	80	4264	$MgO-V_2O_5$	118	4341
K_2O-GeO_2	82	4269	$Y_2O_3-MoO_3$	164	4438
$Li_2O-P_2O_5$	86	4278	HfO_2-SiO_2	165	4443
$Rb_2O-Nb_2O_5$	97	4297	$Ta_2O_5-MoO_3$	177	4465
$BaO-TiO_2$	100	4302	$KF-CrF_3$	355	4804
$BaO-P_2O_5$	100	4303	$NaF-AlF_3$	357	4809
$CaO-Al_2O_3$	103	4308	NiF_2-PbF_2	364	4826
$CaO-Ta_2O_5$	108	4316	Na_2S-Cu_2S	419	4955
$MgO-TiO_2$	116	4336			

1981 Edition

System	Page	Figure	System	Page	Figure
Li_2O-TiO_2	85	5116	$CaO-MoO_3$	97	5146
$Li_2O-V_2O_5$	85, 86	5117	$CdO-V_2O_5$	97	5148
Li_2O-MoO_3	86	5118	$MgO-V_2O_5$	104	5162
$Li_2WO_4-WO_3$	87	5120	$PbO-Al_2O_3$	105	5166
Li_2O-WO_3	87	5121	$PbO-SiO_2$	107, 108, 109	5170, 5171, 5172, 5173
Na_2O-TiO_2	88	5123, 5124			
$Na_2O-P_2O_5$	88	5125	$SrO-V_2O_5$	113	5182, 5183
$Na_2O-V_2O_5$	89	5126	$SrO-WO_3$	114	5185
$BaO-Ge_2O_3$	91	5131	$ZnO-V_2O_5$	116	5188
$BaO-TiO_2$	92, 93	5135	$Al_2O_3-Nb_2O_5$	117	5191

Table 4.2 *(Continued)*

System	Page	Figure	System	Page	Figure
BaO–MoO$_3$	93	5136	Cr$_2$O$_3$–V$_2$O$_5$	123	5206
CaO–Aℓ$_2$O$_3$	95	5141	Y$_2$O$_3$–MoO$_3$	142	5254

c. Dissociation
1964 Edition
Upper temperature limit

Cu$_2$O–PbO	86	163	ZnO–TiO$_2$	120	303
MgO–B$_2$O$_3$	111	261			

Lower temperature limit

Li$_2$O–B$_2$O$_3$	91	180	Aℓ$_2$O$_3$–Y$_2$O$_3$	122	311
BaO–TiO$_2$	98	213	Nb$_2$O$_5$–Ta$_2$O$_5$	144	375
BeO–SiO$_2$	100	222	KF–ThF$_4$	418	1464

1969 Edition
Upper temperature limit

K$_2$O–SiO$_2$	79	2279	KF–LaF$_3$	362	3355
GeO$_2$–P$_2$O$_5$	109	2395	KF–YF$_3$	363	3358
SiO$_2$–ZrO$_2$	110	2400	LiF–ScF$_3$	365	3374
KF–CeF$_3$	361	3353			

Lower temperature limit

La$_2$O$_3$–SiO$_2$	102	2372	Y$_2$O$_3$–SiO$_2$	107	2388
Nb$_2$O$_3$–SiO$_2$	104	2380	Yb$_2$O$_3$–SiO$_2$	108	2391
Sm$_2$O$_3$–SiO$_2$	106	2386	CsF–ThF$_4$	359	3344

1975 Edition
Upper temperature limit

BaO–TiO$_2$	100	4302	SrO–Nd$_2$O$_5$	127	4360
NiO–B$_2$O$_3$	120	4345	B$_2$O$_3$–Eu$_2$O$_3$	138	4383
PbO–SiO$_2$	124	4352	B$_2$O$_3$–Gd$_2$O$_3$	138	4384

Lower temperature limit

CaO–GeO$_2$	104	4309	Aℓ$_2$O$_3$–TiO$_2$	135	4376
CoO–TiO$_2$	111	4324	Ga$_2$O$_3$–GeO$_2$	150	4411
PbO–SiO$_2$	124	4352	TiO$_2$–Nb$_2$O$_5$	170	4453

1981 Edition
Lower temperature limit

Li$_2$O–WO$_3$	87	5121	CaO–MoO$_3$	97	5146
Na$_2$O–SiO$_2$	87	5122			

Table 4.3 Systems With Solid Solution

Systems	Page	Figure	Systems	Page	Figure
		a. Complete s.s.			
		1964 Edition			
MgO–CoO	52	52	UO_2–Y_2O_3	139	353
MgO–FeO	54	63	KF–RbF	416	1454
Fe_3O_4–Mn_3O_4	57	72	ThF_4–UF_4	431	1526
MgO–NiO	110	258	UF_4–ZrF_4	431	1527
Al_2O_3–Cr_2O_3	121	309	BiS–Sb_2S_3	520	1894
UO_2–ThO_2	69, 70	116, 118	CoS–FeS	520	1895
		1969 Edition			
MgO–FeO	36	2154	Na_2CrO_4–K_2CrO_4	116	2418
MgO–MnO	37	2157	RbF–CsF	357	3334
PuO_2–ZrO_2	109	2396	Ag_2S–Cu_2S	525	3988
UO_2–ZrO_2	112	2404			
		1975 Edition			
PuF_6–UF_6	365	4829	Sb_2S_3–Sb_2Se_3	420	4957
		b. Partial s.s.			
		1964 Edition			
Fe–O_2	38, 39	8, 9, 10	Nd_2O_3–ZrO_2	138	350
Ti–O_2	41	22	Y_2O_3–ZrO_2	140	354
Zr–O_2	42	25	CeO_2–ZrO_2	140	355
FeO–Al_2O_3	43	26	GeO_2–SiO_2	140	357
Fe_2O_3–Al_2O_3	43	27	GeO_2–TiO_2	141	358
Fe_2O_3–Cr_2O_3	53	58	SiO_2–ZrO_2	141	361, 362
BaO–Nb_2O_5	99	214	ThO_2–ZrO_2	143	368
BeO–MgO	99	216	TiO_2–ZrO_2	143	369
BeO–Y_2O_3	100	220	TiO_2–Nb_2O_5	143	372
BeO–TiO_2	101	224	ZrO_2–Nb_2O_5	144	373
CaO–CoO	101	228	Nb_2O_5–Ta_2O_5	144	375
CaO–MgO	102	229	KF–MgF_2	417	1458
CaO–NiO	102	230	LiF–MgF_2	420	1471
CaO–ZrO_2	105	243	LiF–ZnF_2	420	1472
MgO–Al_2O_3	110	259, 260	RbF–BeF_2	426	1500
MgO–Cr_2O_3	111	262	RbF–ZrF_4	428	1508
MgO–ZrO_2	113	271	BaF_2–UF_3	429	1512
Al_2O_3–Ga_2O_3	121	310	BeF_2–PbF_2	429	1515

Table 4.3 *(Continued)*

Systems	Page	Figure	Systems	Page	Figure
Ga_2O_3–Sc_2O_3	133	339	FeS–ZnS	521	1897
La_2O_3–ZrO_2	136	346			

1969 Edition

Systems	Page	Figure	Systems	Page	Figure
Ti–O	9	2081, 2082	MgO–ZrO_2	88	2317
$Aℓ_2O_3$–Mn_2O_3	15	2099	ZnO–V_2O_5	94	2338
CaO–FeO	18	2108	$Aℓ_2O_3$–Sc_2O_3	95	2343
CaO–Mn_2O_3	20	2112	Gd_2O_3–TiO_2	101	2368
CaO–Mn_2O_3	21	2113	La_2O_3–ZrO_2	103	2374
CoO–Fe_2O_3	22	2116	Nd_2O_3–ZrO_2	105	2382
Mn_3O_4–Co_3O_4	22	2118	KF–PbF_2	361	3350
Fe_2O_3–Ga_2O_3	34	2149, 2150, 2151	NaF–R.E.F_3	367–370	3385–3400
Fe_2O_3–Mn_2O_3	37	2158	RbF–PbF_2	371	3403
FeO–TiO_2	40	2169	CaF_2–YF_3	373	3412
Fe_3O_4–ZrO_2	45	2178	Ag_2S–Bi_2S_3	525	3990
MgO–MnO	46	2182	Cu_2S–FeS	526	3992
MgO–MnO	47	2183	FeS–ZnS	526	3994
MgO–ZnO	87	2312			

1975 Edition

Systems	Page	Figure	Systems	Page	Figure
Zr–O	17	4172	Gd_2O_3–ZrO_2	152	4417
CuO–Bi_2O_3	20	4178	La_2O_3–Y_2O_3	153	4419
CaO–CoO	21	4179	La_2O_3–ThO_2	154	4420
Fe_2O_3–SnO_2	43	4210	Nd_2O_3–ZrO_2	157	4426
RuO_2–IrO_2	47	4218	Sc_2O_3–TiO_2	158	4429
K_2O–Ta_2O_5	83	4271	Sc_2O_3–ZrO_2	159	4430
Li_2O–Ta_2O_5	87	4279	Sm_2O_3–ZrO_2	160	4433
Na_2O–$Aℓ_2O_3$	88	4282	Y_2O_3–HfO_2	162	4436
Na_2O–Ta_2O_5	92	4289	Y_2O_3–ZrO_2	163	4437
CaO–NiO	102	4307	TiO_2–ZrO_2	169	4452
MgO–ZrO_2	117	4339	Nb_2O_5–P_2O_5	174	4460
Bi_2O_3–MoO_3	142	4393	Nb_2O_5–Ta_2O_5	175	4461
Cr_2O_3–Sc_2O_3	145	4398	CuF_2–PbF_2	362	4823
Gd_2O_3–HfO_2	151	4414	PbF_2–$AℓF_3$	364	4828
Gd_2O_3–ThO_2	151	4415	CdS–MnS	420	4956

1981 Edition

Systems	Page	Figure	Systems	Page	Figure
BaO–Dy_2O_3	90	5129	Gd_2O_3–ZrO_2	129	5224
CaO–La_2O_3	95	5142	Ho_2O_3–ThO_2	131	5227

Table 4.3 (Continued)

Systems	Page	Figure	Systems	Page	Figure
$CaO-Pr_2O_3$	96	5143	$Ho_2O_3-ZrO_2$	131	5228
$CaO-P_2O_5$	96	5145	$La_2O_3-ThO_2$	132	5231
$MgO-Er_2O_3$	100	5155	$La_2O_3-ZrO_2$	133	5232
$MgO-Y_2O_3$	101	5156	$Nd_2O_3-Y_2O_3$	134	5234, 5235
$MgO-Ye_2O_3$	101	5157	$Nd_2O_3-ThO_2$	135	5237
$SrO-Nd_2O_3$	110	5175	$Nd_2O_3-ThO_2$	136	5238
$SrO-Pr_2O_3$	110	5176	$Nd_2O_3-ZrO_2$	136	5239
$SrO-Sm_2O_3$	111	5177	$Sc_2O_3-HfO_2$	137	5242
$SrO-Y_2O_3$	111	5178	$Sm_2O_3-ThO_2$	138	5244
$SrO-Yb_2O_3$	112	5179	$Sm_2O_3-ZrO_2$	139	5245
$Bi_2O_3-Y_2O_3$	118	5195	$Tb_2O_3-ThO_2$	140	5248
$Bi_2O_3-WO_3$	119	5197	$Y_2O_3-ThO_2$	140	5250
$Dy_2O_3-ThO_2$	124	5210	$Y_2O_3-ZrO_2$	141	5251
$Dy_2O_3-ZrO_2$	125	5211	$Y_2O_3-Nb_2O_5$	142	5253
$Er_2O_3-ThO_2$	126	5215	$Yb_2O_3-ThO_2$	143	5256
$Er_2O_3-ZrO_2$	127	5217	$Yb_2O_3-ZrO_2$	143	5257
$Gd_2O_3-ThO_2$	129	5223			

Table 4.4 Immiscible Liquid Diagrams

Systems	Page	Figure	Systems	Page	Figure
			1964 Edition		
$FeO-SiO_2$	59	80	$PbO-B_2O_3$	115	281
$Fe_3O_4-SiO_2$	61	87	$SrO-SiO_2$	119	296
$Mn_3O_4-SiO_2$	66	102	$ZnO-B_2O_3$	119	300
TiO_2-SiO_2	69	113	$ZnO-SiO_2$	120	302
Cu_2O-SiO_2	86	164	$ThO_2-B_2O_3$	125	322
$BaO-B_2O_3$	97	207	$Bi_2O_3-B_2O_3$	125	323
$CaO-B_2O_3$	103	234	$Cr_2O_3-SiO_2$	130	332
$CaO-SiO_2$	104	237	$Ga_2O_3-SiO_2$	134	341
$CdO-B_2O_3$	109	252	ZrO_2-SiO_2	141	362
$CoO-B_2O_3$	109	254	$Nb_2O_5-SiO_2$	142	363
$MgO-B_2O_3$	111	261	$R_2O, RO-SiO_2$	142	365
$MgO-GeO_2$	111	264–265			
			1969 Edition		
$CoO-SiO_2$	23	2120	$Er_2O_3-SiO_2$	100	2365
$CuO-SiO_2$	32	2142	$Gd_2O_3-SiO_2$	101	2367

Table 4.4 (*Continued*)

Systems	Page	Figure	Systems	Page	Figure
CaO–SiO$_2$	85	2302	La$_2$O$_3$–SiO$_2$	102	2372
CoO–SiO$_2$	87	2308	La$_2$O$_3$–TiO$_2$	103	2373
MgO–SiO$_2$	88	2313	Nd$_2$O$_3$–SiO$_2$	104, 105	2380, 2381
MnO–GeO$_2$	90	2322	Sc$_2$O$_3$–SiO$_2$	105	2384
NiO–SiO$_2$	90	2324	Sm$_2$O$_3$–SiO$_2$	106	2386
PbO–SiO$_2$	91	2327	Y$_2$O$_3$–SiO$_2$	107	2388
RO–SiO$_2$	92	2331	Yb$_2$O$_3$–SiO$_2$	108	2391
SrO–B$_2$O$_3$	92	2333	ThO$_2$–SiO$_2$	109	2397
Sc$_2$O$_3$–B$_2$O$_3$	98	2352	UO$_2$–SiO$_2$	110	2399
Nb$_2$O$_5$–B$_2$O$_3$	98	2354	ZrO$_2$–SiO$_2$	110	2400
Dy$_2$O$_3$–SiO$_2$	100	2362	BeF$_2$–ZrF$_4$	372	3410
			1975 Edition		
Si–O	12	4162	Eu$_2$O$_3$–B$_2$O$_3$	138	4383
Th–O	13	4164	Gd$_2$O$_3$–B$_2$O$_3$	138	4384
U–O	14	4166	Ho$_2$O$_3$–B$_2$O$_3$	139	4385
Tℓ$_2$O–B$_2$O$_3$	98	4298	Lu$_2$O$_3$–B$_2$O$_3$	139	4386
CaO–GeO$_2$	104	4309	Nd$_2$O$_3$–B$_2$O$_3$	139	4387
NiO–B$_2$O$_3$	120	4345	Sm$_2$O$_3$–B$_2$O$_3$	140	4388
SmO–SiO$_2$	126	4357	Tm$_2$O$_3$–B$_2$O$_3$	140	4389
Aℓ$_2$O$_3$–SiO$_2$	134	4375	Y$_2$O$_3$–B$_2$O$_3$	140	4390
Dy$_2$O$_3$–B$_2$O$_3$	137	4381	Yb$_2$O$_3$–B$_2$O$_3$	141	4391
Er$_2$O$_3$–B$_2$O$_3$	138	4382			
			1981 Edition		
MgO–B$_2$O$_3$	99, 100	5153, 5154	B$_2$O$_3$–TeO$_2$	118	5193
PbO–B$_2$O$_3$	105	5167	B$_2$O$_3$–V$_2$O$_5$	118	5194
SrO–SiO$_2$	112	5180	Fe$_2$O$_3$–V$_2$O$_5$	128	5220
Aℓ$_2$O$_3$–SiO$_2$	116	5190	La$_2$O$_3$–GeO$_2$	131	5229

Table 4.5 A List of References on Liquid Immiscibility in Binary and Ternary Borate, Silicate, Oxide and Sulfide Systems

1. J. W. Grieg, "Immiscibility in Silicate Melts," Amer. Jour. Sci., Vol. XIII, pp. 134–54, February, 1927.
2. G. W. Morey and Earl Ingerson, "Melting of Danburite; Study of Liquid Immiscibility in the System CaO–B$_2$O$_3$–SiO$_2$," Amer. Mineralogist 22 (1) 37–48, 1937.
3. H. P. Hood and M. E. Nordberg, "Treated Borosilicate Glass," U.S. Pat. 2, 106, 744, February 1, 1938.

Table 4.5 *(Continued)*

4. G. W. Morey, "The Ternary System $Na_2O-B_2O_3-SiO_2$," J. Soc. Glass Tech. 35, 167, 270–83, 1951.

5. E. M. Levin and G. M. Ogrinic, "The System Barium Oxide-Boric Oxide-Silica," J. Research Natl. Bur. Standards 51 (1) 37–56, 1953.

6. R. C. DeVries and R. Roy, and E. F. Osborn, "The System TiO_2-SiO_2," Trans. Brit. Ceram. Soc. 53 (9), 525–40, 1954.

7. R. C. DeVries, R. Roy and E. F. Osborn, "Phase Equilibria in the System $CaO-TiO_2-SiO_2$," J. Am. Ceram. Soc. 38 (5) 158–71, 1955.

8. Ernest M. Levin and Given W. Cleek, "Shape of Liquid Immiscibility Volume in the System Barium Oxide-Boric Oxide-Silica," J. Amer. Cer. Soc. 41 (5) 175–79, May, 1958.

9. Muneo Watanabe, Haruo Noake and Takeshi Aiba, "Electron Micrographs of Some Borosilicate Glasses and Their Internal Structure," J. Am. Ceram. Soc. 42 (12) 593–99, 1959.

10. B. S. R. Sastry and F. A. Hummel, "Studies in Lithium Oxide Systems: III Liquid.Immiscibility in the System $Li_2O-B_2O_3 SiO_2$," J. Am. Ceram. Soc. 42 (2), 81–88, 1959.

11. K. H. Kim and F. A. Hummel, "Studies in Lithium Oxide Systems: VI, Progress Report in the System $Li_2O-SiO_2-TiO_2$," J. Am. Ceram. Soc. 42 (6) 286, June, 1959.

12. B. S. R. Sastry and F. A. Hummel, "Studies in Lithium Oxide Systems: VII, $Li_2O-B_2O_3-SiO_2$," J. Am. Ceram. Soc. 43 (1), 23–33, 1960.

13. F. A. Hummel, T. Y. Tien and K. H. Kim, "Studies in Lithium Oxide Systems: VIII, Application of Silicate Liquid Immiscibility to Development of Opaque Glazes," J. Am. Ceram. Soc. 43 (4) 192, April, 1960.

14. S. M. Ohlberg, H. R. Golob and C. M. Hollabaugh, "Fractography of Glasses Evidencing Liquid-in-Liquid Collodial Immiscibility," J. Amer. Cer. Soc. 45 (1) 1, January, 1962.

15. S. M. Ohlberg, H. R. Golob and D. W. Strickler, "Crystal Nucleation by Glass in Glass Separation" Symposium on Nucleation and Crystallization in Glasses and Melts," The American Ceramic Society, 1962.

16. J. F. Argyle and F. A. Hummel, "Liquid Immiscibility in the System $BaO-SiO_2$," Physics and Chemistry of Glasses, Vol. 4, No. 3, pp. 103–105, June, 1963.

17. Mohammed Ibrahim and N. F. Bright, "The Binary System $Nb_2O_5-SiO_2$," J. Am. Ceram. Soc. 45 (5), 222, 1962. Phase Diagrams for Ceramists, p. 142, Figure 363, 1964 Edition.

18. F. C. Kracek, "Liquidus Relations in the Alkali and Alkaline Earth-Silicate Systems," Phase Diagrams for Ceramists, p. 142, Figure 365, 1964 Edition.

19. S. M. Ohlberg and J. M. Parsons, "The Distribution of Sodium Ions in Soda-Lime-Silicate Glass," Physics of Non-Crystalline Solids, Proceedings of the International Conference, Delft, July 1964.

20. S. M. Ohlberg, J. J. Hammel and H. R. Golob, "Phemonenology of Non-crystalline Microphase Separation in Glass," J. Amer. Cer. Soc. 48 (4) 178–180, 1965.

Table 4.5 *(Continued)*

21. S. M. Ohlberg, H. R. Golob, J. J. Hammel and R. R. Lewihuk, "Noncrystal-
    line Microphase Separation in Soda-Lime-Silica Glass," J. Amer. Cer. Soc.
    48 (6) 331–332, 1965.
22. J. J. Hammel and S. M. Ohlberg, "Light Scattering from Diffusion-Controlled
    Phase Separations in Glass," J. App. Physics 36 (4) 1442–1447, April,
    1965.
23. Werner Vogel, "Struktur Und Kristallisation Der Glasser," VEB Deutscher
    Verlag fur Groundstoffindustrie, Leipzig, 1965.
24. S. M. Ohlberg and J. J. Hammel, "Formation and Structure of Phase Sepa-
    rated Soda-Lime-Silica Glass," Research Into Glass, PPG Industries, p. 128,
    1967.
25. Dibyendu Ganguli, "Macroliquation in the System $Al_2O_3-SiO_2$," Materials
    Research Bulletin, Vol. 2, pp. 25–36, 1967.
26. J. J. Hammel, "Experimental Evidence for Spinodal Decomposition in
    Glasses of the Na_2O-SiO_2 System," Research Into Glass, PPG Industries, p.
    140, 1967.
27. R. J. Charles and F. E. Wagstaff, "Metastable Immiscibility in the $B_2O_3-
    SiO_2$ System," J. Am. Ceram. Soc. 51 (1) 16, January, 1968.
28. David Johnson and F. A. Hummel, "Phase Equilibria and Liquid Immiscibil-
    ity in the System $PbO-B_2O_3-SiO_2$," J. Am. Ceram. Soc. 51 (4) 196, April,
    1968.
29. T. P. Seward, III, D. R. Uhlmann, and David Turnbull, "Phase Separation in
    the System $BaO-SiO_2$," J. Amer. Cer. Soc. 51 (5) 278, May, 1968.
30. R. R. Shaw and D. R. Uhlmann, "Subliquidus Immiscibility in Binary Alkali
    Borates," J. Amer. Cer. Soc. 51 (7) 377, July, 1968.
31. E. F. Riebling, "Nonideal Mixing in Binary GeO_2-SiO_2 Glasses," J. Amer.
    Cer. Soc. 51 (7) 406, July, 1968.
32. E. R. Plumat, "New Sulfide and Selenide Glasses: Preparation, Structure
    and Properties," J. Am. Ceram. Soc. 51 (9) 499, September, 1968.
33. T. P. Seward, III, D. R. Uhlmann, and David Turnbull, "Development of
    Two-Phase Structure in Glasses, with Special Reference to the System $BaO-
    SiO_2$ System," J. Am. Ceram. Soc. 51 (11) 634, November, 1968.
34. J. F. MacDowell and G. H. Beal, "Immiscibility and Crystallization in
    $Al_2O_3-SiO_2$ Glasses," J. Am. Ceram. Soc. 52 (1) 17, January, 1969.
35. Rare Earth–SiO_2 Systems, p. 100, Figure 2362, Figure 2365, p. 101,
    Figure 2367, p. 102, Figure 2372, p. 105, Figure 2381, Figure 2384, p. 106,
    Figure 2386, p. 107, Figure 2388, p. 108, Figure 2391 (1969 Edition, Phase
    Diagrams for Ceramists).
36. W. Haller, D. H. Blackburn, F. E. Wagstaff and R. J. Charles, "Metastable
    Immiscibility Surface in the System $Na_2O-B_2O_3-SiO_2$," J. Amer. Ceram.
    Soc. 53 (1) p. 34, January, 1970.
37. Akio Makashima and Teruo Sakaino, "Scanning Electron Micrographs of
    Phase-Separated Glasses in the System $Na_2O-B_2O_3-SiO_2$," J. Amer. Ceram.
    Soc. 53 (1) p. 64, January, 1970.
38. T. H. Elmer, M. E. Nordberg, G. B. Carrier, and E. J. Korda, "Phase Separa-

Table 4.5 (*Continued*)

tion in Borosilicate Glasses as Seen by Electron Microscopy and Scanning Electron Microscopy," J. Amer. Ceram. Soc. 53 (4) p. 171, April, 1970.

39. Paul D. Calvert and Robert R. Shaw, "Liquidus Behavior in the Silica-Rich Region of the System $PbO-SiO_2$," J. Amer. Ceram. Soc. 53 (6) p. 350, June, 1970.

40. Edwin Roedder and Paul W. Weiblen, "Silicate Immiscibility in Lunar Rocks," Geotimes, V 15, No. 3, p. 10, March, 1970.

41. Yoji Kawamoto and Shoji Tsuchihashi, "Properties and Structure of Glasses in the System Ge-S," J. Amer. Cer. Soc. 54 (3) 131, March, 1971.

42. J. P. DeLuca and C. G. Bergeron, "Structural Interpretations of Dielectric Measurements in a Lead Borate Melt," J. Amer. Cer. Soc. 54 (4) 191, April, 1971.

43. R. R. Shaw and J. F. Breedis, "Secondary Phase Separation in Lead Borate Glasses," J. Amer. Cer. Soc. 55 (8) 422, August, 1972.

44. J. A. Topping and M. K. Murthy, "Effect of Small Additions of Al_2O_3 and Ga_2O_3 on the Immiscibility Temperature of Na_2O-SiO_2 Glasses," J. Amer. Cer. Soc. 56 (5) 270, May, 1973.

45. Joseph A. Simmons, "Miscibility Gap in the System $PbO-B_2O_3$," J. Amer. Cer. Soc. 56 (5) 284, May, 1973.

46. S. M. Ohlberg and Helen R. Golob, "Undulating Structures in Complex Silicate Glasses," J. Amer. Cer. Soc. 56 (6) 300, June, 1973.

47. E. F. Riebling, "Depolymerization of GeO_2 and $B_2O_3-GeO_2$ Network Glasses by Sb_2O_3," J. Amer. Cer. Soc. 56 (6) 303, June, 1973.

48. R. W. Haskell, "Introduction to the Thermodynamics of Spinodal Decomposition," J. Amer. Cer. Soc. 56 (7) 355, July, 1973.

49. Minoru Tomozawa and Richard A. Obara, "Effect of Minor Third Components on Metastable Immiscibility Boundaries," J. Amer. Cer. Soc. 56 (7) 378, July, 1973.

50. Wolfgang Haller, Douglas H. Blackburn and Joseph H. Simmons, "Miscibility Gaps in Alkali-Silicate Binaries—Data and Thermodynamic Interpretation," J. Amer. Cer. Soc. 57 (3) 120, 1974.

51. J. A. Topping, I. T. Harrower, and M. K. Murthy, "Properties and Structure of Glasses in the System $PbO-GeO_2$," J. Amer. Cer. Soc. 57 (5) 209, 1974.

52. J. A. Topping, N. Cameron, and M. K. Murthy, "Properties and Structure of Glasses in the System $Bi_2O_3-SiO_2-GeO_2$," J. Amer. Cer. Soc. 57 (12) 519, 1974.

53. I. A. Aksay and J. A. Pask, "Stable and Metastable Equilibria in the System $Al_2O_3-SiO_2$," J. Amer. Cer. Soc. 5 (11–12) 507, 1975.

54. Richard Baylor, Jr. and Jesse J. Brown, Jr., "Phase Separation of Glasses in the System $SrO-B_2O_3-SiO_2$," J. Amer. Cer. Soc. 59 (3–4) 131, 1976.

55. Minoru Tomozawa and Takeshi Takamori, "Effect of Phase Separation on HF Etch Rate of Borosilicate Glasses," J. Amer. Cer. Soc. 60 (7–8) 301, 1977.

56. C. E. Vallet and J. Braunstein, "Thermodynamically Predicted Miscibility Gap in the System BeF_2-LiF," J. Amer. Cer. Soc. 60 (7–8) 315, 1977.

Table 4.5 (Continued)

57. Z. Strnad and P. Strnad, "Calculation of Metastable Two-Liquid Tie Lines in Ternary Glass-Forming Systems," J. Amer. Cer. Soc. 61 (7–8) 283, 1978.

58. Takeshi Takamori and Minoru Tomozawa, "HCl Leaching Rate and Micro-structure of Phase-Separated Borosilicate Glasses," J. Amer. Cer. Soc. 61 (11–12) 509, 1978.

59. Subhash H. Risbad and Joseph A. Pask, "On the Location of Metastable Immiscibility in the System $SiO_2-Al_2O_3$," J. Amer. Cer. Soc. 62 (3–4) 214, 1979.

60. Takeshi Takamori and Minoru Tomozawa, "Viscosity and Microstructure of Phase-Separated Borosilicate Glasses, J. Amer. Cer. Soc. 62 (7–8) 370, 1979.

61. J. P. Guha, "Liquid Immiscibility in the System $BaTiO_3-CeO_2$," J. Amer. Cer. Soc. 62 (11–12) 627, 1979.

62. R. McPherson, "Evidence for a Metastable Miscibility Gap in the System Mullite-Alumina," J. Amer. Cer. Soc. 63 (1–2) 110, 1980.

63. Takeshi Takamori and Minoru Tomozawa, "Morphology of Creep Fracture of a Phase-Separated Borosilicate Glass," J. Amer. Cer. Soc. 63 (3–4) 126, 1980.

64. Joanne H. Markis, Kevin Clemens, and Minoru Tomozawa, "Effect of Fluorine on the Phase Separation of Na_2O-SiO_2 Glasses," J. Amer. Cer. Soc. 64 (1) C-20, 1981.

65. H. D. Jarnnek and D. E. Day, "Sodium Motion in Phase-Separated Sodium Silicate Glasses," J. Amer. Cer. Soc. 64 (4) 227, 1981.

66. Y. Kawamoto and M. Tomozawa, "Prediction of Immiscibility Boundaries of the Systems K_2O-SiO_2, $K_2O-Li_2O-SiO_2$, $K_2O-Na_2O-SiO_2$, and $K_2O-BaO-SiO_2$," J. Amer. Cer. Soc. 64 (5) 289, 1981.

67. Y. Kawamoto, K. Clemens and M. Tomozawa, "Effects of MoO_3 on Phase Separation of $Na_2O-B_2O_3-SiO_2$ Glasses," J. Amer. Cer. Soc. 64 (5) 292, 1981.

68. Peter Taylor and Derrick G. Owen, "Liquid Immiscibility in the System $Na_2O-ZnO-B_2O_3-SiO_2$," J. Amer. Cer. Soc. 64 (6) 360, 1981.

69. K. Clemens, M. Yoshiyagawa and M. Tomozawa, "Liquid-Liquid Immiscibility in $BaO-B_2O_3$," J. Amer. Cer. Soc. 64 (6) C-91, 1981.

70. R. J. Charles, "Immiscibility and Its Role in Glass Processing," Bull. Amer. Cer. Soc. 52 (9) 673, 1973.

APPENDIX B   REFERENCES ON HIGH PRESSURE METHODS AND MATERIALS

Table 4.6   References on High Pressure Methods

1. William Paul and Douglas M. Warschauer, "Solids Under Pressure", McGraw-Hill Book Co., 1963 (Dedicated to P. W. Bridgman).

2. Francis Birch, Eugene C. Robertson, and Sydney P. Clark, Jr., "Apparatus for Pressures of 27,000 Bars and Temperatures of 1400°", Ind. Eng. Chem., 49, 1965-6 (1957).

Table 4.6 *(Continued)*

3. John C. Jamieson, "Introductory Studies of High-Pressure Polymorphism to 24,000 Bars by X-ray Diffraction with Some Comments on Calcite II", J. Geol., 65, 334–42 (1957).
4. H. Tracy Hall, "Chemistry at High Pressures and High Temperatures", J. Wash. Acad. Sci., 47, 300–4 (1957).
5. W. S. Fyfe, J. Geol., 68, 553, 1960.
6. F. Dachille and R. Roy, "Encyclopedia of Science and Technology", Vol. 6, pp. 443–444, McGraw-Hill, N.Y., 1960.
7. J. R. Goldsmith and H. D. Heard, J. Geol., 69, 45, 1961.
8. D. C. Munro, "High Pressure Physics and Chemistry", R. S. Bradley, editor, Vol. 1, p. 11, Academic Press, N.Y., 1963.
9. F. Dachille and R. Roy, "Modern Very High Pressure Techniques", R. H. Wentorf, editor, Butterworths, Washington, D.C., 1962.
10. A. Neuhaus, "Synthesis, Structural Behavior, and Valence State of Inorganic Matter in the Realm of Higher and Highest Pressure", Chimia 18(3), 93–103, (1964).
11. E. A. Perez-Albuerne, K. F. Forsgren, and H. G. Drickamer, "Apparatus for X-ray Measurements at Very High Pressure", Rev. Sci. Instr., 35 (1), 29–33, (1964).
12. H. J. Van Hook, "Apparatus for High Pressure High Temperature Studies of Oxide Materials", Rev. Sci. Instr., 36 (8), 1119–20, (1965).
13. William Klement, Jr. and Aiyasami Jayaraman, "Phase Relations and Structures of Solids at High Pressures", Progr. Solid State Chem., 3, 289–376, 1967.
14. R. H. Wentorf, Jr., "Modern Very-High-Pressure Research", Brit. J. Appl. Phys., 18(7), 865–82, (1967).
15. Josef Klimovic, "Phase Transitions and Behavior of Solids Under Very High Static Pressures", Cesk. Cas. Fys., 17 (4), 344–65 (1967).
16. Massimo Marezio, "Oxides at High Pressure", Trans. Amer. Crystallogr. Assn., 5, 29–37, 1969.
17. Will Kleber and K. T. Wilke, "Synthesis and Crystal Chemistry of Inorganic Compounds at High Pressures and Temperatures", Krist. Tech. 4(2), 165–99, 1969.
18. D. J. J. Van Resburg, "Limits in Science. II. Pressure Limits", Spectrum, 7(2), 86–8, 1969.
19. Mario D. Banus, "X-ray Diffraction at High Pressure", High Temp-High Pressures, 1(5), 483–515, 1969.
20. Robert D. Shannon and Charles T. Prewitt, "Coordination and Volume Changes Accompanying High-Pressure Phase Transformations of Oxides", Mater. Res. Bull., 4(1), 57–62, 1969.
21. John Beynon, "What Do We Mean by Pressure?" Vaccum, 20(10), 443–4, 1970.

Table 4.6 *(Continued)*

22. Noboru Nakayama and Hiroshi Hayashi, "High-Pressure Technology", Nippon Kinzoku Gakkai Kaiho, 10(6), 402-4, 1971.
23. Shunicki Akimoto, "Phases of Materials Under High Pressure", Nippon Kinzoku Gakkai Haiho 10(6), 364-72, 1971.
24. Hisao Mitsui, "Fixed Points of Super Pressure", Nippon Kinzoku Gakkai Kaiho, 10(6), 382-4, 1971.
25. Hiroshi Iwasaki, "Sodium Chloride-Type Metal Oxides Under High Pressure", Nippon Kinzoku Gakkai Kaiho, 10(6), 397-9, 1971.
26. Akira Sawaoka, "Physics Research and High Pressure", Nippon Kinzoku Gakkai Kaiho, 10(6), 358-63, 1971.
27. Shigeru Minomura, "Structure of Metals and Their Electronic Structure Under High Pressure", Nippon Kinzoku Gakkai Kaiho 10(6), 385-8, 1971.
28. Tsunesaburo Asada, "Cubic Pressing", Nippon Kinzoku Gakkai Kaish, 10(6), 400-2, 1971.
29. Robert H. Wentorf, "High Pressure Phenomena", Phys. Chem., 1, 571-611, 1971.

MATERIALS

1. A. E. Ringwood and Merren Seabrook, "Some High-Pressure Transformations in Pyroxenes", Nature, 196, 883-4, (1962).
2. R. D. Shannon, "New High Pressure Phases Having the Corundum Structure", Solid State Commun., 4(12), 629-30 (1966).
3. Peter M. Bell, William W. Atkinson, Jr., Bobby G. Cahoon, and C. S. Hurlbut, Jr., "Research on Chemistry and Physics of Inorganic Systems Under Extreme High Pressure and Temperature", U.S. Dept. Comm. AD 633148.
4. Massimo Marezio, Joseph P. Remeika, P. D. Dernier, "High Pressure Synthesis of $YGaO_3$, $GdGaO_3$, and $YbGaO_3$", Mater. Res. Bull., 1(4), 247-55, 1966.
5. K. F. Seifert, "Pressure Crystal Chemistry of AX_2 Compounds", Fortschr. Mineral. (Pub. 1968), 45(2), 214-80, 1967.
6. Allen Forrest Reid and Alfred E. Ringwood, "Newly Observed High Pressure Transformations in Hausmannite, Calcium Aluminate, and Zirconium Silicate (Zircon)", Earth Planet. Sci. Lett., 6(3), 205-8, 1969.
7. Karl A. Wilhelmi, Lena Jahnberg, Sten Andersson, "High Pressure Synthesis of Nb_3O_7F with U_3O_8-type Structure", Acta Chem. Scand., 24(4), 1472-3, 1970.
8. Robert D. Shannon and Charles T. Prewitt, "Effective Ionic Radii and Crystal Chemistry", Inorg. Nucl. Chem., 32(5), 1427-41, 1970.
9. A. Jayaraman and L. H. Cohen, Chapter VI, "Phase Diagrams in High Pressure Research", pp. 245-287, Phase Diagrams, Materials Science and Technology, Vol. 6-I, Academic Press, 1970.

SUGGESTED CLASS DISCUSSIONS WHICH CAN ENLARGE THE UNDERSTANDING OF EXPERIMENTAL TECHNIQUES

1. Discuss in detail the chemical purity and chemical characteristics of starting materials for certain specific systems.
2. Discuss in detail the physical characteristics of starting materials for certain specific systems.
3. Discuss in detail the time and temperature characteristics which are necessary for the determination of equilibria in certain specific systems.
4. Discuss in detail the manner in which optical and X-ray diffraction is used for determination of polymorphism, melting, dissociation, solid solution, partial solid solution, unmixing and immiscibility in certain specific systems.
5. Discuss in detail the difference between "sluggish" glass-forming systems and extremely rapid non-glass forming systems, the quenching technique and the difference between equilibrium at room temperature vs. equilibrium at high temperature in certain specific systems.
6. Discuss in detail the use of high pressure, controlled oxygen pressure or hydrothermal methods in specific systems.
7. Discuss in detail the reasons for the revision of ZrO_2 binary systems, other systems which have undergone revisions (see section VCii or the system $Al_2O_3-SiO_2$.

REFERENCES

1. F. de Körösy, Colloidal Mixtures As Batches for Glass Melting, Bull. Amer. Cer. Soc. 20 (5) p. 162, May, 1941.
2. Rustum Roy, Aids in Hydrothermal Experimentation: II, J. Amer. Cer. Soc. 39 (4) 145–46, 1956.
3. Rustum Roy, Gel Route to Homogeneous Glass Preparation, J. Amer. Cer. Soc. 52 (6) p. 344, June, 1969.
4. Gregory J. McCarthy and Rustum Roy, Gel Route to Homogeneous Glass Preparation: II Gelling and Desiccation, J. Amer. Cer. Soc. 54 (12), p. 639, 1971.
5. J. F. Schairer and N. L. Bowen, The System $K_2O-Al_2O_3-SiO_2$, Amer. J. Sci. 253, 681–746, December, 1955.
6. W. Hume-Rothery, J. W. Christian and W. B. Pearson, Metallurgical Equilibrium Diagrams, The Institute of Physics, London, 1952, Chapman and Hall, Ltd.
7. Frederick N. Rhines, Phase Diagrams in Metallurgy, McGraw-Hill Book Company, 1956.

8. R. C. Mackenzie, Differential Thermal Analysis: Vol. I, Fundamental Aspects, Academic Press, New York, 1970.

9. R. C. Mackenzie, Differential Thermal Analysis: Vol. II, Applications, Academic Press, New York, 1972.

10. W. J. Smothers, Differential Thermal Analysis, Chemical Publishing Co., New York, 1958.

11. W. J. Smothers and C. Yao, Handbook of Differential Thermal Analysis, Chemical Publishing Co., N.Y., 1966.

12. W. W. Wendlandt, Thermal Methods of Analysis, Interscience Publisher, New York, 1964.

13. J. B. MacChesney and P. E. Rosenberg, Chapter III, The Methods of Phase Equilibria Determination and Their Associated Problems, Vol. 6-I, Phase Diagrams, Materials Science and Technology, Edited by Allen M. Alper, Academic Press, N.Y.

14. H. Rawson, Inorganic Glass-Forming Systems, Academic Press, New York, 1967.

15. S. R. Scholes and C. H. Greene, Modern Glass Practice, Cahners Publishing Co., 1975.

16. L. D. Pye, V. D. Frechette and N. J. Kreidl, Borate Glasses-Structure, Properties, Applications, Materials Science Research, Vol. 12, Plenum Press, New York.

17. R. H. Doremus, Glass Science, Wiley-Interscience, N.Y., 1973.

18. Ernest M. Levin, Carl R. Robbins and Howard F. McMurdie, Phase Diagrams for Ceramists, American Ceramic Society, Inc., 1964.

19. P. D. S. St. Pierre, Constitution of Bone China: I, High Temperature Phase Equilibrium Studies in the System Tricalcium Phosphate-Alumina-Silica, J. Amer. Cer. Soc., 37, 6, 243-58, 1954.

20. Takeshi Takamori and Rustum Roy, Rapid Crystallization of SiO_2-Al_2O_3 Glasses, J. Amer. Cer. Soc. 56(12), 639, 1973.

21. State of the Art Review of Rapid Solidifcation of Technology (RST), Metals and Ceramic Information Center 81-45, October 1981.

22. Arnulf Muan, Phase Equilibria At High Temperatures in Oxide Systems Involving Changes in Oxidation States, Amer. J. Sci. 256, p. 171-207, March, 1958.

23. Arnulf Muan and E. F. Osborn, Phase Equilibria Among Oxides in Steelmaking, Addison-Wesley Publishing Company, Inc., 1965.

24. Ernest G. Ehlers, The Interpretation of Geological Phase Diagrams, W. H. Freeman and Company, 1972.

25. Donald H. Lindsley, Chapter 2, Experimental Studies of Oxide Minerals, Page 67-68, Mineralogical Society of America, Oxide Minerals, Vol. 3, November, 1976, Short Course Notes.

26. P. H. Ribbe, Editor, Sulfide Mineralogy, Mineralogical Society of America, Short Course Notes, Vol. 1, November, 1974.

27. G. Kullerud, Sulfide Studies, P. H. Abelson, Editor, Researches in Geo-chemistry, Vol. 2, p. 286-321, John Wiley and Sons, 1967b.

28. G. Kullerud, Experimental Techniques in Dry Sulfide Research, G. C. Ulmer, Editor, Research Techniques for High Pressure and High Temperature, p. 288-315, Springer-Verlag, N.Y., 1971.

29. R. W. Haskell and R. C. DeVries, Estimate of Free Energy Formation of Kyanite, J. Amer. Cer. Soc., 47, (4), 203, 1964.

30. Wilhelm Eitel, Silicate Melt Equilibria, paragraph 30, Figure 6, Rutgers University Press, New Brunswick, New Jersey, 1951.

31. Manfred Berretz and Smith L. Holt, Optical and Magnetic Properties of Some Transition Metal Ions in Barium Phosphate Glass, J. Amer. Cer. Soc., 61, (3-4), 136, 1978.

32. F. A. Hummel, Glass Formation, J. Amer. Cer. Soc., 62, (5-6), 312, 1979.

33. D. K. Smith and C. F. Cline, Verification of Existence of Cubic Zirconia at High Temperature, J. Amer. Cer. Soc., 45, (5), p. 249-50, 1962.

34. V. S. Stubican and J. R. Hellman, Phase Equilibria in Some Zirconia Systems, Chapter II, p. 25, Advances in Ceramics; Science and Technology of Zirconia, American Ceramic Society, Columbus, Ohio, 43214, 1981.

35. E. S. Shepard, G. A. Rankin and F. E. Wright, The Binary Systems of Alumina with Silica, Lime and Magnesia, Am. J. Sci., 28 (4), 292-333, 1909.

36. N. L. Bowen and J. W. Grieg, The System Al_2O_3-SiO_2, J. Amer. Cer. Soc., 7 (4), 238-254, 410, 1924.

37. R. B. Sosman, A Pilgrimage to Mull, Bull. Amer. Cer. Soc., 35 (3), 130-131, 1956.

38. S. Aramaki and R. Roy, Revised Phase Diagram for the System Al_2O_3-SiO_2, J. Amer. Cer. Soc., 45 (5), 229-242, 1962.

39. K. Konopiky, Diskussionsbemerkung Ziem Schmelzdiagram Al_2O_3-SiO_2, Ber. Deut. Keram. Ges., 40 (5), 286-288, 1963.

40. Dibyendu Ganguli and P. Saha, Macroliquation in the System Al_2O_3-SiO_2, Mat. Res. Bull., 2, 25-36, 1967.

41. J. F. MacDowell and G. H. Beall, Immiscibility and Crystallization in Al_2O_3-SiO_2 Glasses, J. Amer. Cer. Soc., 52 (1), 17, 1969.

42. Takeshi Takamori and Rustum Roy, Rapid Crystallization of SiO_2-Al_2O_3 Glasses, J. Amer. Cer. Soc., 56 (12), 639-643, 1973.

43. F. A. Hummel and W. F. Horn, The Quaternary System B_2O_3-Al_2O_3-SiO_2-P_2O_5: Literature Review and Exploratory Data on the Ternary Subsystems, J. Australian Cer. Soc., 17 (1), 25-32, 1981.

44. Paul Gordon, Principles of Phase Diagrams in Materials Systems, McGraw-Hill Book Co., Inc., 330 West 42nd St., New York, N.Y., 10036, 1968.

45. Arnold Reisman, Phase Equilibria, Academic Press, 111 Fifth Ave., New York, N.Y., 10003, 1970.

46. A. Prince, Alloy Phase Equilibria, American Elsevier Publishing Co., Inc., 52 Vanderbilt Avenue, New York, N.Y., 1966.

5

TERNARY SYSTEMS WITHOUT SOLID SOLUTION

I. INTRODUCTION

Systems containing three components are called ternary systems. The phase rule becomes $F = 3 - P + 2 = 5 - P$, and an invariant point is a position where five phases coexist. If pressure is constant (a "condensed" system), the phase rule for the system is $F = 4 - P$, and an invariant point involves equilibrium between four phases. At constant temperature AND pressure ($F = 3 - P$), three phases can coexist (three liquids, three solids or a combination of solid(s) and liquid(s)).

A condensed ternary system can be bounded by any type of binary system discussed in Chapter 3. For example, a metallurgist might regard a ternary system which was bounded by three binary systems showing continuous solid solution without a maximum or minimum in the liquidus as being the simplest possible type of configuration. If the components melted congruently and underwent no polymorphic changes, the solid model would consist of only three simple regions; a single phase subsolidus region of continuous solid solution (no unmixing), a lens shaped region of coexistence of liquids and solid solutions and a single phase homogeneous liquid region (no unmixing). However, crystallization paths, isothermal sections, vertical sections and other fundamental features of such a system are not of the simplest type and it is customary in ceramics to begin a discussion of solid models of ternary systems by using the so-called simple eutectic model (Figure 5.1), similar to the start of the discussion in Chapter 3 on binary systems.

159

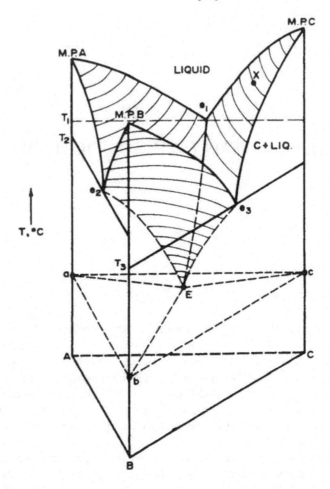

Figure 5.1. Simple Eutectic Ternary System, p = Constant, Usually 1 atm

A. The Three-Dimensional Solid Model; The Simple Eutectic Type Ternary System; Isothermal and Vertical Sections

In this case, the end members melt congruently, have no inversions and form simple binary eutectic type systems. The ternary system is also a simple eutectic type where boundary lines from the binary eutectic invariant points meet in the interior of the system at the ternary eutectic invariant point and outline three liquidus surfaces of primary crystallization of the "end-member" components. A two-dimensional projection of the three-dimensional ternary liquidus surface is shown in Figure 5.2. (See boundary lines and isotherms)

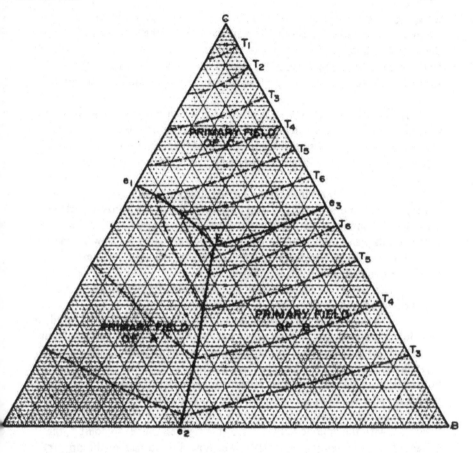

Figure 5.2. Isothermal Section at Temperature T_E showing Primary Fields of Crystalline A, B, and C and Projected Isotherms

The base consists of an equilateral triangle the apices of which represent the composition of the three components. The temperature axis is perpendicular to this base. The top surface describes the topography of the liquidus which is a curved surface. The sides of this solid represent the limiting or bounding binary systems. In Figure 5.1 the limiting binary systems, A–B, B–C, C–A, each contain one binary eutectic, shown by the points e_2, e_3, and e_1, respectively.

The liquidus surface is the locus of all temperature-composition points representing the maximum solubility (saturation) of a solid phase in a ternary liquid phase. At temperatures above the liquidus surface the equilibrium phase is homogeneous liquid (F=4–1=3). Points on the liquidus surface represent equilibrium between a ternary liquid and a solid phase (F=4–2=2). For example, point

X located on the liquidus surface M.P.C-e_1-E-e_3 represents equilibrium between a ternary liquid and crystals of C. The intersections of the liquidus surfaces form the boundary lines e_1-E, e_2-E, and e_3-E. Points on the boundary lines represent equilibrium between two crystalline phases and a ternary liquid (F=4-3=1). Along the boundary line e_1-E, crystals of A and C are in equilibrium with a liquid.

The boundary lines intersect at point E which is the ternary eutectic. Crystals of A, B, and C are in equilibrium with a liquid at this point, and the system is invariant. (F=4-4=0.) The eutectic point represents the lowest melting mixture of crystals of A, B, and C. No liquid exists at temperatures below that of point E.

In Figure 5.1 the solidus lines for the limiting binary systems of the ternary have been drawn at T_1, T_2 and T_3. In addition, the horizontal plane a-b-c which contains the eutectic point E has been constructed; lines have been drawn in this plane from E to the points a, b, and c. Figure 5.1 now contains all of the boundary lines necessary to define the eight "volumes" of "spaces" of which the figure is composed, as follows:

a. The *liquid space* is the region above the liquidus surface and is composed of homogeneous liquid only (F=4-1=3).

b. The *three spaces of primary crystallization*, A + Liquid, B + Liquid, and C + Liquid, are bounded on their upper surfaces by the liquidus surface. Their lower surfaces are generated in the following manner: Consider the primary crystallization space of B + Liquid. Place a slender rod, such as a pencil, in a horizontal position with one end contacting the vertical line B-M.P.B at T_3 and the other end contacting the boundary line e_3-E at point e_3. If the left end of the rod is permitted to slide from T_3 to b while being maintained horizontal and in contact with the boundary line e_3-E as the right end slides to point E, the surface thus generated is part of the lower bounding surface of the primary crystallization space of B. The remainder of the lower surface is generated by sliding the rod horizontally from T_2 to b while maintaining contact with boundary line e_2-E. The space thus generated is shown in Figure 5.3. All composition-temperature points within this space represent equilibrium between crystals of B and ternary liquids whose compositions lie on the liquidus surface M.P.B-e_2-E-e_3 (F=4-2=2). The lower surfaces of the primary crystallization spaces of A + Liquid and C + Liquid are generated in a manner similar to that described for B + Liquid.

c. The *three spaces of binary crystallization*, A + B + Liquid, A + C + Liquid, and B + C + Liquid, are bounded by the lower surfaces of the spaces of primary crystallization and the horizontal plane a-b-c which contains the eutectic point E. A "cave-like" space will be generated as shown in Figure 5.4 for A + B + Liquid (F=4-3=1). All composition-temperature points within this type of space (A+C+Liquid, A+B+Liquid or B+C+Liquid) represent equilibrium

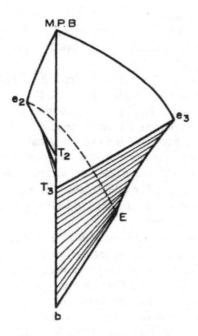

Figure 5.3. Space of Primary Crystallization of Component B

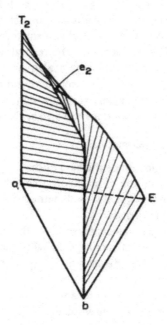

Figure 5.4. Space of Primary Crystallization of Components A and B

between two crystalline phases and a ternary liquid, the composition of which will lie on the appropriate boundary line (either $e_1 E$, $e_2 E$, or $e_3 E$). For example, in the binary crystallization space of A + C + Liquid, all liquid compositions will lie on the boundary line $e_1 E$.

    d. The *space containing crystalline phases only* is that space below the plane a–b–c and is composed of crystals of A, B, and C.

    At this point it should be emphasized that ternary systems could not be easily diagrammed in three dimensions unless they are "condensed" systems at constant pressure. The constancy of the pressure variable permits the three-dimensional composition vs. temperature diagram shown in Figure 5.1. If the pressure variable was included, the geometrical diagramming of five variables would be exceedingly complex. The four variable diagram using an equilateral triangle for composition variables permits considerable quantitative calculations to be made as will be shown later.

    e. For practical purposes, *Isothermal and Vertical Sections*, the simple eutectic model may be divided into two regions, one involving liquid or liquid and solid(s) which is called the liquidus region and the other involving only solid phases, called the sub-solidus region. The model may be sliced horizontally at liquidus or sub-solidus temperatures, giving rise to "isothermal" sections. In collections of phase diagrams it is very common to show a projection of the liquidus surface on a two dimensional surface (Figure 5.2) which contains boundary lines, isotherms and joins connecting composition points in the system. Later, the technique of generating isothermal sections at liquidus temperatures or below the temperature of liquid formation will be discussed.

    The model may also be sliced vertically in any manner whatever, giving rise to simple or complicated vertical sections. The vertical section taken at random in a ternary system may be extremely complicated, but if the section is judiciously selected along certain joins, the section may turn out to be relatively simple and in some cases may be a true binary system within the ternary system. The technique of generating vertical sections will also be discussed later.

B. Methods of Representing Compositions

The equilateral triangle which forms the base of the three-dimensional ternary figure permits the representation of all possible combinations of the three components. There are two frequently used methods for determining the composition of a point with the triangle:

    a. The length of perpendiculars drawn from the sides of the equilateral composition triangle to the point are proportional to the quantities of A, B and C in the sample. For example, in Figure 5.5a, composition X is composed of 40% C, 20% A, and 40% B.

    The composition can be plotted on a weight or mole percent basis, whichever is most convenient or useful. The amount of any component is always given

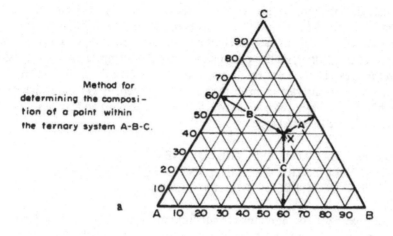

Method for determining the composition of a point within the ternary system A-B-C.

a

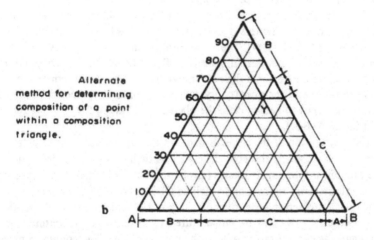

Alternate method for determining composition of a point within a composition triangle.

b

Figure 5.5a, b. Methods of Representing Compositions on an Equilateral Triangle in a Ternary System

by A=100-B-C, etc. The equilateral triangle has other useful properties. For example, if a straight line is drawn from any apex, say C, to the opposite side of the triangle (A-B), the ratio of A:B in the composition will be constant on this line as the amount of C varies.

b. Lines are drawn through the composition point Y, parallel to the sides of the equilateral triangle. The intersection of these lines with any side of the triangle gives the proportions of A, B and C represented by the point. For example, in Figure 5.5b, the point Y is composed of 10% A, 30% B, and 60% C. (See side AB or BC) This method is often used for determining liquid and crystalline percentages or crystalline percentages at constant temperature in composition triangles which are not equilateral types. This will be demonstrated later when isoplethal analyses are made in many types of ternary systems which contain binary or ternary compounds which generate composition triangles which are not of the equilateral type.

II. ISOPLETHAL ANALYSIS; THE EQUILIBRIUM CRYSTALLIZATION PATH IN A SIMPLE EUTECTIC TYPE SYSTEM; USE OF THE LEVER RULE

The changes which occur during the equilibrium cooling of a melt of a given composition and the quantitative isoplethal analysis are usually worked out through the use of a plane projection of the liquidus surface. However, the rules of construction which are used on the plane projection have been derived from the three-dimensional solid representing the ternary system. Therefore, it is desirable to understand the correlation between the three dimensional figure and its plane projection. In Figure 5.6 an isopleth (XX') has been constructed for the composition X (A+B+C=100). The plane projection of the liquidus surface is shown at the base of the three-dimensional figure (A–B–C). The temperature scale on the three-dimensional figure is shown by a series of isothermal lines on the liquidus surface of the primary crystallization space of component C. These isotherms have been projected onto the plane at the base of the figure (equilateral triangle A–B–C).

The intersection of the isopleth with the liquidus surface occurs at temperature T_3 and is shown by the point labeled 3. A tie line, T_3-3, has been constructed to join the conjugate phases (essentially 100% of a melt of composition 3 and an infinitesimal amount of crystalline C). The corresponding projection onto the plane projection is denoted by the point X' and the line C–X'. As the melt of composition X cools to temperature T_4, crystals of C precipitate from the melt and the concentration of C in the melt decreases. The conjugate phases are again determined by drawing the tie line (in a direction away from the crystallizing component, C) from T_4 through the isopleth at 4 and on to its intersection with the liquidus surface 4' (A+B+C=100). The position of 4' within the composition triangle A–B–C gives the composition of the melt at temperature T_4. Further cooling of the sample of overall composition X to temperature T_5 results in additional precipitation of crystals of C and a corresponding change in the melt composition to that given by point 5' (A+B+C=100).

At temperature T_6, the composition of the melt corresponds to that given by point 6' which is located on the boundary line e_1 E. The isopleth is in contact

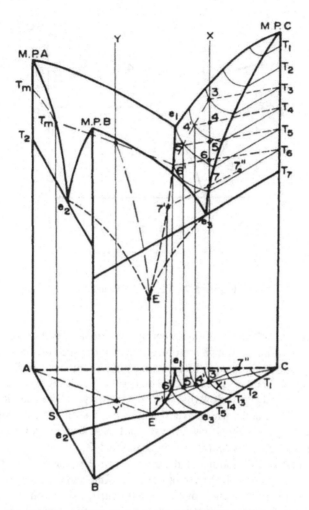

Figure 5.6. Cooling Path for Liquid of Composition X

with one of the lower surfaces of the primary crystallization space of C (point 6). The vertical section C–S is shown in Figure 5.7 and aids in the understanding of the crystallization path and the isoplethal analysis. Further cooling now places the composition into the region of binary crystallization, C+A+Liquid. At temperature T_7 the composition of the melt is given by the point at 7′, and the crystalline portion of the sample is composed of A and C (point 7″.

As the sample is cooled toward the ternary eutectic temperature E, the liquid composition moves along the boundary line (in the direction of decreasing

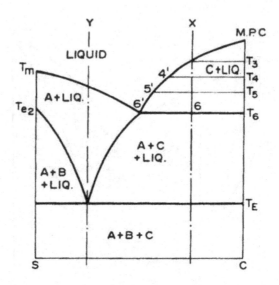

Figure 5.7. Vertical Section C–S

temperature) from 7' to E. Simultaneously, additional quantities of C and A precipitate from the melt. At a temperature just above the eutectic temperature, crystals of C and A are in equilibrium with a melt of composition E. At temperature T_E, liquid E and crystalline A, B and C are in equilibrium. As the temperature is decreased slightly below that of E, the melt crystallizes completely. That is, crystals of A, B, and C precipitate simultaneously until all of the liquid has been solidified. In this simple case, the amount of A, B and C in the crystalline assemblage is the same as that in the original liquid, X.

Quantitative calculations of the amounts of the phases present at each temperature are made with the use of the Lever Rule and are most easily followed on the plane projection of the liquidus surface. For convenience, this plane projection has been reproduced in Figure 5.8 and the composition of the original melt (X) will be taken as 20% A, 10% B and 70% C. At temperature T_3 the sample is composed of the original melt plus an infinitesimal quantity of crystals of C. On cooling to temperature T_4, additional quantities of C are precipitated. The quantities are determined by drawing a tie line from C through point X and terminating on the T_4 isotherm at point 4'.

The percentages of melt and crystal are determined from the lever rule, as follows:

$$\frac{CX}{C4'} \times 100 = \% \text{ melt of composition } 4' \ (23.5\% \text{ A}, 11\% \text{ B}, 65.5\% \text{ C})$$

$$\frac{X4'}{C4'} \times 100 = \% \text{ crystals of C}$$

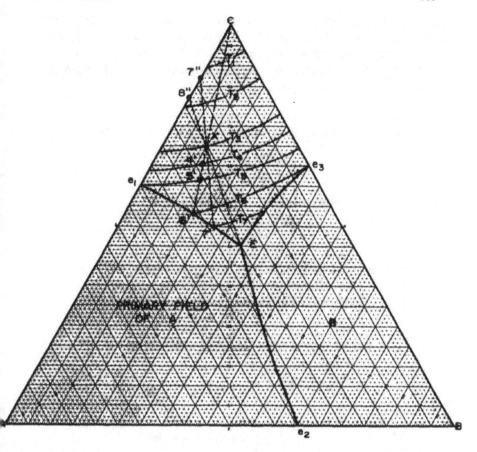

Figure 5.8. The Lever Rule and Isoplethal Analyses at Various Temperature Levels

At temperature T_5:

$$\frac{CX}{C5'} \times 100 = \% \text{ melt of composition } 5' \ (26\% \text{ A, } 12.5\% \text{ B, } 61.5\% \text{ C})$$

$$\frac{X5'}{C5'} \times 100 = \% \text{ crystals of C}$$

At temperature T_6:

$$\frac{CX}{C6'} \times 100 = \% \text{ melt of composition } 6' \ (32\% \text{ A, } 15\% \text{ B, } 53\% \text{ C})$$

$$\frac{X6'}{C6'} \times 100 = \% \text{ crystals of C}$$

At T_6 an infinitesimal quantity of crystals of A appear, and on cooling from T_6 to T_7, the quantity of A increases. The new tie line is drawn from $7'$ through X and on to its intersection with the line AC, point $7''$. This intersection divides the line AC into two lengths which are used to determine the relative proportions of A and C crystals in the sample at temperature T_7:

$$\frac{7''X}{7''7'} \times 100 = \% \text{ melt of composition } 7' \ (29\% \text{ A}, 21\% \text{ B}, 50\% \text{ C})$$

$$\frac{X7'}{7''7'} \times 100 = \% \text{ crystals} \begin{cases} \dfrac{7''A}{CA} \times 100 = \% \text{ C} = 87\% \\[2ex] \dfrac{C7''}{CA} \times 100 = \% \text{ A} = 13\% \end{cases}$$

On cooling to a temperature just above the eutectic temperature, additional precipitation of A and C occurs and the quantity of melt decreases.

At T_E^+: (Temperature slightly above the eutectic temperature)

$$\frac{8''X}{8''E} \times 100 = \% \text{ melt of composition E } (25\% \text{ A}, 30\% \text{ B}, 45\% \text{ C})$$
$$= \text{approx. } 33.3\%$$

$$\frac{XE}{8''E} \times 100 = \% \text{ crystals} \begin{cases} \dfrac{8''A}{CA} \times 100 = \% \text{ C} = 82\% \\[2ex] \dfrac{8''C}{CA} \times 100 = \% \text{ A} = 18\% \end{cases} = \text{approx. } 66.7\%$$

As heat is withdrawn at the ternary eutectic, the temperature remains constant while all the liquid disappears and A, B, and C crystallize. In this simple case, all three phases crystallize during the eutectic reaction and A, B, and C are present in the same proportions as they were in the original melt (20% A, 10% B, and 70% C). In later examples we shall see that the events which occur during final crystallization at other types of invariant points are not always so simple. That is, a crystalline phase may be resorbed (taken into solution) during final crystallization, rather than being precipitated. Furthermore, binary or ternary crystalline phases may be present in the final crystalline assemblage rather than the simple end member crystals. The total bulk composition of the crystalline assemblage in terms of A, B, and C is always the same as that of the original melt, however.

For example, at T_E^+, the approximate percentages of A, B, and C in the melt and crystal assemblage are:

Liquid	Crystals	Approximate Total %
%A = 25 X 33.3 = 8.32	%A = 18 X 66.7 = 12.00	20.32
%B = 30 X 33.3 = 9.99		9.99
%C = 45 X 33.3 = 14.98	%C = 82 X 66.7 = 54.69	69.67
		99.98

To summarize, the cooling path for any melt in a simple eutectic type system will terminate at the eutectic invariant point E. The crystals which precipitate as the melt cools will depend upon the location of the liquid composition within the ternary diagram. Melt compositions will generally fall within the primary field (also called primary phase field or field of primary crystallization) of A, B, or C. It is possible that the original melt composition would fall on a boundary line such as $e_1 E$, $e_2 E$, or $e_3 E$ in which case two crystalline phases would appear simultaneously and continue to precipitate as the melt composition moved down the boundary line to E. A melt of the eutectic composition would precipitate all three end members simultaneously and change from 100% liquid to 100% solid (A, B and C) as heat is withdrawn.

It should be pointed out that isoplethal analyses can be carried out by heating as well as cooling. In other words, by starting with a 100% crystalline assemblage and heating until the composition is entirely liquid. Under perfect equilibrium conditions, the analysis of the heating path will be precisely the reverse of the cooling path and quantitative calculation at any temperature level can be made in the same manner using the Lever Rule.

III. THE INFLUENCE OF BINARY COMPOUNDS ON THE TERNARY SYSTEM; JOINS AND ALKEMADES LINES; COMPOSITION OR COMPATIBILITY TRIANGLES; CRYSTALLIZATION PATHS

A. A Congruently Melting Binary Compound

i. The True Binary Join; Isoplethal Analyses

The fundamental configuration is shown in Figure 5.9. The binary system A-B contains a congruently melting compound AB (40% A, 60% B) and A-C and B-C are simple eutectic type systems. The binary compound AB has a primary field in the ternary system bounded by $e_2 E_1$, $E_1 E_2$ and $e_3 E_2$. It is a characteristic of the congruently melting binary compound that its composition point

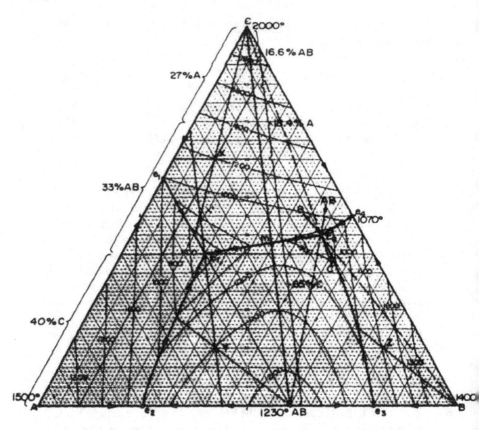

Figure 5.9. Congruently Melting Binary Compound AB; True Binary Join C–AB; Composition Triangles A–AB–C and B–AB–C; Ternary Eutectics E_1 and E_2

always falls within its primary field of stability in the ternary system. The arrows on the boundary lines indicate the direction of falling temperature in the system. A most important feature is the introduction of the join between C and AB which creates two composition or compatibility triangles, A–AB–C and B–AB–C. The join C–AB is known as a true binary join (Figure 5.10) within the ternary system. It is also known as an Alkemades line which is defined as a line connecting two composition points whose primary fields are adjacent.

A theorem by Alkemades states that the intersection of a boundary line (or its extension) with its corresponding Alkemades line (or its extension) represents a temperature maximum on that boundary line and a temperature

minimum on the Alkemades line. Point m is thus a temperature maximum on
the boundary line E_1-E_2 in the system A-B-C and a temperature minimum (in
this case a eutectic) in the vertical section C-AB. This section is a true binary
section because the composition of all of the liquid or solid phases appearing in
the system can be expressed in terms of the end members C and AB.

The concept of the composition triangle is extremely important because
under equilibrium conditions, all liquids having compositions within the composition triangle A-AB-C will terminate their crystallization path at eutectic E_1,
and all those having compositions within the triangle B-AB-C will finally crystallize at E_2. Thus, in this case, there is a 1:1 correspondence between the invariant points E_1 and E_2 and the composition triangles. Note that the original
liquid composition in A-AB-C may lie in the primary field of A, AB or C. Regardless of the original position in a primary field, the final crystallization will
take place at E_1. Likewise for B-AB-C, a liquid in the primary field of B, AB or
C will undergo final crystallization at E_2. The two composition triangles
A-AB-C and B-AB-C are called ternary "subsystems", similar to the binary
"subsystems", A-AB and B-AB in the system A-B.

It will be instructive to follow the equilibrium crystallization paths of
melts of composition X (25A, 10B, 65C), Y (50A, 35B, 15C) and Z (10A, 75B
and 15C). A ternary liquid of composition X will first precipitate an infinitesimal
amount of crystals of C. Greater and greater amounts of C will be precipitated as
the liquid moves to the boundary line $e_1 E_1$. For example, at X′, approximately
30.5% C and 69.5% liquid (36% A, 14% B and 50% C) are in equilibrium (approximate total, 69.5 X 36 = 25.02% A, 69.5% X 14 = 9.73% B, 69.5 X 50 = 34.75%
C (34.75 + 30.5) = 65.25% C). After reaching the boundary line $e_1 E_1$, A and C
then crystallize from the liquid until the eutectic E_1 is reached, when A, C and
AB then crystallize. The amounts of A, C, and AB in the final assemblage are
18.4, 65, and 16.6%, respectively. (See construction on diagram). Compound AB
contains 40% A and 60% B, therefore the (approximate) final amounts of A, B
and C are 18.4 + 6.6 = 25, 9.96, and 65%, respectively, corresponding to the
original liquid composition, 25A, 10B, 65C.

Liquid Y will crystallize AB to the boundary $e_2 E_1$, A and AB along the
boundary to E_1, and finally A, C and AB at the E_1 eutectic. Liquid Z will
crystallize B to the boundary $e_3 E_2$, B and AB along the boundary to E_2, and
finally B, AB and C at the eutectic E_2. (The student should make quantitative
calculations of intermediate states while the liquid is present and of the final
state when all the liquid has disappeared. Calculations should also be made to
show that the amounts of A, B and C in any mixture of phases is always the
same as they are in the original liquid.)

It should be emphasized at this time that when two crystalline phases
precipitate from a ternary liquid when going down boundary lines, for example,

Boundary Line	Crystalline Phases
$e_1 E_1$	A and C
$e_2 E_1$	A and AB
$e_3 E_2$	B and AB
$e_4 E_2$	C and B
$m E_1$	C and AB
$m E_2$	C and AB

they do so because all tangents to any of the boundary lines fall in between the joins A-C, A-AB, B-AB, B-C, or C-AB. In many cases in ternary systems, the boundary line tangents do not fall in between the end members of the corresponding join. In such a case, one crystalline phase will resorb (dissolve) while the second one precipitates. Such cases will be discussed later when appropriate.

Also, when three crystalline phases precipitate at the E_1 and E_2 invariant points while the ternary liquid is disappearing, they do so because lines from A, C and AB to E_1 (or from B, C and AB to E_2, see Figure 5.9) are within a 360° range and not within a 180° range. Like the case of boundary lines, if lines from the primary phases through the invariant points are within a 180° range, one of the phases will resorb (dissolve) rather than crystallize during the four phase (F=4-4=0) equilibrium session.

Not all liquid compositions fall within the area of primary crystallization of A, B, C or AB. If a liquid composition falls on a boundary line, two crystalline phases will immediately precipitate from the ternary liquid as it proceeds toward a ternary eutectic. If a melt has the composition of a ternary eutectic, it will solidify to three crystalline phases, whose amounts are derived by drawing lines through E_1 or E_2 parallel to any two sides of the appropriate composition triangle. The intersections of these parallels with the third side of the composition triangle will subdivide it into three segments which are proportional to the amounts of the three solid phases in the assemblage. See eutectic E_1 in Figure 5.9 and the corresponding amounts of A (27%), AB (33%) and C (40%).

Special cases involve points which are located on the join C-AB. Liquids will lie in the primary field of C or AB but they all finish their crystallization at m, as is the case for any simple binary eutectic system (Figure 5.10). Calculations of the amount and composition of coexisting phases are made according to those previously described for a simple binary eutectic type system. The major point to be made is that all liquids on the join C-AB finish their crystallization at m and do not proceed to E_1 or E_2, as do all other ternary liquids in the system.

ii. Isothermal Sections

The type of diagram shown in Figure 5.9 lends itself to an introduction to the generation of simple isothermal sections. Using the isotherms shown on Figure 5.9, a sequence of isothermal sections at temperatures of 1200°, 1000°, 900° and 600° is shown as Figures 5.11 to 5.14.

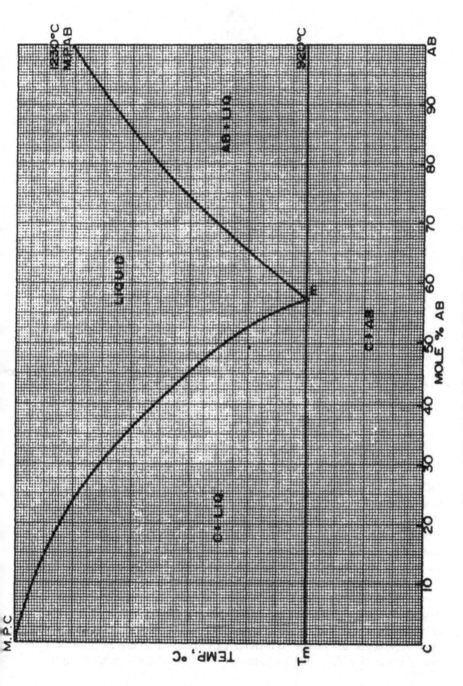

Figure 5.10. Vertical Section C–AB; Binary Eutectic m

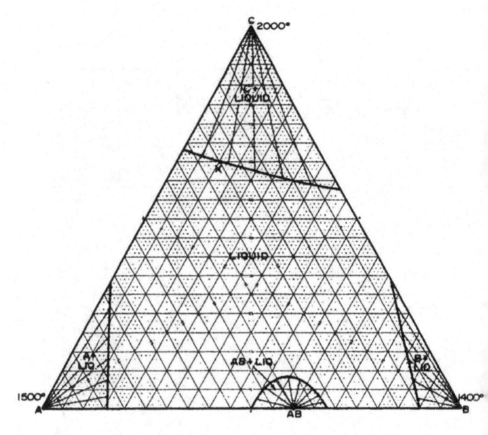

Figure 5.11. 1200°C Isothermal Section

iii. The Non-Binary Join; The Ternary Peritectic Point

An important variation of Figure 5.9 is shown as Figure 5.15 to illustrate several new definitions and principles. The binary system A–B has not been fundamentally changed, but the binary compound AB is now 35A and 65B and e_2 is 80% A, 20% B. Eutectic e_4 is now 30% B, 70% C and ternary eutectic E is now 45A, 15B and 40C. The boundary lines originating from e_3 and e_4 have been redrawn in such a way as to position both ternary invariant points E and P in the composition triangle A–AB–C. Such a change illustrates the generation of the ternary PERITECTIC point P, which falls outside of its associated composition triangle B–AB–C. All liquid compositions in this triangle now undergo final crystallization at P, but as heat is withdrawn and the liquid disappears, two phases crystallize and one (B) is PARTIALLY resorbed during the invariant reaction. For example, for liquid X containing 10A, 40B and 50C, which is in the

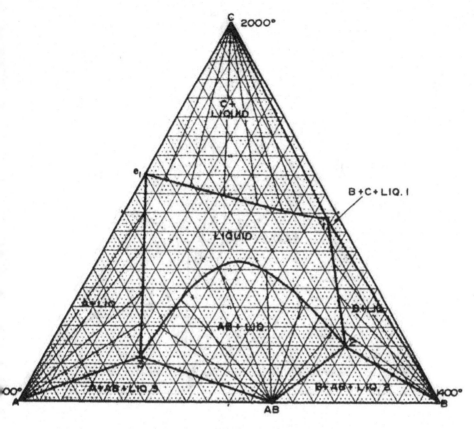

Figure 5.12. 1000°C Isothermal Section

primary field of B and in the composition triangle B–AB–C, B will crystallize to the boundary line e_4 P, C and B will crystallize as the liquid moves toward P and finally, SOME B is resorbed at P as C and AB crystallize and the liquid disappears. A quantitative calculation of the equilibrium just before and just after final crystallization will prove that B is partially resorbed during the invariant (F=4–4=0) reaction. Using the lever P-k, it is seen that just before final crystallization, about 67% solid (B and C) and 33% liquid (30% A, 25% B, 45% C) are in equilibrium. The solid contains 48% B and 52% C, hence the amount of B present before final crystallization is approximately 0.67 × 48 = 32%. Just after final crystallization (see construction on Figure 5.15) about 21.5% will be present in the assemblage, so approximately 32–21.5 ≃ 10.5% B has been resorbed. This resorbtion of B at the peritectic will be characteristic of all liquids in the composition triangle B–AB–C.

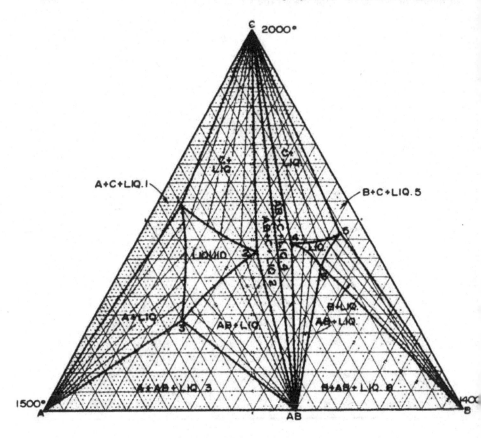

Figure 5.13. 900°C Isothermal Section

It should be further noted that a small portion of the primary field of B lies in the left-hand composition triangle A–AB–C. This means that final crystallization under equilibrium conditions must take place at E. Therefore, a liquid composition such as Y will first crystallize B, then B + C as the path moves toward P. At P, the peritectic reaction L+B+C $\rightleftarrows$ L+AB+C will occur (F=4–4=0), all of B will disappear and crystallization (Liq + AB + C) will proceed to E, giving A, AB and C as the final mixture. E, of course, is the point of final crystallization of ALL liquids in the triangle A–AB–C.

The line C–AB is now a non-binary join due to the fact that the primary field of B overlaps the join. There is no way in which composition B can be expressed in terms of the compositions of the end members of the join, C and AB, therefore the join is said to be non-binary in the region where it overlaps the primary field of B.

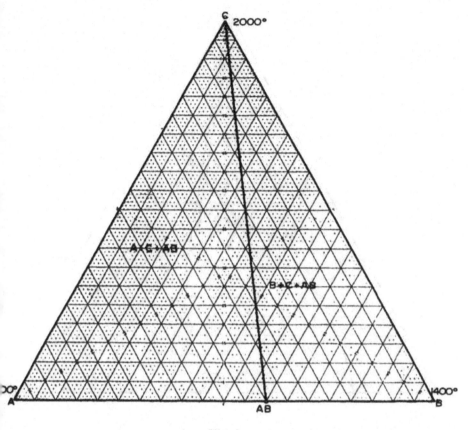

Figure 5.14. 600°C Isothermal Section

It should be noted that certain liquids in the primary fields of C or AB whose paths encounter the boundary lines e_4P or e_3P, respectively, will undergo the peritectic reaction at P, causing the disappearance of previously crystallized B, before proceeding to E.

To summarize, when an isoplethal analysis of a ternary liquid composition is made, it is always necessary to ask the following questions before proceeding:

1. The composition is in what primary field? Or is it on a boundary line or join?
2. The composition is in which composition triangle? Or is it on a join which is one side of a composition triangle?
3. What invariant point is associated with a particular composition triangle?

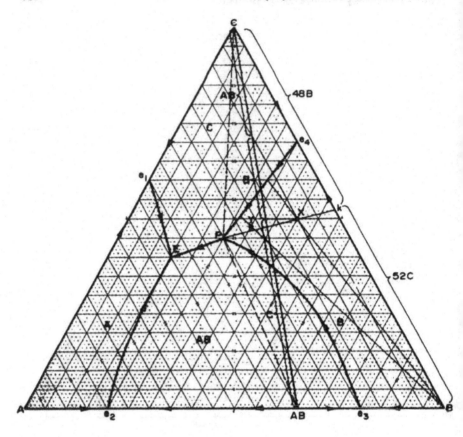

Figure 5.15. The Non-binary Join; The Ternary Peritectic Point

In this section it has been shown that ternary eutectic points and ternary peritectic points exist, similar to their binary system counterparts.

In the case of invariant equilibrium at eutectic points, three phases crystallize from the eutectic liquid composition. At peritectic points, two phases crystallize and one is partially resorbed during disappearance of the eutectic liquid as heat is withdrawn under equilibrium conditions.

B. An Incongruently Melting Binary Compound; Isoplethal Analysis; Recurrent Crystallization; Composition on a Join

Figure 5.16 shows the fundamental configuration and its relationship to the binary system A–B and its incongruently melting compound, AB. Note that the composition point of the incongruently melting binary compound falls outside

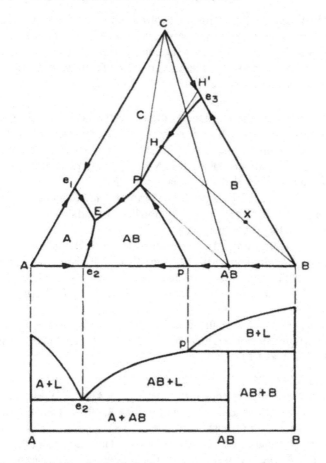

Figure 5.16. Ternary System with a Binary Incongruently Melting Compound AB. Binary System A–B Shown Below

of its primary field in the ternary system. Boundary lines leading into the ternary system from e_2 and p encounter the ternary eutectic E and the ternary peritectic P, respectively. The ternary peritectic falls outside of its associated composition triangle B–AB–C. Note carefully the arrows indicating the direction of falling temperature on ALL boundary lines.

The line C–AB does not intersect the boundary line E–P separating the primary fields of C and AB and it has a large overlap of the primary field of B, thus making it a non-binary join.

As usual, two questions should always be asked when considering the crystallization path of any ternary liquid:

1. What composition triangle is involved and where is the associated invariant point at which final crystallization will take place?
2. In which primary field does the composition lie; where is the associated composition point for the primary field and in what direction does crystallization proceed?

Compositions in the left hand triangle A–AB–C will crystallize under equilibrium conditions at E. Note that these include compositions in the primary fields of A and AB and in part of the primary fields of B and C. Compositions in the triangle B–AB–C will crystallize at P. Note that only parts of the primary fields of B and C are included in the triangle B–AB–C.

Tangent lines like H–H′ indicate the ratio of B and C crystals which are precipitating as temperatures decrease on the P–e_3 boundary line (so-called "instantaneous" crystalline composition).

When crystallization paths are traveling along the boundary line P–E, tangents to the boundary line will end at the C–AB join. When calculations of liquid and crystalline percentages are being made as liquids cool down boundary lines P–E, p–P, e_2–E and e_1–E, corresponding joins are C–AB, B–AB, A–AB and A–C, respectively.

Several new rules can be illustrated in this particular system by following the crystallization paths of certain specially selected starting liquids, as shown in Figure 5.17. (X, Y and Z)

i. Composition X (10% A, 70% B, 20% C)
Composition X is located in composition triangle B–AB–C, the primary field of B and the final crystalline assemblage will be 40% B, 40% AB and 20% C. (See Figure 5.17, lines drawn through X, parallel to B–C and B–AB and intersecting C–AB) Table 5.1. Final crystallization will occur at invariant point P.

As liquid X cools, it will intersect the B liquidus and an infinitesimal amount of crystalline B will coexist with the ternary liquid X. (See Table 5.1). As the temperature lowers, the liquid will proceed on the B–X extention and more and more crystalline B will precipitate. For example, at X′, 50% liquid (20% A, 40% B, 40% C) and 50% B coexist (Total 10A, 70B, 20C). At X″, an infinitesimal amount of C appears and $\frac{30}{65}$ = 46.1% liquid (22A, 35B, 43C), 53.9% B, and C coexist. (Table 5.1)

Along the boundary line e_3P, B and C crystallize from the liquid and at point X′‴, 32.3% Liquid (32% A, 32% B, 36% C) and 67.8% crystals coexist. See tie line X′‴–X–m and Table 5.1 for calculations.

At a temperature slightly above the invariant peritectic point, the tie line P–n will give the percentage of coexisting phases as shown in Table 5.1. Note that 62.5% B coexists with the peritectic liquid. During the peritectic reaction, AB and C precipitate, B is partially resorbed, and the liquid disappears (See

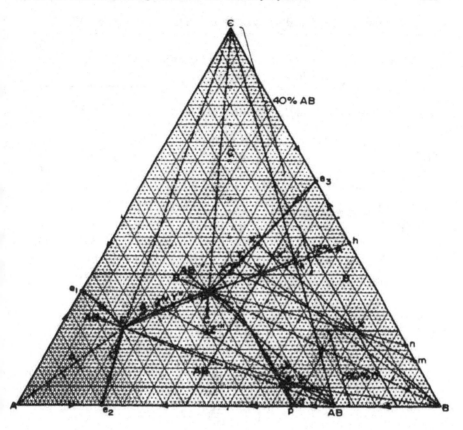

Figure 5.17. Ternary System with a Binary Incongruently Melting Compound AB, Ternary Eutectic and Peritectic; Isoplethal Analyses

Table 5.1). The final percentage of B is 40, indicating that 62.5 − 40 = 22.5% B was resorbed during the peritectic reaction. This is qualitatively indicated in Figure 5.17 by the arrows on peritectic point P. The arrows indicate in which direction C, AB, and B would continue from point P. The arrows are all within 180°, therefore it would be necessary to flip arrow B in the opposite direction to have them exist in a 360° region. This is the qualitative indication that crystalline B will be partially resorbed during the invariant peritectic reaction. In contrast, note the direction of the C, AB, and A arrows at E (360° region). At ternary eutectic points, three crystalline phases always precipitate from the eutectic liquid as it disappears under perfect equilibrium conditions.

Note that that the amount of crystalline B increased from infinitesimal (at X) to 62.5% before final crystallization and was finally 40% of the crystalline

Table 5.1 Isoplethal Analysis for Composition X, Figure 5.17

	$A = 10$		$B = 40$
Original Liquid	$B = 70$	Final Crystalline Assemblage	$AB = 40$
	$C = 20$		$C = 20$

Temp.	Proportions and Compositions of Phases		Analysis A	B	C
1 (Pt. X)	$\frac{40}{40}$ Melt = 100%	$\begin{cases} A = 10 \\ B = 70 \\ C = 20 \end{cases}$	10	70	20
	Infinitesimal Amount of Crystals of B				
2 (Pt. X")	$\frac{30}{65}$ Melt = 46.1%	$\begin{cases} A = 22 \\ B = 35 \\ C = 43 \end{cases}$	10	16.1	19.8
				$\underline{53.9}$	
	$\frac{35}{65}$ Xtals = 53.9% B			70.0	
	Infinitesimal Amount of Crystals of C				
3 (Pt. X"')	$\frac{18}{56}$ Melt = 32.2%	$\begin{cases} A = 32 \\ B = 32 \\ C = 36 \end{cases}$	10.3	10.3	11.6
				$\underline{59.7}$	$\underline{8.1}$
				70.0	19.7
	$\frac{38}{56}$ Xtals = 67.8%	$\begin{cases} B = \frac{88}{100} = 88 \\ C = \frac{12}{100} = 12 \end{cases}$			
4 slightly above Pt. P	$\frac{13.5}{53.5}$ Melt = 25.2%	$\begin{cases} A = 40 \\ B = 30 \\ C = 30 \end{cases}$	10	7.5	7.5
				$\underline{62.5}$	$\underline{12.3}$
				70.0	19.8
	$\frac{40}{53.5}$ Xtals = 74.8%	$\begin{cases} B = \frac{83.5}{100} = 83.5 \\ C = \frac{16.5}{100} = 16.5 \end{cases}$			
5 slightly below Pt. P	Peritectic reaction occurs: $B(S)+C(S)+A,B,C,(\ell) \rightleftarrows AB(S)+B(S)+C(S)$ yielding final Xtals $\begin{cases} B = 40 \\ AB = 40\ (25A, 75B) \\ C = 20 \end{cases}$		10	30 $\underline{40}$ 70	20

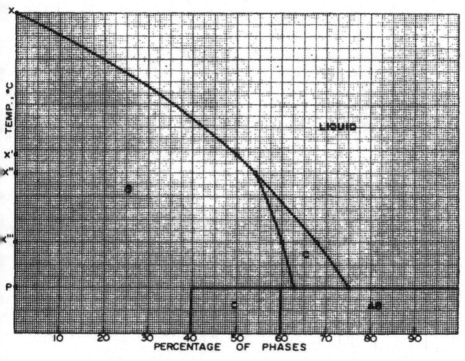

Figure 5.18. Phase Analysis Diagram for Composition X

assemblage. Crystalline C increased from an infinitesimal amount at X″ to 12.3% before final crystallization and finally became 20% of the assemblage. 40% AB crystallized at the invariant point to become that portion of the final assemblage. Figure 5.18 is a "phase analysis" diagram based on the calculations in Table 5.1.

ii. Composition Y (25% A, 40% B, 35% C)
Composition Y is located in composition triangle A–AB–C, the primary field of B and the final crystalline assemblage will be 12% A, 53% AB and 35% C. (See Figure 5.17, lines drawn through Y, parallel to A–C and A–AB, and intersecting C–AB) Table 5.2. Final crystallization will occur at invariant point E (eutectic), but prior to this, an invariant reaction will take place at point P (peritectic).

Starting at Y, an infinitesimal amount of B is in equilibrium with liquid Y. At Y′, an infinitesimal amount of C appears in equilibrium with liquid Y′ and crystalline B. B and C crystallize from Y′ to P and at P the peritectic reaction L+B+C ⇌ L+AB+C continues as the temperature lowers and B disappears. Note tic line P-Y-h and Table 5.2 calculations. As the temperature decreases the liquid proceeds from P to E, C and AB crystallize. At Y″, note conjugation line

Table 5.2 Isoplethal Analysis for Composition Y, Figure 5.17

	Original Liquid $A = 25$ $B = 40$ $C = 35$	Final Crystalline Assemblage $A = 12$ $AB = 53$ $C = 35$		
		Analysis		
Temp.	Proportions and Composition of Phases	A	B	C
Y	Original Liquid Melt = 100% $\begin{cases} A = 25 \\ B = 40 \\ C = 35 \end{cases}$	25	40	35
	Infinitesimal amount of crystals of B			
Y'	$\frac{60}{67}$ Melt = 89.6% $\begin{cases} A = 28 \\ B = 33 \\ C = 39 \end{cases}$	25	29.6	35
	$\frac{7}{67}$ Xtals = 10.4% $\begin{cases} B = 100 \\ C = \text{Infinitesimal Amount} \end{cases}$		$\underline{10.4}$	
		25	40.0	35
	$\frac{25}{40}$ Melt = 62.5% $\begin{cases} A = 40 \\ B = 30 \\ C = 30 \end{cases}$	25	18.75	18.75
P$^+$	$\frac{15}{40}$ Xtals, B, C = 37.5% $\begin{cases} \frac{56.5}{100} \ B = 56.5 \\ \frac{43.5}{100} \ C = 43.5 \end{cases}$		$\underline{21.2}$	$\underline{16.3}$
			39.95	35.05

P⁻

$\dfrac{9.5}{24.5}$ Melt = 38.7% $\left\{\begin{array}{l}A = 40\\B = 30\\C = 30\end{array}\right.$ 15.5 11.6 11.6

$\dfrac{15}{24.5}$ Xtals, AB, C = 61.2% $\dfrac{61.5}{100}$ AB = 37.6 $\begin{array}{l}A = 25\\B = 75\end{array}$ $\dfrac{9.4}{24.9}$ $\dfrac{28.2}{39.8}$ $\dfrac{23.4}{35.0}$

$\dfrac{38.5}{100}$ C = 23.4

Y″

$\dfrac{9.5}{34.5}$ Melt = 27.5% $\left\{\begin{array}{l}A = 50\\B = 23.5\\C = 26.5\end{array}\right.$ 13.75 6.5 7.3

$\dfrac{25}{34.5}$ Xtals, AB, C = 72.5% $\dfrac{61.5}{100}$ AB = 44.5 $\begin{array}{l}A = 25\\B = 75\end{array}$ $\dfrac{11.25}{25.00}$ $\dfrac{33.5}{40.0}$ $\dfrac{28}{35.3}$

.5% $\dfrac{38.5}{100}$ C = 28

E⁺

$\dfrac{10}{50}$ Melt = 20% $\left\{\begin{array}{l}A = 65\\B = 15\\C = 20\end{array}\right.$ 13.0 3 4

$\dfrac{40}{50}$ Xtals, C, AB = 80% $\dfrac{61}{100}$ AB = 48.8 $\begin{array}{l}A = 25\\B = 75\end{array}$ $\dfrac{12.2}{25.2}$ $\dfrac{36.6}{39.6}$ $\dfrac{31.2}{35.2}$

$\dfrac{39}{100}$ C = 31.2

E⁻

$\begin{array}{l}A\ \ 12\\AB\ 53\ \left\{\begin{array}{l}25\ A\\75\ B\end{array}\right.\\C\ \ 35\end{array}$ $\dfrac{12}{13.25}$ $\dfrac{36.6}{39.6}$ $\dfrac{31.2}{35.2}$

25.25 39.75 35

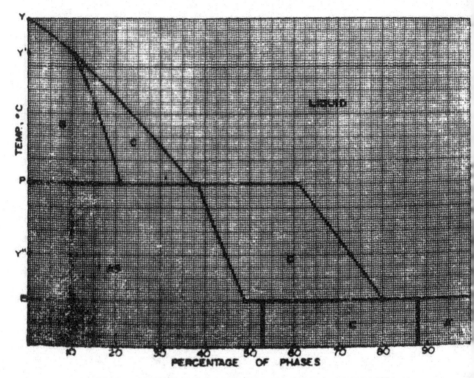

Figure 5.19. Phase Analysis Diagram for Composition Y

Y''-Y-k and Table 5.2. Join C-AB is now the one which corresponds to crystallization of C and AB on boundary line P-E. At an infinitesimal temperature above eutectic E, note tie line E-Y-ℓ, Table 5.2 calculations. At E, the eutectic reaction (F=4-4=0) occurs, L ⇄ C+AB+A and finally 12% A, 53% B and 35% C coexist. The phase analysis diagram based on calculations in Table 5.2 is shown in Figure 5.19.

iii. Composition Z (30% A, 65% B, 5% C)
Composition Z is located in composition triangle A-AB-C, the primary field of B and the final crystalline assemblage will be 8.5% A, 86.5% AB and 5% C. (See Figure 5.17, lines drawn through Z, parallel to A-C and A-AB and intersecting C-AB) Table 5.3. Final crystallization will occur at invariant point E (eutectic). but prior to this, a new rule will be demonstrated during isoplethal analysis.

As liquid Z cools, it will intersect the liquidus surface B and an infinitesimal amount of crystalline B will coexist with the ternary liquid. When the liquid reaches the boundary line p-P, an infinitesimal amount of AB will coexist with ternary liquid Z' and B. (Table 5.3). If a tangent to p-P is now drawn at Z', it

Table 5.3 Isoplethal Analysis for Composition Z, Figure 5.17

	Original Liquid $A = 30$, $B = 65$, $C = 5$	Final Crystalline Assemblage $A = 8.5$, $AB = 86.5$, $C = 5$		
			Analysis	
Temp.	Proportions and Composition of Phases	A	B	C
Z	Melt = 100% { A = 30, B = 65, C = 5 }	30	65	5
	Infinitesimal Amount of Crystals of B			
Z'	$\frac{35}{40}$ Melt = 87.5% { A = 34.5, B = 60, C = 5.5 }	30	52.5	5
	$\frac{5}{40}$ Xtals, B, AB = 12.5% { B = 100 }		12.5	
			65.0	
	Infinitesimal Amount of Crystals of AB			
Z"	$\frac{10}{18.5}$ Melt = 54.2% { A = 34.5, B = 56.5, C = 9.5 }	18.7	30.6	5.1
	B has been totally resorbed			
	$\frac{8.5}{18.5}$ Xtals, B, AB = 45.8% AB = 45.8 A = 25, B = 75	11.4	34.4	
		30.1	65.0	
	$\frac{10}{40}$ Melt = 25% { A = 46, B = 35, C = 19 }	11.5	8.8	4.75

Table 5.3 (Continued)

Temp.	Proportions and Composition of Phases		Analysis		
	Original Liquid $A = 30$, $B = 65$, $C = 5$	Final Crystalline Assemblage $A = 8.5$, $AB = 86.5$, $C = 5$	A	B	C
Z''	$\dfrac{30}{40}$ AB = 75%	$\Big\{$ A = 25 B = 75	$\underline{18.75}$ 30.25	$\underline{56.2}$ 65.0	
	$\dfrac{10}{53.5}$ Melt = 18.7%	$\Big\{$ A = 53 B = 21.5 C = 25.5	9.9	4.0	4.8
Z'ᵛ	$\dfrac{43.5}{53.5}$ Xtals, AB, C = 81.3%	AB = 81.3% $\Big\{$ A = 25 B = 75 C = Infinitesimal Amount	$\underline{20.3}$ 30.2	$\underline{61.0}$ 65.0	
	$\dfrac{8.3}{58.3}$ Melt = 14.2%	$\Big\{$ A = 65 B = 15 C = 20	9.2	2.1	2.8
E⁺	$\dfrac{50}{58.3}$ Xtals, AB, C = 85.7%	$\dfrac{97.5}{100}$ AB = 97.5 $\Big\{$ A = 25 B = 75 $\dfrac{2.5}{100}$ C = 2.5	$\underline{20.8}$ 30.0	$\underline{62.8}$ 64.9	$\underline{2.14}$ 4.94
E⁻		A = 8.5 AB = 86.5 $\Big\{$ 25 A 75 B C = 5	8.5 $\underline{21.5}$ 30.0	65	5

will intersect the EXTENSION of the join AB-B, indicating that B will resorb as
the temperature drops and the ternary liquid moves down the boundary line p-P.
In all previous discussions of boundary line crystallization, the tangents to the
boundary lines always fell within the two end members of the corresponding
join and two crystalline phases precipitated from the ternary liquid. In this case,
B resorbs and AB crystallizes as the liquid moves down p-P. When the liquid
reaches Z'' and when the conjugation line includes Z''-Z-AB, all of B has been
resorbed and the path will proceed as the extension of Z''-Z-AB across the
primary field of AB until the boundary line P-E is intercepted at Z'^v (See calculations in Table 5.3 for points Z'', Z''' and Z'^v). From Z'^v to E, C and AB crystallize and the join C-AB must be used for quantitative calculations of the amounts
of liquid, C and AB. At E, crystalline A appears and the calculation at temperature E^+ is shown in Table 5.3. Finally the eutectic liquid disappears as heat is
withdrawn and the final assemblage of crystals contains 8.5% A, 86.5% AB and
5% C. The phase analysis diagram based on calculations in Table 5.3 is shown in
Figure 5.20.

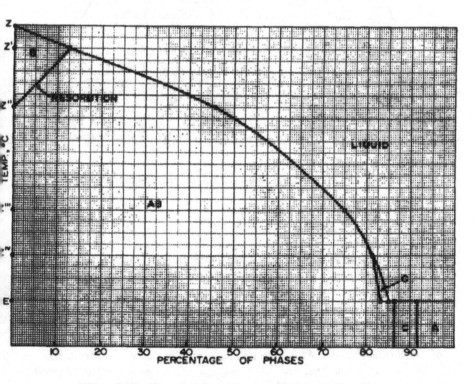

Figure 5.20. Phase Analysis Diagram for Composition Z

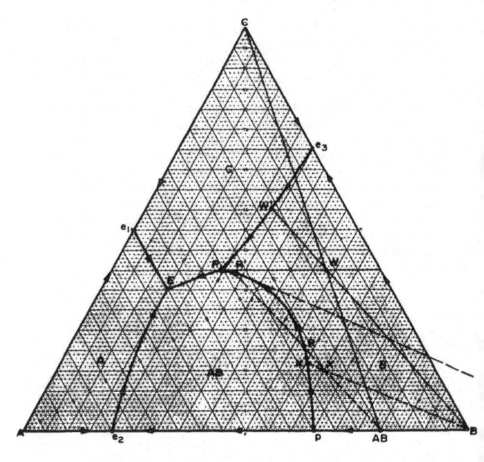

Figure 5.21. Diagram Illustrating Recurrent Crystallization

iv. Recurrent Crystallization

A variation on the example just given occurs when the boundary line P–p has a strong curvature as shown in Figure 5.21. Composition X is in the composition triangle A–AB–C and in the primary field of B. Under equilibrium conditions, the final crystalline assemblage will be composed of A, AB and C. B crystallizes from the liquid from X to K. B is resorbed from K to R while AB crystallizes. All of B has been resorbed when the liquid reaches point R and the melt follows a path across the primary field of AB as only AB crystallizes (the prolongation of X–AB from R to R'). At R', B once again crystallizes (recurrent crystallization) while AB is resorbed. Note that the tangent to the p–P boundary line at R' would intersect the EXTENSION of the join AB–B on the B side, indicating resorbtion of AB from R' to P. At P, B is completely resorbed and C and AB crystallize from the liquid as it moves toward E. Thus B has alternately crystallized, dissolved, crystallized and dissolved as the liquid has moved from X to E.

v. Composition on a Join (W)
In Figure 5.21 the composition W appears on the join C–AB which does not intersect its boundary line P–E and, therefore, makes the join C–AB non-binary. Nevertheless, the melt will completely solidify to form C and AB in the proportions determined by the position of the point W on the C–AB join. (40% C, 60% AB) The sequence of reactions is different from any described previously because W is located on a join and is not associated with either of the two composition triangles. (Table 5.4) Crystalline B will precipitate from liquid W to the point W' at which crystals of C and AB are in equilibrium. As the melt changes composition along the boundary line from W' to point P, crystals of B and C precipitate. At point P the peritectic reaction occurs in which B reacts with the melt to form crystals of C and AB:

$$B(s) + C(s) + A_xB_yC_z(\ell) \rightleftarrows C(s) + AB(s) + A_xB_yC_z(\ell)$$

As heat is withdrawn, B is completely resorbed and the melt crystallizes to form 40% C and 60% AB.

vi. Variation of Figures 5.17 and 5.21; Ternary Eutectics E_1 and E_2
Figure 5.22 shows a possible variation of the case of a binary incongruently melting compound with a primary field in the ternary system. The binary system A–B is basically the same as previously described, having a eutectic e_2, a peritectic p, and incongruently melting AB. However, the boundary lines in the ternary system lead to a ternary eutectic E_1, where all liquids in the composition triangle A–C–AB terminate their equilibrium crystallization and a ternary eutectic E_2 which is the terminus of crystallization for all liquids whose compositions are originally in composition triangle B–C–AB. The join C–AB determines the temperature maximum on E_1–E_2. The portion of the join C–AB which crosses the primary field of B is responsible for the fact that it is a non-binary join. Whatever the composition of C and AB might be, composition B could not be expressed in terms of C and AB.

vii. Isothermal and Vertical Sections
The sequence of reactions may be better understood by considering isothermal and vertical sections through the solid model of the ternary system. Figure 5.23 shows a projection of the liquidus surface on which the isotherms have been drawn. The cooling path of liquid X from 1100° to 800° is shown on the diagram by the line X–X'. An isothermal section at 800° is shown in Figure 5.24. Note that the point X is located in the region of B plus liquid and is on the tie

Table 5.4 Isoplethal Analysis for Composition W

		Final Crystalline Assemblage		
	Melt $\begin{array}{l}A = 12\\ B = 48\\ C = 40\end{array}$	$\begin{array}{l}AB = 60\\ C = 40\end{array}$ $=$		
Temp.	Proportions and Composition of Phases	A	B	C
			Analysis	
		A	B	C
W	Melt $= 100\%$ $\begin{array}{l}A = 12\\ B = 48\\ C = 40\end{array}$	12	48	40
	Xtals = Infinitesimal Amount of Crystals of B			
W'	$\frac{52}{72}$ Melt $= 72.2\%$ $\begin{array}{l}A = 16.5\\ B = 28\\ C = 55.5\end{array}$	11.9	20.2	40
	$\frac{20}{72}$ Xtals, C, B $= 27.8\%$ $\begin{array}{l}B = 100\\ C = \text{Infinitesimal Amount of C}\end{array}$		$\dfrac{27.8}{48.0}$	
P⁺	$\frac{12}{35}$ Melt $= 34.2\%$ $\begin{array}{l}A = 35\\ B = 25\\ C = 40\end{array}$	12	8.55	13.6
	$\frac{23}{35}$ Xtals, C, B $= 65.8\%$ $\begin{array}{l}B = 60\\ C = 40\end{array}$		$\dfrac{39.5}{48.05}$	$\dfrac{26.3}{39.9}$
P⁻	0 Melt $= 0\%$			
	All Xtals $= 100\%$ $AB = 60$ $\begin{array}{l}A = 20\\ B = 80\\ C = 40\end{array}$	12	48	40

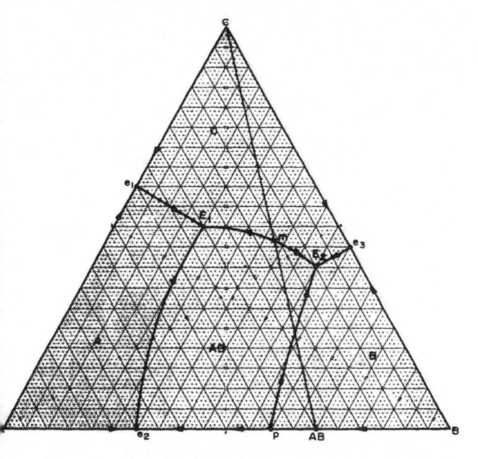

Figure 5.22. Variation of Figures 5.17 and 5.21; Ternary Eutectic E_1 and E_2

line B-X-X'. At 730°, which is the peritectic temperature and the temperature of final solidification, the isothermal section appears as shown in Figure 5.25. The point X now appears on the vertical plane C-AB (Figure 5.26) which separates the two three-phase spaces, C + AB + Liquid P and B + AB + C in Figure 5.25. Note that the isopleth passes sequentially through the spaces B + L, C + B + L, and finally AB + C.

C. Dissociating Binary Compounds

i. Upper Temperature Limit of Stability
In Figure 5.27, the binary compound AB (65% A, 35% B) decomposes to A and B at T_R. The end members form a eutectic at Te_2. Figure 5.28 shows the

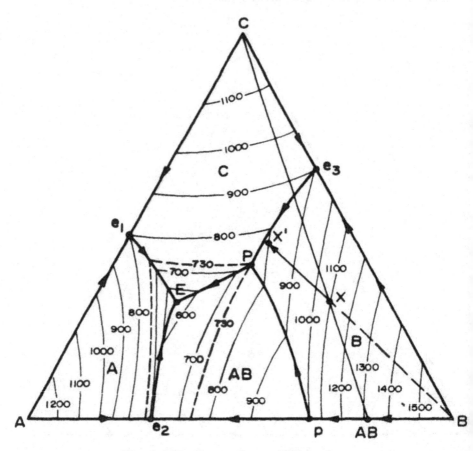

Figure 5.23. Ternary System With Isotherms

primary field of compound AB which will appear in the ternary system when liquidus temperatures become low enough (less than T_R) to allow AB to appear on the liquidus surface. Temperatures E_1 and E_2 are both lower than T_R. The join C-AB is non-binary due to the substantial intrusion of the primary field of A across the join. However, the intersection of the join C-AB with the boundary line E_1-E_2 determines that a temperature maximum exists at m. The boundary line E_1-E_2 separates the primary fields of C and AB. Note the direction of falling temperature on all other boundary lines. The two ternary eutectics fall within their respective composition triangles A-C-AB and B-C-AB. The point R is a invariant (reaction) point where liquid and three crystalline phases coexist at equilibrium (F=4-4=0), but there is no corresponding composition triangle (A-B-AB) and no crystallization paths terminate at R.

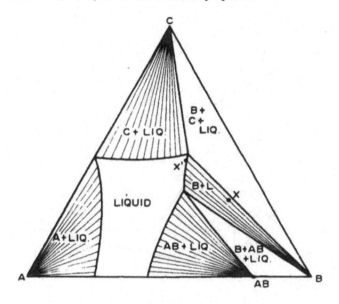

Figure 5.24. 800° Isothermal Section of Figure 5.23

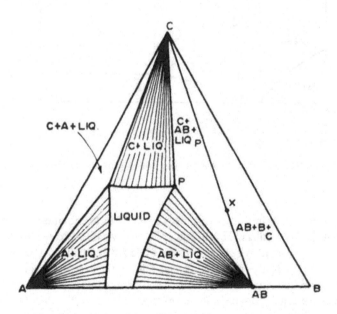

Figure 5.25. 730° Isothermal Section of Figure 5.23

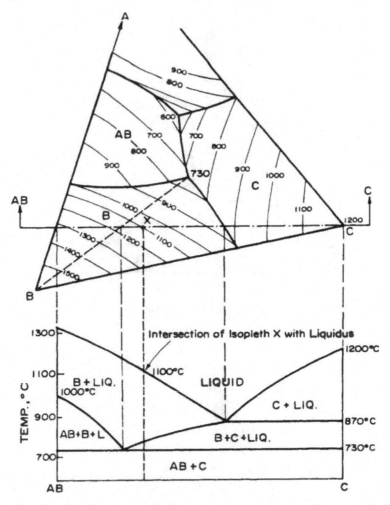

Figure 5.26. Vertical Section Along Join C-AB (not a true binary section)

1. The point X is in the primary field of A and in the composition triangle A-C-AB and final crystallization will end at E_1. A crystallizes to the boundary line R-E_1, at which point it begins to resorb while AB crystallizes. The tangent to the boundary line R-E_1 (see X'-k) always falls on the extension of the join A-AB, indicating that A will be resorbed. A continues to be resorbed and AB crystallizes as the liquid composition moves to E_1. At E_1, A, C and AB crystallize and the liquid disappears as heat is withdrawn at constant temperature.

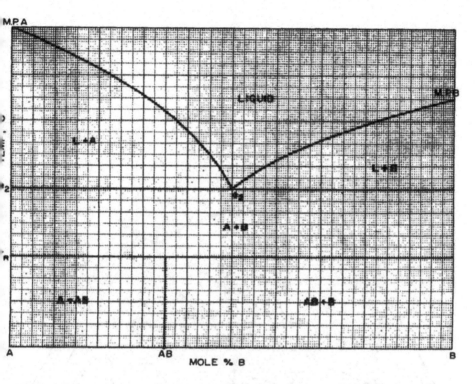

Figure 5.27. Dissociating Binary Compound with a Lower Temperature Limit of Stability

2. The path of composition Y is unusual. A will crystallize to the boundary line e_2-R, at which point A and B crystallize together into point R. The liquid path would appear to have two choices at R, boundary line R-E_1 or boundary line R-E_2. It follows R-E_1, while A is resorbed and AB crystallizes. (If the liquid followed the boundary R-E_2, the tie line from any point on the boundary projected back through Y would intersect the join A-AB, which leads to an absurdity because R-E_2 separates the primary fields of B and AB and no quantitative analysis could be made.) At point X′, A has been completely resorbed and AB crystallizes as the ternary liquid path moves to point ℓ. C and AB crystallize from ℓ to E_2 and at E_2, C, AB and B crystallize as the eutectic liquid disappears.

3. Composition Z is in the primary field of A, but in the composition triangle B-C-AB and hence finishes equilibrium crystallization at E_2. A precipitates from the liquid from Z to the boundary line. At the boundary line, A

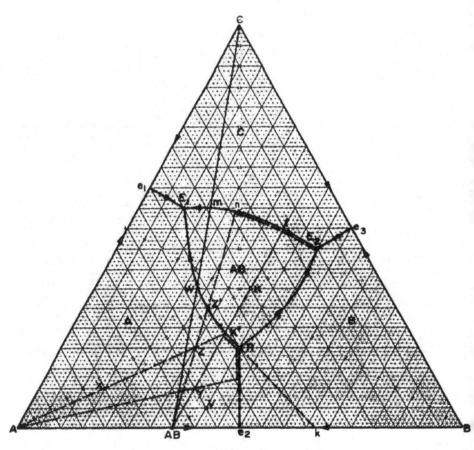

Figure 5.28. Ternary System with a Dissociating Binary Compound (Upper Temperature Limit of Stability)

resorbs while AB crystallizes and at point Z′, all of A has been resorbed. The liquid moves across the primary field of AB, crystallizing only AB, until the boundary line E_1-E_2 at point n is encountered. C and AB crystallize together to E_2, where final solidification of C, AB, and B takes place.

    4. Liquid V would precipitate A, move to the boundary e_2-R, crystallize A and B to R, resorb A completely at R, move along R-E_2 while AB and B crystallize and terminate at E_2 while C, AB, and B crystallize.

    5. Liquid K will crystallize AB as the path moves away from composition point AB to the boundary line E_1-E_2, AB + C from ℓ to E_2 and at E_2, eutectic crystallization takes place.

6. Composition W is on the boundary line R-E$_1$ and on the join line C-AB, therefore only these two phases will be present after final crystallization. AB will crystallize as the path of the liquid moves across the primary field of AB to m, at which point C and AB will crystallize as the liquid disappears at constant temperature. Any composition on the join will finish its crystallization at m, but all those liquid compositions on the join which are in the primary field of A will reach m via boundary lines e$_2$-R and R-W or R-W only.

Quantitative analyses of the amount and composition of liquid and solid(s) can be made by the standard method at any point during crystallization of X, Y, Z, V, K or W. The composition of invariant points e$_1$, e$_2$, e$_3$, E$_1$, E$_2$ and R are:

e$_1$	E$_1$	E$_2$	R
40 A, 60 C	35A	10A	40 A
e$_2$			
50 A, 50 B	10 B	45 B	40 B
e$_3$			
50 B, 50 C	55 C	45 C	20 C

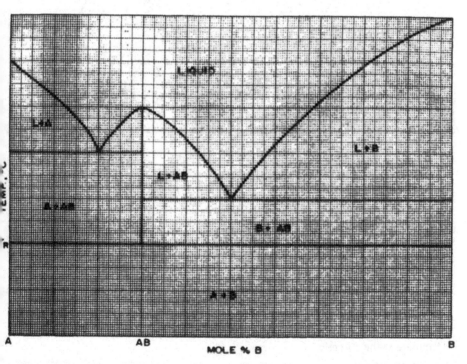

Figure 5.29. Dissociating Binary Compound with a Lower Temperature Limit of Stability

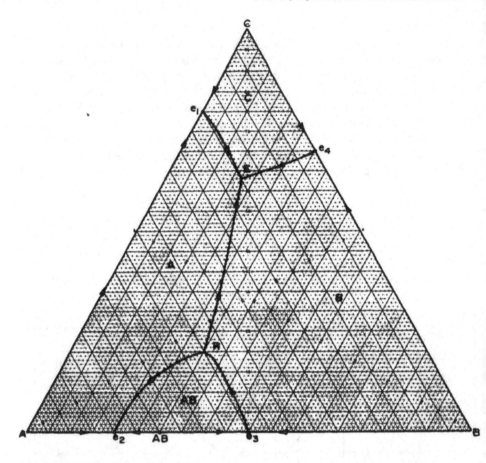

Figure 5.30. Ternary System with Dissociating Binary Compound (Lower Temperature Limit of Stability)

The joins which must be used for quantitative calculations are A–C, B–C A–AB, B–AB and C–AB, as the ternary liquid moves along the corresponding boundary lines.

ii. Lower Temperature Limit of Stability

Figures 5.29 and 5.30 show the binary system A–B with the binary compound AB which dissociates at T_R, and the associated ternary system A–B–C, where R is again an invariant reaction point (F=4-4=0). All crystallization paths will terminate at the eutectic, E.

The binary compound AB need not melt congruently as shown in Figure 5.29. It may melt incongruently or dissociate at some higher temperature. In the latter case, the compound is said to have both an upper and lower temperature limit of stability.

In this system, it should be noticed that the only composition triangle is A-B-C, corresponding to the invariant point E, and that no composition triangle corresponding to invariant point R (A, AB, B) exists. Ternary liquids in all primary fields (A, B, C, AB) finally crystallize under perfect equilibrium conditions at E. Certain compositions will undergo the invariant reaction (L, A, B, AB) at point R before final crystallization at the ternary eutectic.

IV. THE INFLUENCE OF TERNARY COMPOUNDS ON THE TERNARY SYSTEM; CRYSTALLIZATION PATHS

A. A Congruently Melting Ternary Compound

Figure 5.31 shows the fundamental configuration. The composition point of the ternary compound ABC falls within its own primary field and joins between ABC and the end members create three composition triangles. All liquids in composition triangle A-ABC-C finish crystallization at E_1; those in the triangle A-ABC-B finish at E_2 and those in the triangle B-ABC-C finish at E_3. Temperature maxima on the boundary lines at m_1, m_2 and m_3 are created by the intersection of the joins with E_1-E_3, E_1-E_2 and E_2-E_3, respectively. The system A-B-C is divided into three true ternary subsystems and crystallization paths follow the course outlined in section II of this chapter. The composition triangles are no longer equilateral triangles, however, and the final equilibrium assemblage of crystals must be found by drawing lines through the original liquid composition, parallel to any two sides of the composition triangle. For example, composition x will consist of 27.5 parts (42.3%) ABC, 27.5 parts (42.3%) A, and 10 parts (15.4%) C after final crystallization. As always, original liquids may be located in primary fields, on joins, on boundary lines or at ternary invariant points. The ternary compound ABC is presumed to be stable from room temperature to its congruent melting point, which is the highest temperature point in the primary field of ABC.

A combination of a congruently melting ternary compound ABC and a congruently melting binary compound AB would give the configuration shown in Figure 5.32. Note the four ternary eutectics, E_1, E_2, E_3 and E_4, the corresponding triangles A-C-ABC, A-AB-ABC, B-AB-ABC and B-C-ABC, and temperature maxima m_1, m_2, m_3 and m_4. A portion of the primary field of ABC is in each of the four composition triangles.

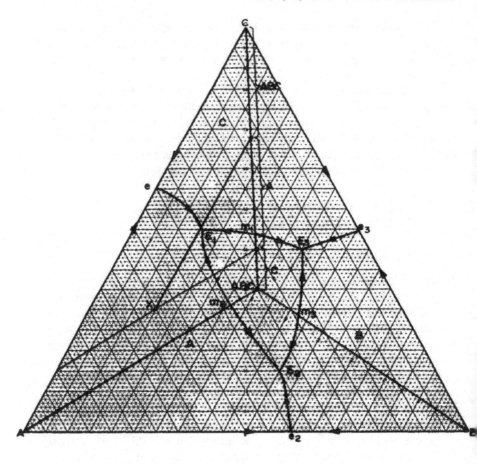

Figure 5.31. Congruently Melting Ternary Compound

B. An Incongruently Melting Ternary Compound

i. Single Incongruency

The fundamental configuration is shown in Figure 5.33. The ternary compound ABC falls outside its primary field and joins connecting ABC with the end members create the composition triangles A–ABC–C, B–ABC–C, and A–ABC–B, associated with their respective invariant points P_1, E, and P_2. The important concept here is the "covered temperature maximum" (m) on the boundary line P_1–P_2, created by extending the join A–ABC (dashed lines) until it intersects P_1–P_2. A melt of the composition ABC will crystallize A as it moves toward P_1–P_2. At m, the peritectic reaction, Liq+A $\rightleftarrows$ ABC, occurs and the temperature remains constant while all of the liquid reacts with A to yield 100% ABC.

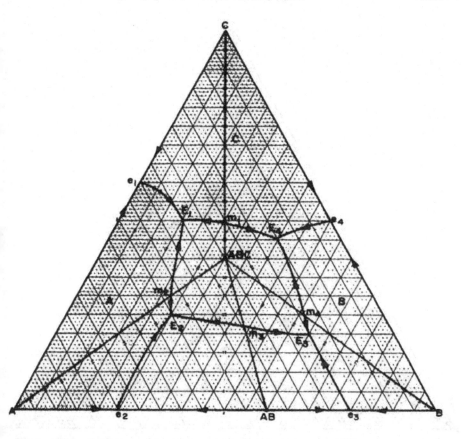

Figure 5.32. A System with a Congruently Melting Binary and Ternary Compound

This reaction is similar to the incongruent melting of a binary compound (Chapter 3, Section III-B, Figure 3.5), except that the liquid and ABC are now ternary compositions containing certain proportions of all three end members.

Conversely, if the ternary compound ABC is heated under equilibrium conditions, the first liquid will appear at m and ABC will completely react to form crystals of A. As the temperature is increased, A dissolves and the liquid moves to the point ABC, the temperature of complete liquifaction.

A liquid of composition X will crystallize A to P_1-P_2. At X′, ABC crystallizes while A is resorbed, until A is completely resorbed at X″. The ternary compound crystallizes in its own primary field until the boundary line P_1-E is encountered at X′″. C and ABC crystallize together to E, where eutectic crystallization occurs. The path of liquid Y is similar, proceeding from Y′ to Y″ while A

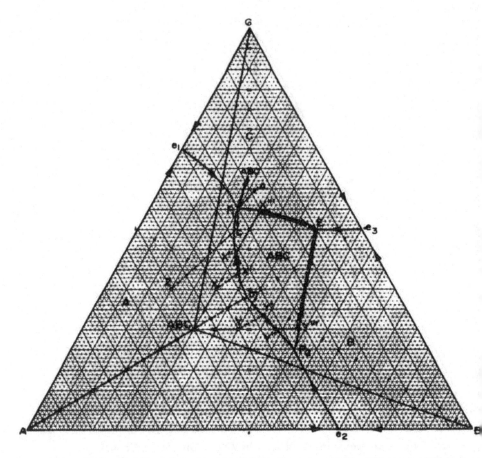

Figure 5.33. Ternary Incongruently Melting Compound (Single Incongruency)

is resorbing and ABC crystallizing, crossing the primary field of ABC to Y''' and approaching E as B and ABC crystallize.

For a point (liquid composition) such as Z (primary field A, composition triangle A-C-ABC), note that A is resorbing on the boundary P_1-P_2 and that it will continue to be (partially) resorbed at P_1, as C and ABC crystallize. Note A, C and ABC arrows on peritectic point P_1.

Some variations of Figure 5.33 are possible. For example, the primary field of ABC could overlap the join C-ABC, creating a eutectic within the composition triangle C-ABC-A. Likewise, the primary field of ABC could overlap the join B-ABC, creating a eutectic within the composition triangle B-ABC-A. Finally, the primary field of ABC could overlap both C-ABC and

B–ABC. Although the reactions at the invariant points would differ, depending on what kind of invariant point was involved (E or P), the fundamental concepts previously described for liquids X, Y and Z would not change as the liquids moved down the boundary from either side of m.

ii. Double Incongruency

The fundamental configuration is shown in Figure 5.34. A liquid of composition ABC will crystallize A to the intersection of the boundary line e_1-P at K. A and C will crystallize from the liquid from K to P. At P, the invariant reaction, L+A+C $\rightleftarrows$ ABC will occur to give 100% ABC as the temperature remains constant while heat is withdrawn. Conversely, if ABC is heated under equilibrium conditions, the first liquid will form at P, ABC will disappear in favor of A and C and the liquid will move from P to K as A and C dissolve. At K, C completely

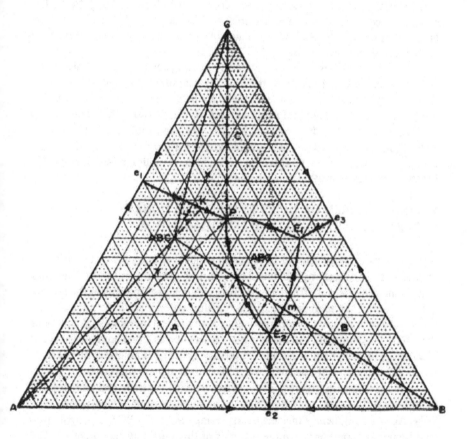

Figure 5.34. Ternary Incongruently Melting Compound (Double Incongruency)

disappears and A dissolves as the temperature increases to ABC where the composition becomes completely liquid.

It should be noted that other liquids near ABC such as X, Y and Z, which are in composition triangle C-ABC-B or A-ABC-B, will undergo reaction at P on their way to final crystallization at E_1 or E_2, respectively. These liquids may be located in the primary field of A or C at the outset of crystallization.

Note that the composition triangles A-ABC-C, B-ABC-C and A-ABC-B have corresponding invariant points P, E_1 and E_2, respectively.

iii. Dissociating Ternary Compounds

The ternary compound ABC may have an upper, lower, or upper and lower temperature limit of stability, analogous to the binary dissociating compound discussed in Chapter 3, Section III-C under Dissociation. Two very simple cases of dissociation would have the ternary compound ABC yielding end members A, B and C when heated or cooled (assume congruent melting of ABC). In the first case, if the compound had an upper limit of dissociation to A, B and C, at some higher temperature, the end members would be involved in a eutectic reaction in the ternary system as in Figure 5.1, 5.2 and 5.6.

In the second case, if the congruently melting compound ABC dissociated at some lower temperature limit, the liquidus relationships would be similar to those in Figure 5.31, but at some temperature below the lowest temperature ternary eutectic (the temperature at which ABC dissociates to A, B and C) only one composition triangle involving the end members would exist. (Triangle A-B-C). Many complicated cases would exist if the ternary compound dissociated at an upper or lower temperature to some combination of end members, binary compounds or other ternary compounds.

V. POLYMORPHISM OR PHASE TRANSFORMATIONS IN TERNARY SYSTEMS WITHOUT SOLID SOLUTION

If phase transformations occur in the end members or in binary or ternary compounds in ternary systems without solid solution, the temperatures of the phase transformations are not affected by a change in composition within the ternary system. That is, the temperature of the transformation will be the same as it is in the pure one, two or three component phase and the inversions take place as they do if the component or compound was isolated as a one-component system. The change in composition within the ternary system has no effect on the inversion temperatures of end members, binary or ternary compounds.

For example, in Figure 5.35, a system which is partially real and partially hypothetical illustrates polymorphism or phase transitions. Component C is SiO_2, AC is a congruently melting binary compound and ABC is a congruently melting ternary compound. In the real part of the ternary system involving SiO_2

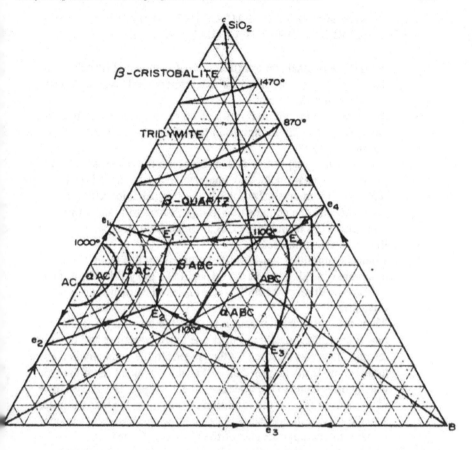

Figure 5.35. Phase Transitions of End Members, Binary or Ternary Compounds in Ternary Systems Without Solid Solution

and components A and B, the tridymite-cristobalite and β-quartz-tridymite transitions occur at 1470° C and 870° C, temperatures identical with the transitions in pure SiO_2.

In the hypothetical binary and ternary compounds AC and ABC, the $\alpha \rightleftarrows \beta$ transitions occur at constant temperatures 1000° and 1100°, respectively, in spite of the fact that component B is involved in the AC compound inversion and that the inversion of compound ABC occurs at constant temperature in the ternary system in positions which are variable with respect to the amounts of components A, B and C.

VI. REVIEW AND SUMMARY OF FACTORS WHICH MUST
BE CONSIDERED DURING ISOPLETHAL ANALYSES

1. What is the original composition? (A-B-C equilateral triangle)
2. The original composition is in which primary field? Is it on a boundary line or invariant point?
3. The original composition is in which composition triangle? Is it on a particular side of the composition triangle? Is it on the apex of a composition triangle?
4. If the original composition is in a composition triangle, where is the associated invariant point and where will it finally crystallize after perfect equilibrium crystallization? If the original composition is on a particular side of a composition triangle, where will final crystallization occur? If it is on the apex of a composition triangle, where will final crystallization occur?
5. When a composition moves toward a boundary line from its position in a primary field, it moves on a straight line away from the primary field composition point and the lever rule can be used for quantitative calculations.
6. When the crystallization path intersects the boundary line, do two phases crystallize together or does one phase crystallize and the other resorb? The answer is given by the tangent to the boundary line at the intersection. If the tangent falls within the end members of the join which corresponds to the boundary line, both phases will crystallize. If the tangent falls on the extension of the join, one phase will resorb and the other will crystallize. The phase which resorbs is the one which is on the opposite side of the extension of the join.
7. When liquids are proceeding on boundary lines, it is very important to recognize the corresponding join which will be used for quantitative calculations by the use of conjugation lines and the lever rule.
8. One must be alert for cases where one phase is completely resorbed as the liquid moves down the boundary line and when the liquid then moves across a primary field to a second boundary line.
9. One must be alert for cases where the crystallization path goes through a reaction point R (F=4-4=0) before going to a final eutectic or peritectic invariant point. Or for cases where the crystallization path goes through a peritectic point P (F=4-4=0) before going to the final eutectic point.
10. During final crystallization at invariant points (P or E), the composition of the ternary liquid, the conjugation line which gives the relative amount of liquid and solids, and the appropriate join which gives the relative amount of solids must be used for a quantitative calculation at

an infinitesimal temperature *above* the temperature of the invariant point.

11. To qualitatively determine what phases crystallize or resorb at a peritectic point, draw arrows from the composition point of each crystalline phase through the invariant point. The arrow which has to be directionally reversed to change the group of arrows from a 180° to a 360° configuration will indicate which phase will resorb. However, the quantitative proof of crystallize and absorbtion is given by examination of the quantitative calculations of the phases which exist just before and just after final crystallization. (See items 10 and 12).

12. At an infinitesimal temperature *below* the temperature of the invariant point during final crystallization, the final amounts of three crystalline phases are determined by drawing lines through the original liquid composition, parallel to any two sides of the composition triangle which then intersect the third side of the triangle.

13. After making several quantitative calculations between the liquidus temperature and the temperature immediately below final crystallization, a "phase analysis" diagram may be constructed to quantitatively present the entire isoplethal analysis. The phase analysis diagram indicates what happens during cooling or heating under perfect equilibrium conditions.

VII. LIQUID IMMISCIBILITY

A. General; Liquid Region Only

In all isoplethal analyses so far discussed, it has been assumed that crystallization took place from a single phase, homogeneous liquid. In condensed ternary systems it is possible for two or three liquids to coexist and for crystallization to take place from conjugate liquids rather than a single liquid. Furthermore, in ternary systems, it is possible for immiscible liquids to coexist in three different ways. In the first way, the immiscible liquids coexist only in the liquid region and there are no relationships with crystalline phases. In the second way, immiscible liquids are on the conventional ternary liquidus surfaces and coexist at certain temperatures with crystalline phases. Thirdly, ternary immiscible liquids may be metastable, coexisting at temperatures below ordinary liquidus temperatures or even below the lowest ternary eutectic temperature in the system (where only crystalline phases should exist under perfect equilibrium conditions). The approach to an understanding of liquid immiscibility in ternary systems might be facilitated by first discussing low temperature chemical systems or systems in which only the liquid region is concerned. It should be remembered from the discussion of liquid immiscibility in binary systems in Chapter III,

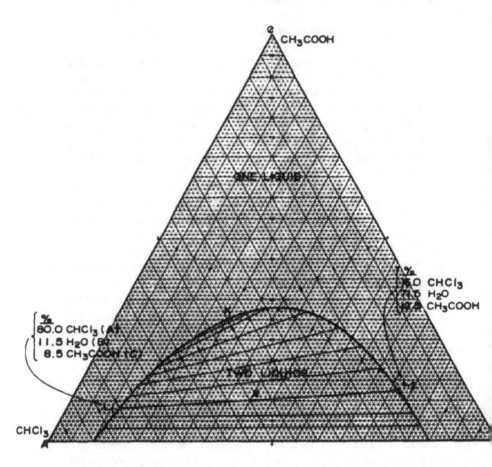

Figure 5.36. Room Temperature Immiscible Liquid Region in the Ternary System $CH_3COOH-CHCl_3-H_2O$

Section X that it is possible to have coexisting liquids with an upper or lower consolute point or both. This concept also applies to ternary liquid immiscibility.

Figure 5.36 shows the isothermal section for a system which is completely liquid at room temperature. Chloroform and water form an immiscible system. If acetic acid is added as a third component, the immiscibility gap narrows in the ternary system and the ternary two liquid region is formed. Note that the tie lines change position as they move from the binary system to the ternary critical or "plait" point K. In the general case the tie lines are not parallel to the binary system $A(CHCl_3)-B(H_2O)$, and they rotate in some manner as dictated by the position of the ternary critical point. Composition X, consisting of 10% CH_3COOH, 40% H_2O and 50% $CHCl_3$ will disproportionate to $\dfrac{XL_2}{L_1L_2} \times 100\%$ of L_1 and $\dfrac{XL_1}{L_1L_2} \times 100\%$ of L_2, whose compositions can be read directly from the diagram.

Figure 5.37a shows the case of two pairs of immiscible liquids at T, and at some lower temperature T_1 (Figure 5.37b), where the two regions have merged. Figure 5.37c shows three pairs of immiscible liquids at T and at some lower temperature T_1 (5.37d), where the regions have merged. Figure 5.37d consists of three homogeneous liquid regions labeled 1, three regions of two liquid coexistence labeled 2, and one region of three coexisting liquids whose compositions are given by L_1, L_2 and L_3. The amount and composition of the two liquids can be determined using the tie lines in each of the three two phase regions. For example, composition ℓ consists of $\dfrac{28}{60} \times 100 = 46.66\%$ ℓ_1 (22A, 6B, 72C) and $\dfrac{32}{60} \times 100 = 53.33\%$ ℓ_2 (74A, 14B, 12C).

The region where three liquids coexist is a composition triangle and for any composition within this triangle, such as X, the amounts of the three coexisting liquids of compositions L_1, L_2 and L_3 can be determined by drawing lines through the composition point, parallel to any two sides of the triangle. The intersection on the third leg will give the proportion of L_1, L_2 and L_3 in the three liquid mixture.

In oxide and silicate systems, the simplest kind of immiscible liquid region might be that which did not intersect the liquidus surface or the binary systems. Such a region might have an upper and lower ternary consolute point and "float" above the liquidus surface like a "bubble", although the shape might actually be quite irregular, not a spheroid or ellipsoid. Tie lines in such a three-dimensional region would connect the two or three ternary liquid compositions in equilibrium at various temperatures. Various other combinations are possible if the two liquids originate in one or more of the bounding binary systems, and of course, three conjugate liquid systems which originate in binary systems are possible (Figure 5.37d).

B. Immiscibility Originating in Binary Systems

In real oxide and silicate systems, certain types of liquid immiscibility occur more frequently than others. In Figure 5.38a, a two liquid region extends from the binary system C–B into the primary field of C. The binary systems are all simple eutectic types, as is the ternary system. The intersection of the dome-like space with the liquidus surface is emphasized in Figure 5.38b, where the dome has been removed to clearly reveal the tie lines connecting conjugate phases at several temperature levels. In Figure 5.38a, the composition of the upper

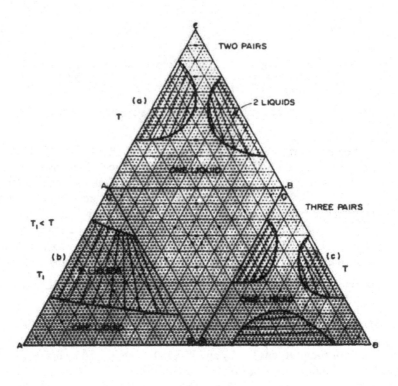

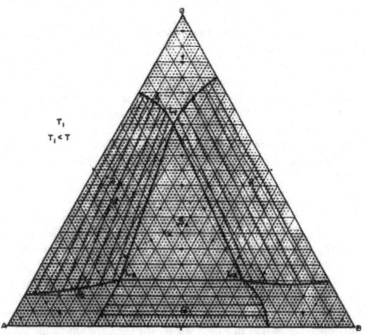

Figure 5.37a, b, c, d. Immiscible Liquid Regions in Ternary Systems at Constant Temperature

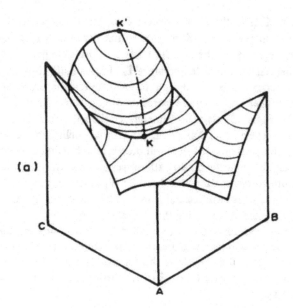

Figure 5.38a. Liquid Immiscibility Gap in Ternary System A–B–C. Isotherms

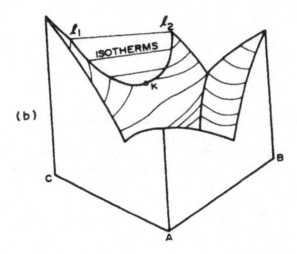

Figure 5.38b. Two Liquid Region Removed to Show Intersection with Liquidus Surface

consolute point K' is in the binary system C–B, but in certain systems, the composition of consolute point K' may be within the ternary system.

A plane projection of the liquidus surface of this system is shown in Figure 5.39. An isothermal section at temperature T_4, Figure 5.40, shows the triangular-shaped space of coexistence of $C + L_1 + L_2$. An isoplethal analysis of a composition which undergoes liquid-liquid separation during cooling is as follows: Liquid composition Y (Figure 5.39) will ultimately cool to the eutectic and finally crystallize to a mixture of A, B and C. During initial cooling to the liquidus surface, an infinitesimal amount of C will precipitate. Further cooling will cause more C to precipitate and the liquid path moves along a straight line from Y to a. At a temperature corresponding to a (between T_2 and T_3), a separation into two liquids will occur and an infinitesimal amount of L_2 will be formed.

For convenience in discussing the crystallization path, the C-rich liquids on one side of the immiscibility gap will be called L_1 type liquids and those on the B-rich side of the gap will be designated L_2 type liquids. At a temperature corresponding to T_3, one liquid will have the composition of b' (L_1 type) and the other a composition of b'' (L_2 type). The proportions are calculated as follows:

At Temp T_3,

$$\% \text{ Solid C} = \frac{Yb}{Cb} \times 100$$

$$\% \text{ Liquids} = \frac{CY}{Cb} \times 100$$

$$\text{but } \% L_1 = \frac{bb''}{b'b''} \times 100$$

$$\text{and } \% L_2 = \frac{bb'}{b'b''} \times 100$$

Each liquid, of course, contains a definite amount of A, B and C as given by its location on a ruled grid on the ternary diagram.

As the composition cools to T_4, L_1 changes in composition from b' to c' and L_2 changes from b'' to c''. The relative amounts in the mixture are as follows:

$$\% \text{ Solid C} = \frac{Yc}{Cc} \times 100$$

$$\% \text{ Liquids} = \frac{CY}{Cc} \times 100$$

$$\text{but } \% L_1 = \frac{cc''}{c'c''} \times 100$$

$$\text{and } \% L_2 = \frac{cc'}{c'c''} \times 100$$

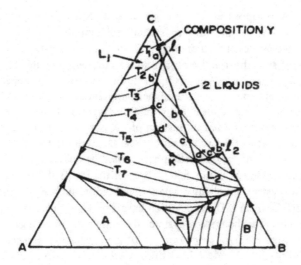

Figure 5.39. Plane Projection of Liquidus Surface from Figure 5.38

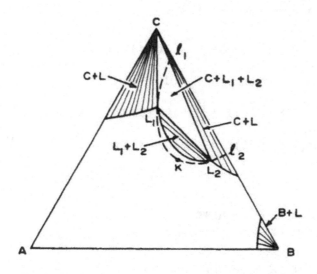

Figure 5.40. Isothermal Section at Temperature T_4

At T_5, the composition of L_1 is given by d', but only an infinitesimal amount now remains and the mixture is essentially composed of C and L_2. C continues to precipitate on further cooling from d'' to 9. At point 9, B and C crystallize together as the liquid moves from 9 to the ternary eutectic.

If the original composition was b, inside the immiscibility gap, and if the immiscibility gap closed over at some higher temperature, as shown in Figure 5.38a, an original homogeneous liquid would separate into two liquids during cooling (at the temperature of intersection of the isopleth with the dome) and at some lower temperature C would begin to separate from the two liquids. After the appearance of C, the crystallization path would be qualitatively of the same type as for composition Y, which falls outside of the immiscibility gap.

In silicate systems involving divalent metal oxides such MgO, MnO, FeO, CoO, NiO, CaO, ZnO, or SrO, all of which have a binary immiscibility gap with SiO_2, it is common for the ternary system to assume a two-liquid gap such as that shown in Figure 5.37b. For examples of real systems, see especially the systems MgO-"FeO"-SiO_2, MgO-CaO-SiO_2, CaO-"FeO"-SiO_2, and CaO-ZnO-SiO_2 in the 1964 edition of "Phase Diagrams for Ceramists".

It is possible for an immiscibility gap to lie in two primary fields of crystallization and overlay the boundary line separating these fields as shown in Figure 5.41. In such a case, the boundary line $e_1 E$ is isothermal between k' and k'' and is the location of the invariant reaction, $L_1 + L_2 + C \rightleftarrows A$ (P=4, F=0). Composition X (30A, 25B, 45C) will at some temperature involve an equilibrium between L_1, L_2 and C.

For example, at X', the analysis is:

$$\% C = \frac{11.5}{66.5} \times 100 = 17.3\%$$

$$\% \text{Liquid} = \frac{55}{66.5} \times 100 = 82.7\%$$

$$(46.5A, 14B, 39.5C)\ \ell' = \frac{19}{35} \times 100 = 54.3\%$$

$$(24.5A, 49B, 26.5C)\ \ell'' = \frac{16}{35} \times 100 = 45.7\%$$

Totals,

82.7 × 54.3 × 46.5 = 20.8% A
82.7 × 45.7 × 24.5 = 9.2% A
                    30.0% A
82.7 × 54.3 × 14   = 6.4% B
82.7 × 45.7 × 49   = 18.5% B
                    24.9% B

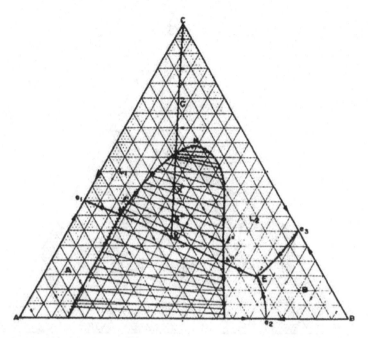

Figure 5.41. Immiscible Liquid Region in Two Primary Fields, Originating in the Binary System A–B

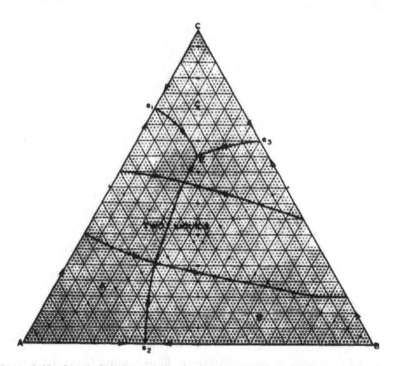

Figure 5.42. Immiscible Liquid Region in Two Primary Fields, Originating in Two Binary Systems A–C and B–C

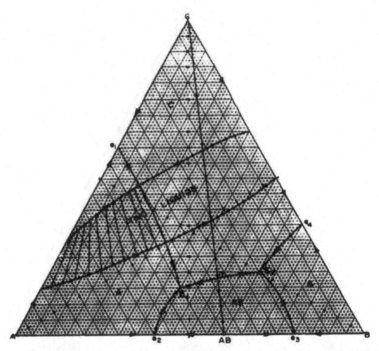

Figure 5.43. Immiscible Liquid Region in a System with a Congruently Melting Binary Compound

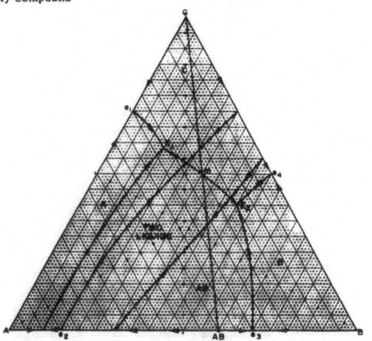

Figure 5.44. Immiscible Liquid Region in a System with a Congruently Melting Binary Compound, Crossing the Primary Fields of C and AB

$$82.7 \times 54.3 \times 39.5 = 17.5\% \, C$$
$$82.7 \times 45.7 \times 26.5 = 10.2\% \, C$$
$$\underline{17.3\% \, C}$$
$$45.0\% \, C$$

As heat is withdrawn, more C crystallizes and the composition of the two liquids change as indicated by the boundary of the gap. When the temperature is lowered to that which corresponds to point k, a four phase equilibrium between L_1, L_2, A and C is maintained as the temperature remains constant and as all of L_1 disappears as heat is withdrawn. When L_1 completely disappears, A and C crystallize from L_2 as it moves from k″ into the eutectic E. Composition points located in the primary field of either A or C, which encounter the immiscibility gap, will all undergo the invariant reaction during cooling, before final crystallization at E.

In Figures 5.37b, 5.38a the immiscibility gap is in ONE primary field. Extensions of Figures 5.37b and 5.41 are shown in Figures 5.42, 5.43 and 5.44. In Figure 5.42, immiscibility occurs in binary systems A–C and B–C and crosses the boundary line between the primary fields of A and B. Figures 5.43 and 5.44 contain a congruently melting binary compound AB. In Figure 5.43, the two liquid area originates in systems A–C and B–C and crosses the primary fields of A and C. In Figure 5.44 the two liquid areas originate in systems A–B and B–C and cross the primary fields of AB and C and boundary line E_1–E_2. In each case shown in Figures 5.41 to 5.44 it is possible that the immiscibility would not originate in any of the binary systems. The boundary lines would then be within the ternary system and they would contain TWO ternary critical or "plait" points, rather than only one as shown in Figure 5.41.

C. Three Liquid Phases

Figure 5.45 represents a simple eutectic type ternary system with a three liquid region. The immiscible liquid regions originate in the simple eutectic type binary systems and at a certain constant temperature the three liquids will be saturated with crystalline C.

Compositions within the triangle CL_2L_3 may first crystallize C from a homogeneous liquid, a two liquid or a three liquid combination. Regardless of the origin of the composition, at the temperature of L_2L_3, the isothermal section will exist as shown in Figure 5.46. At constant temperature, as heat is withdrawn, the invariant reaction $L_1 \rightleftarrows C + L_2 + L_3$ will occur (F=4-4=0). At this temperature L_1 will disappear and C will continue to precipitate from two immiscible liquids in the primary field of C as it approaches the boundary line e_1 E. At the temperature of $L_2'L_3'$, the usual $C + L_2' + L_3' \rightleftarrows A$ invariant reaction (F=4-4=0) occurs, L_2' disappears and $A + C + \ell_3$ continues to coexist from L_3'

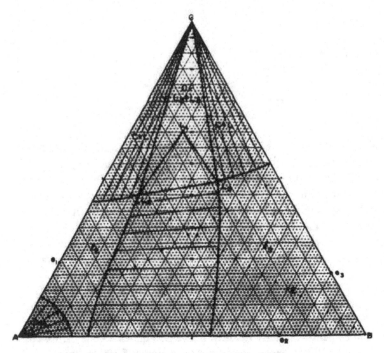

Figure 5.45. System with Three Liquid Phases

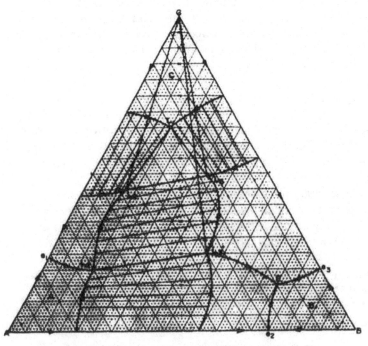

Figure 5.46. Isothermal Section at Temperature $L_2 L_3$

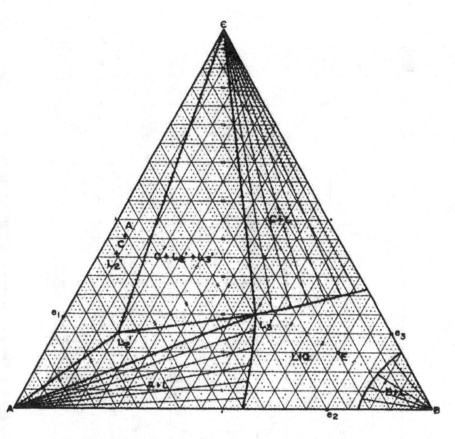

Figure 5.47. Isothermal Section at Temperature $L_2' L_3'$

to E. The isothermal section at temperature $L_2' L_3'$ is shown in Figure 5.47. If the original composition was in triangle $CL_3 L_3'$ as well as triangle $CL_2 L_3$, the crystallization or isoplethal analysis would pass through the first invariant reaction $C + L_2 + L_3 \rightleftarrows L_1$, but not the second invariant reaction. A great variety of isoplethal sequences will occur in such a system during cooling of compositions in the primary field of C, in addition to those compositions in the triangle $CL_2 L_3$ which were briefly and qualitatively discussed.

D. Relationships Between Binary and Ternary Compounds and Two or Three Liquid Phases

Although some of the more common types of immiscible liquid ternary systems were discussed, many more types exist. There can be relationships between

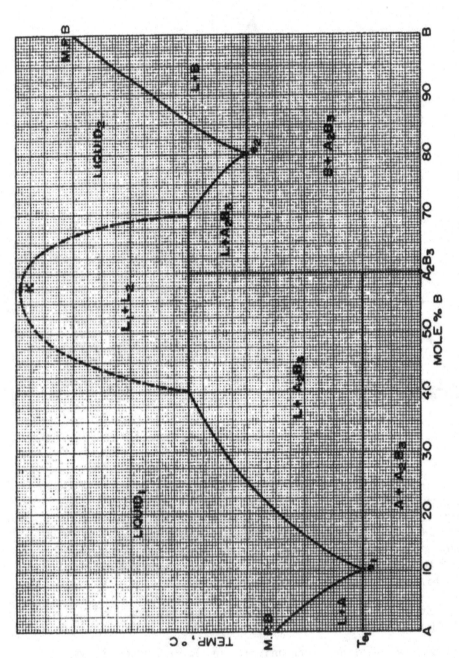

Figure 5.48. Decomposition of Binary Compound to Immiscible Liquids

congruently melting, incongruently melting or dissociating binary or ternary compounds and two or three immiscible liquids in condensed ternary systems.

In a condensed ternary system, it must be realized that only four phases may coexist at an invariant temperature (F=4-4=0) and that only the following immiscible liquid situations may take place:

1. $L_1 + L_2 + L_3 \rightleftarrows$ Solid
2. $L_1 + L_2 \rightleftarrows S_1 + S_2$

In a condensed binary system, a binary compound may melt to two immiscible liquids, Figure 5.48. In a condensed ternary system, it is possible for a ternary compound to melt to three immiscible liquids, $L_1 + L_2 + L_3 \rightleftarrows ABC$, similar to the situation in section VII-C. (Figure 5.49)

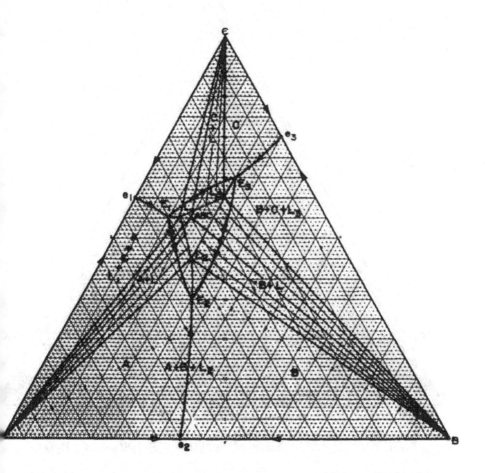

Figure 5.49. Decomposition of Ternary Compound to Three Immiscible Liquids

VIII. ISOTHERMAL SECTIONS IN AQUEOUS SYSTEMS; ISOTHERMAL EVAPORATION

To facilitate the understanding of isothermal sections in high temperature ceramic systems, it might be appropriate to start with a brief discussion of isothermal sections of aqueous systems (constant pressure), involving the equilibria between well known salts and H_2O at or near room temperature.

In Figure 5.50, point a represents the solubility of $(NH_4)_2SO_4$ in H_2O and point b the solubility of NH_4Cl in H_2O at 30°C. The influence of NH_4Cl on the solubility of $(NH_4)_2SO_4$ is shown by the curve a-P which is an isobaric, isothermal saturation curve. Likewise, the curve b-P shows the influence of

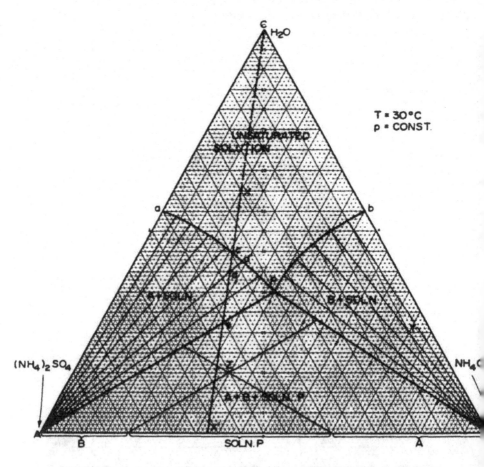

Figure 5.50. Isothermal Section of the System $(NH_4)_2SO_4$–NaCl–H_2O

$(NH_4)_2SO_4$ on the solubility of NH_4Cl. If point P is connected with A and B, the system is divided into three types of regions, 1) unsaturated solution (one phase) 2) A + Solns and B + Solns (two phase regions containing tie lines) and 3) a three phase region (a compatibility triangle) where A, B and solution P coexist in various proportions.

At 30°, composition X is an unsaturated solution containing 60.0% H_2O, 25% $(NH_4)_2SO_4$ and 15% NH_4Cl. Composition Y is a two phase mixture of 50% $NH_4Cl(B)$ and 50% of a saturated liquid solution containing 50% H_2O, 10% $(NH_4)_2SO_4$ and 40% NH_4Cl. Composition Z contains 35% $(NH_4)_2SO_4$ (A), 20% NH_4Cl (B) and 45% of a solution (P) saturated with both salts (35% H_2O, 30% A and 35% B). These aqueous diagrams are only schematic and the actual diagram for the system may be somewhat different from that shown in Figure 5.50. However, Figure 5.50 is appropriate to illustrate the principles involved.

Isothermal evaporation of composition X at 30°C will involve movement along X-X' as follows: As H_2O is removed, the solution becomes more concentrated and at c, an infinitesimal amount of $(NH_4)_2SO_4$ appears. At point d, $(NH_4)_2SO_4$ is in equilibrium with a solution containing 42.5% H_2O, 33.5% $(NH_4)_2SO_4$ and 24% NH_4Cl (tie line d'-A). At point e, an infinitesimal amount of NH_4Cl will appear and from e to X' various proportions of A, B and saturated solution P will coexist. (See analysis of point Z). Removal of the last trace of H_2O will result in a mixture of dry salts containing 38% NH_4Cl and 62% $(NH_4)_2SO_4$. During all stages of isothermal evaporation, the ratio of A to B will remain constant.

Figures 5.51a-f show isothermal sections of various kinds of aqueous systems each of which illustrates additional principles or concepts. Figure 5.51a is referred to as a system containing a double salt which is not decomposed by water. The join D-H_2O is a true binary system, since the addition of H_2O to D results in various amounts of a saturated solution R in equilibrium with various amounts of D. Additions of H_2O to D beyond R results in homogeneous solutions.

Note the three regions of two coexisting phases (solid plus various saturated solutions) and the two regions of three coexisting phases, each involving a very special solution phase (P or Q) which is saturated with respect to two salts.

Figure 5.51b is said to be characterized by a double salt, D, which is decomposed by water. That is, as soon as H_2O is added to D, the composition enters a three phase region where D, B and liquid Q coexist. This could also be described as incongruent solubility. Note the existence of two and three phase regions as before and the importance of special saturated solutions P and Q.

Figure 5.51c contains the binary intermediate compound $Na_2SO_4 \cdot 10H_2O$, which has a characteristic solubility in H_2O at 15°C and whose solubility is affected by the presence of NaCl as indicated by k-P. This system is described as one in which the hydrate is not decomposed by the presence of the second salt.

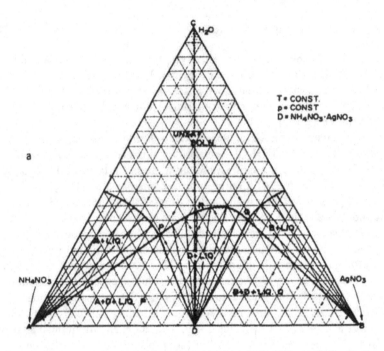

Figure 5.51a. Isothermal Section of the System NH_4NO_3–$AgNO_3$–H_2O

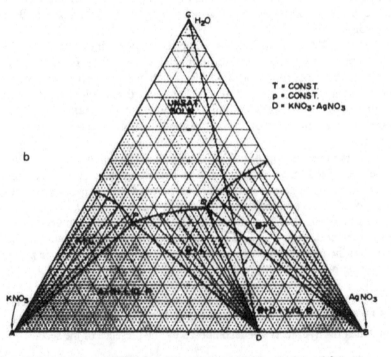

Figure 5.51b. Isothermal Section of the System KNO_3–$AgNO_3$–H_2O. Binary Compound $KNO_3 \cdot AgNO_3$ (D) Decomposed by Water

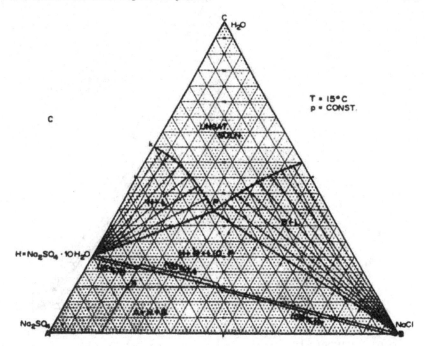

Figure 5.51c. Isothermal Section of the System Na_2SO_4–$NaC\ell$–H_2O at $15^{\circ}C$.
Binary Compound $Na_2SO_4 \cdot 10H_2O$

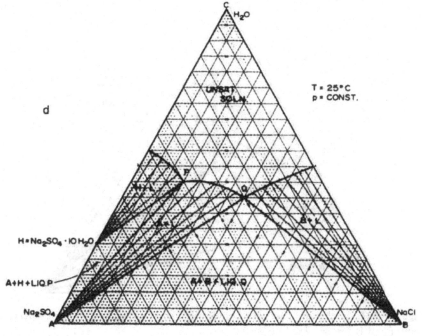

Figure 5.51d. Isothermal Section of the System Na_2SO_4–$NaC\ell$–H_2O at $25^{\circ}C$

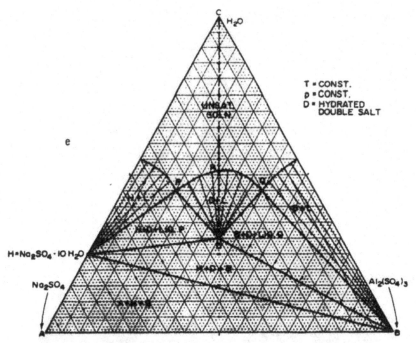

Figure 5.51e. Isothermal Section of the System $Na_2SO_4-Al_2(SO_4)_3-H_2O$. Ternary Compound D

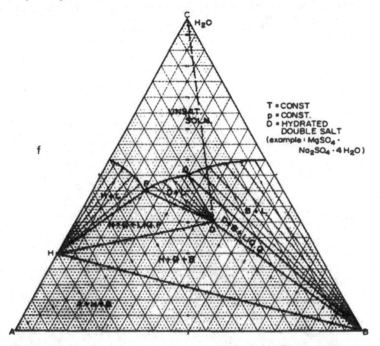

Figure 5.51f. Isothermal Section of the System $MgSO_4-Na_2SO_4-H_2O$. Ternary Compound D Decomposed by Water

That is, if NaCl (B) is added to $Na_2SO_4 \cdot 10H_2O$, this simply results in mixtures of the two salts as indicated by the binary join H-B. Note that this join creates two composition triangles 1) H-B-Soln P and 2) A+H+B. The proportion of each coexisting phase in the composition triangle can be determined by the method previously described. For example, composition X consists of approximately 59% $Na_2SO_4 \cdot 10H_2O$, 26% Na_2SO_4 and 15% NaCl.

Figure 5.51d illustrates the decomposition of the binary compound $Na_2SO_4 \cdot 10H_2O$ by NaCl. As soon as NaCl is added to the hydrate, the composition falls in the field of A+H+liquid P and is thus "decomposed" (partially) to some Na_2SO_4 and some liquid P.

Figure 5.51e contains a ternary compound (a hydrate of a double salt) which is not decomposed by H_2O. Addition of H_2O results in various amounts of saturated solution R in coexistence with D, up to point R. Addition of more H_2O results in homogeneous unsaturated solutions. Note the four composition triangles, H+D+liq P, B+D+liq Q, H+D+B AND A+H+B.

Finally, Figure 5.51f illustrates the case of a hydrated double salt which is decomposed by the addition of water. As soon as H_2O is added to D, the composition enters the triangle D+B+liq Q, another case of incongruent solubility.

The purpose of these aqueous diagrams is to illustrate one, two and three phase regions, saturation curves, mutual saturation points such as P and Q, quantitative calculations and isothermal evaporation. The three phase regions are composition triangles. The basic concepts are all related to isothermal sections in ceramic systems, except for isothermal evaporation which is usually not a consideration in high temperature systems.

IX. ISOTHERMAL SECTIONS IN TERNARY OXIDE SYSTEMS

A. A Congruently Melting Binary Compound;
Ternary Eutectics E_1 and E_2

A few isothermal sections have already been shown in connection with discussions of congruently melting binary compounds (Figures 5.11 to 5.14), incongruently melting binary compounds (Figures 5.23, 5.24 and 5.25) and liquid immiscibility (Figures 5.40, 5.46 and 5.47).

However, it is now appropriate to show in some detail how the isothermal sections are generated in both abstract and real systems. Let us assume the general temperature relationships shown in Figure 5.52, where D is a binary congruently melting compound (AB), the vertical section C-AB is a simple binary eutectic system with the eutectic located at m, and m is a temperature maximum on the boundary line between ternary eutectics E_1 and E_2. It is assumed that temperatures decrease in the order B, D, e_1, A, C, e_2, e_3, e_4, m, E_1, E_2. Horizontal isothermal sections through the solid model will then have the appearance shown in Figures 5.53a-k at the temperature levels indicated.

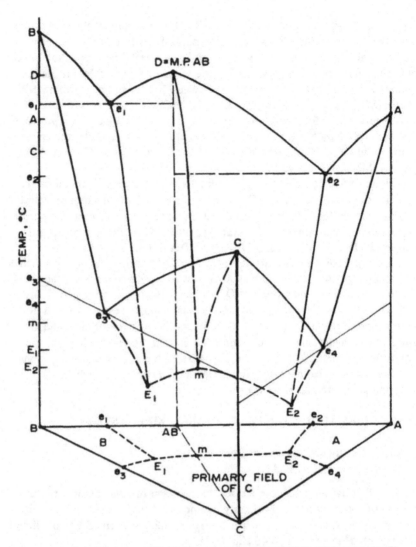

Figure 5.52. System with a Binary Congruently Melting Compound AB

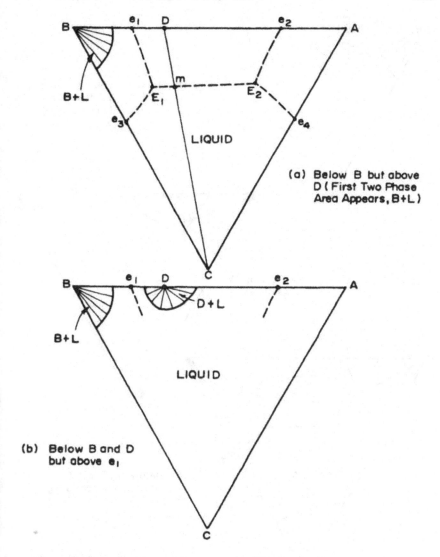

(a) Below B but above D (First Two Phase Area Appears, B+L)

(b) Below B and D but above e_1

Figure 5.53a-k. Isothermal Sections of Solid Model in Figure 5.52 from Temperature B to E_2

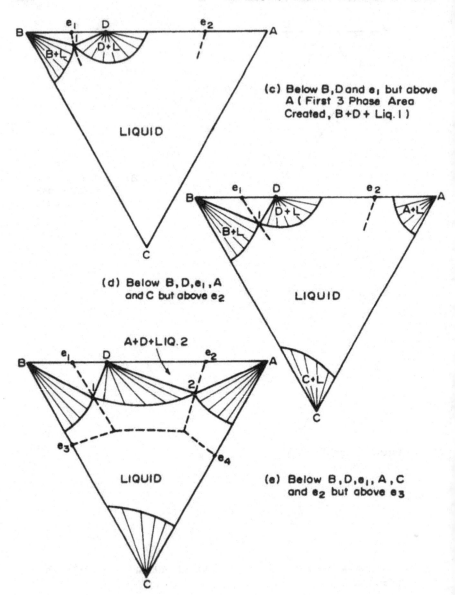

Figure 5.53c,d,e.

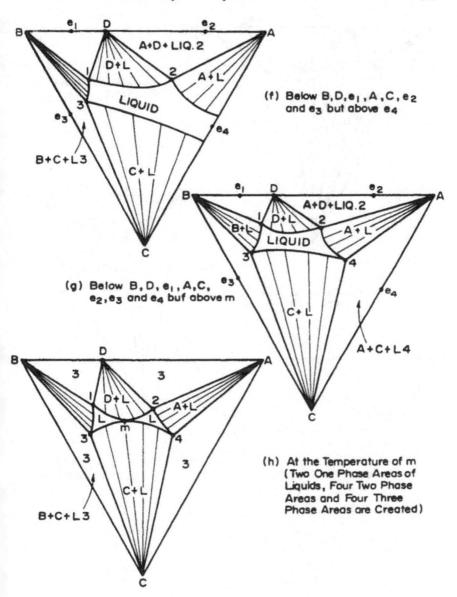

(f) Below B,D,e₁,A,C,e₂ and e₃ but above e₄

(g) Below B, D, e₁, A,C, e₂,e₃ and e₄ but above m

(h) At the Temperature of m (Two One Phase Areas of Liquids, Four Two Phase Areas and Four Three Phase Areas are Created)

Figure 5.53f,g,h.

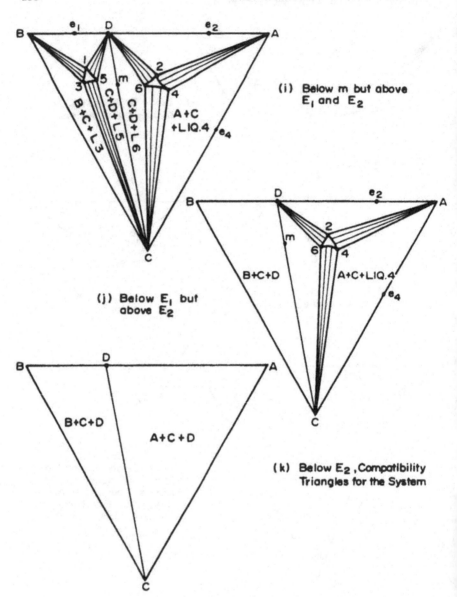

(i) Below m but above E_1 and E_2

(j) Below E_1 but above E_2

(k) Below E_2, Compatibility Triangles for the System

Figure 5.53i,j,k.

The case just discussed for Figures 5.52, 5.53a-k, is actually a very simple one because only one binary congruently melting compound D was involved in conjunction with two ternary eutectics which divided the system into two true ternary subsystems A-C-AB and B-C-AB. The only case which would be less complicated is one where all binary systems were of the simple eutectic type with boundary lines leading into a single ternary eutectic point. The generation of isothermal sections is in all other cases more complicated when incongruently melting or dissociating binary or ternary compounds are involved or when liquid immiscibility is present in the system. A few examples will illustrate this point.

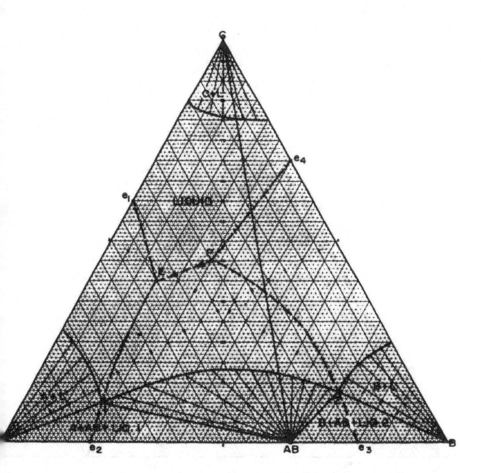

Figure 5.54. Isothermal Section at a Temperature Above e_1, e_4, P and E

B. A Congruently Melting Binary Compound; Ternary Eutectic and Peritectic

Figures 5.54, 5.55 and 5.56 show a simple variation of the relationships shown in Figures 5.9, 5.10 and 5.52. A ternary peritectic is present and the primary field of B overlaps the join C–AB. At a temperature below the melting points of A, B, C and AB and the binary eutectics e_2 and e_3, but above the eutectics e_1 and e_4 and the ternary invariant points E and P, the isothermal section has the configuration shown in Figure 5.54. At a lower temperature, below that of A, B, C, AB, e_2, e_3, and e_4 but above e_1, E and P, the section takes the form shown in Figure 5.55. Figure 5.56 gives the section at a temperature below all the invari-

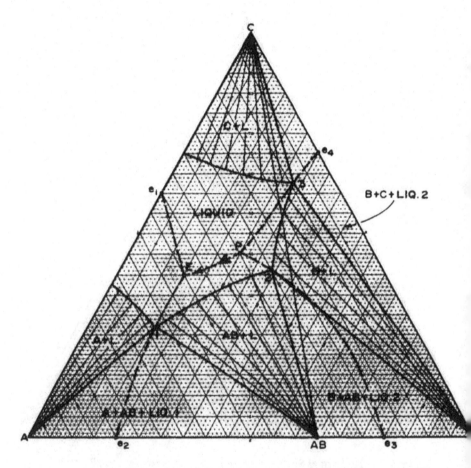

Figure 5.55. Isothermal Section at a Temperature Above e_1, P and E

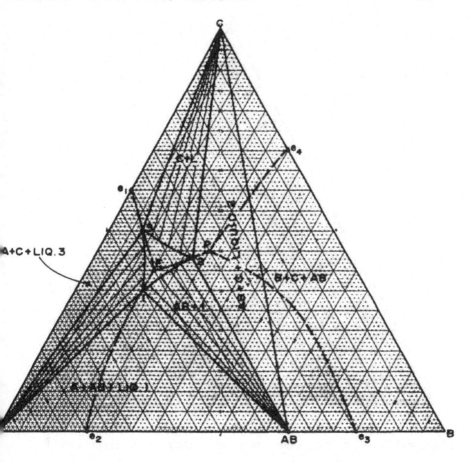

Figure 5.56. Isothermal Section at a Temperature Above E

ant points except E. Note that a small region of liquid exists in the neighbor-hood of the eutectic, while only solid phases coexist in the composition triangle B-C-AB. Below E, the section consists of two composition triangles, A-C-AB and B-C-AB.

C. Immiscible Liquid System

In the case of an immiscible liquid system as shown in Figure 5.41 where the two liquid region is in the primary fields of A and C and overlays the boundary line e_1-E, three isothermal sections are shown in Figures 5.57, 5.58 and 5.59. Figure 5.57 is an isothermal section above the temperature of e_1, e_2, e_3, and E.

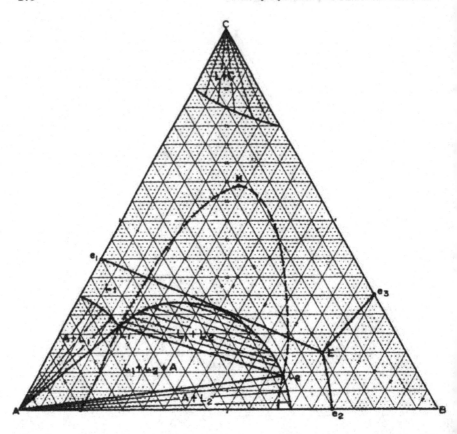

Figure 5.57. Isothermal Section Above the Temperature of e_1, e_2, e_3 and E

In Figure 5.58, the temperature is below e_1, but above e_2, e_3 and E. Finally, at a temperature below e_1 and e_3, above e_2 and E, but after the disappearance of L_1, Figure 5.59 gives the isothermal relationships.

D. A Real Oxide System

The real ternary system $CaO-TiO_2-SiO_2$ is shown in Figure 5.60 and the isothermal section at 1395°C is shown as Figure 5.61. The latter diagram may be used to calculate the amout and composition of coexisting phases in two and three phase regions of the diagram. Note the liquid region, the four regions where SiO_2, TiO_2, $CaTiO_3$ and $CaSiO_3$ coexist with liquids saturated with respect to these compounds at 1395°C, and the five composition triangles located in the CaO-rich region of the system. Also note the four regions where special

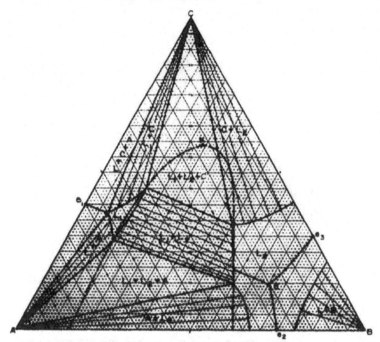

Figure 5.58. Isothermal Section Above the Temperature of e_2, e_3, and E, but below e_1

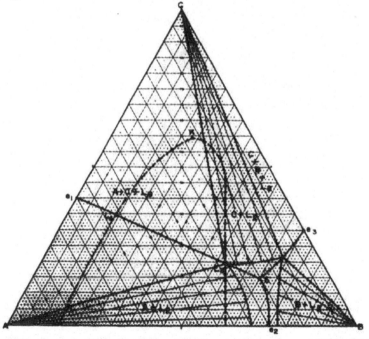

Figure 5.59. Isothermal Section Above the Temperature of e_2 and E, but below e_1 and e_3

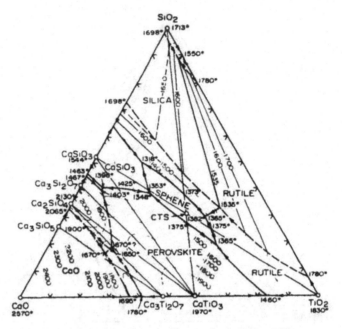

Figure 5.60. The Ternary System CaO–TiO₂–SiO₂

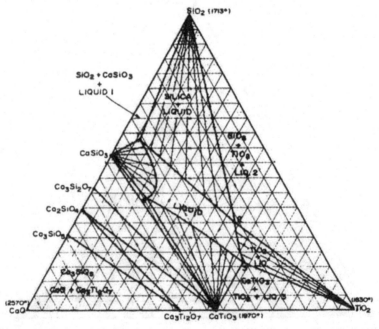

Figure 5.61. The 1395°C Isothermal Section of the System CaO–TiO₂–SiO₂

liquids 1, 2, 3 and 4 coexist with two solid phases. Figure 5.61 was constructed by the use of the 1400°C isotherm in Figure 5.60 and by lowering 5°C to avoid the liquid region which would be generated by the 1398°C ternary eutectic involving $CaSiO_3$, $Ca_3 Si_2 O_7$ and $CaTiO_3$.

X. VERTICAL SECTIONS IN TERNARY OXIDE SYSTEMS

A. General; Simple Ternary Eutectic System

It has already been shown that some vertical sections may be true binary systems within the ternary system, as for example, Figure 5.10. In general, however,

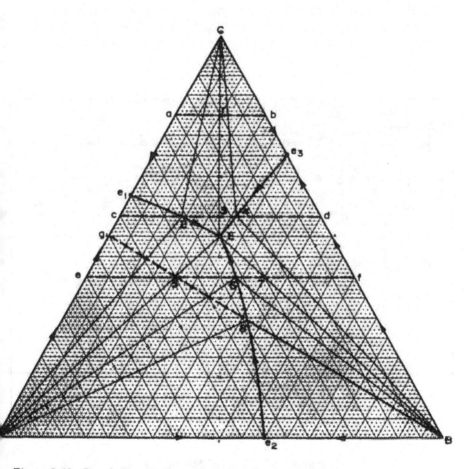

Figure 5.62. Simple Ternary Eutectic System; Vertical Sections a–b, c–d, e–f and B–g

most vertical sections are non-binary, at least in part. This is true in most cases even for vertical sections which run between two composition points in a ternary system and it is most often true when the vertical section is chosen at random. Construction of a vertical section is a good test of the understanding of crystallization paths, since the path of ALL points on the section must be traced in order to present a complete picture of the relationships. The following examples will illustrate some of the rules which govern the construction of the vertical section.

Some relative temperature relationships must be assumed for the simple ternary eutectic system shown in Figure 5.62. For the first vertical section a-b, let us assume that the temperature of $T_a > T_b$ and $T_{e_1} > T_{e_3}$. The crystallization paths are of two types, those which crystallize A and C and those which crystallize B and C as the liquid path moves toward the ternary eutectic E on boundary lines $e_1 E$ and $e_3 E$, respectively. Composition 1 is unique inasmuch as it precipitates C from the liquid as the liquid path moves directly toward the ternary

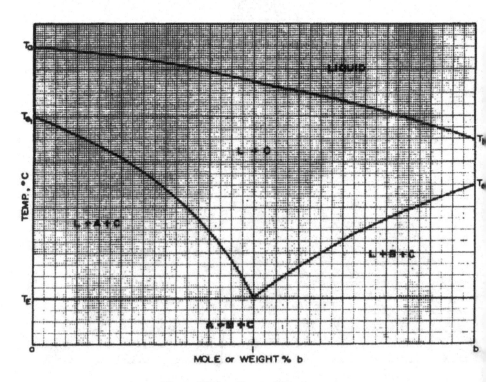

Figure 5.63a. Vertical Section a-b

eutectic, at which point A, B and C crystallize together. The section will take the form shown in Figure 5.63a.

For the second section c-d, it is assumed that $Tc>T_d$, $Te_1>Te_3$ and $T_2>T_3>T_4$. Crystallization paths are of four types, those between c and 2 which crystallize A, then A and C and finally A, B and C; those between 2 and 3 which crystallize in order, C, A+C, and A, B and C; those between 3 and 4 which crystallize in order C, B+C, and A, B and C; and those between 4 and 5 which crystallize B, B+C and finally A, B and C. The configuration is shown in Figure 5.63b.

The third section e-f, shown in Figure 5.63c assumes that $Te>T_f$, $Te_1>Te_3$ and some reasonable temperature for T_6. Points on the section are divided into four types from e-5, 5-6, 6-7 and 7-f. Compositions 5 and 7 crystallize A or B, respectively, as the liquid path moves directly toward E.

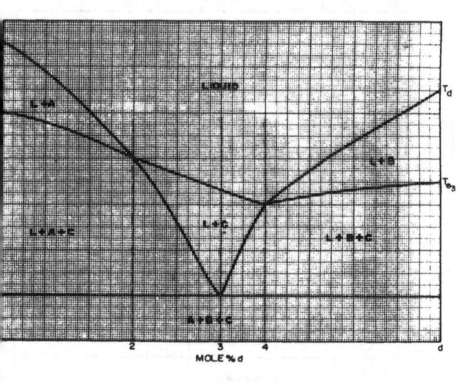

Figure 5.63b. Vertical Section c–d

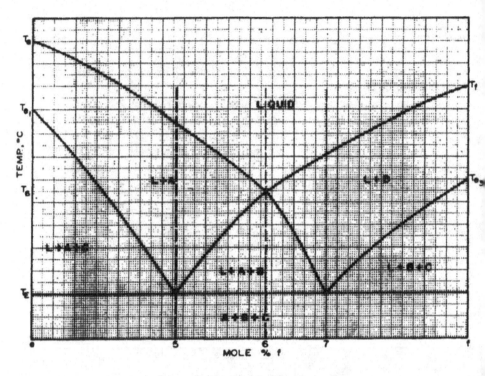

Figure 5.63c. Vertical Section e-f

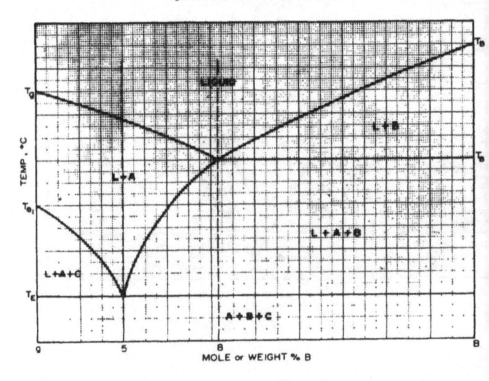

Figure 5.63d. Vertical Section B-g

Points on the fourth vertical section B–g are of three types as shown in Figures 5.62 and 5.63d. All liquids between B and 8 crystallize B to a temperature of T_8, then A and B from 8 to E. Liquids between 8–5 crystallize A, then A+B to E. Liquids from 5–g crystallize A, then A+C to E. Liquid 5 is unique, crystallizing A to the eutectic, where A, B and C precipitate simultaneously.

B. Ternary System With a Binary Congruently Melting Compound

The presence of a congruently melting binary compound AB as shown in Figure 5.64 makes the random vertical section quite complicated as shown in Figure 5.65. The section crosses two composition triangles and three primary fields of crystallization, A, AB and B. Eight groups of liquids are involved in eight different types of crystallization. Liquids of composition 1, 3, 5 and 7 crystallize

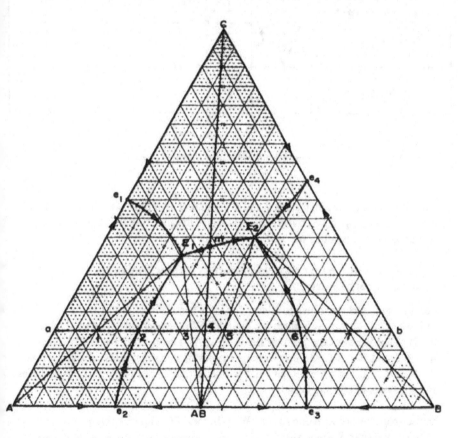

Figure 5.64. Congruently Melting Binary Compound in Ternary System

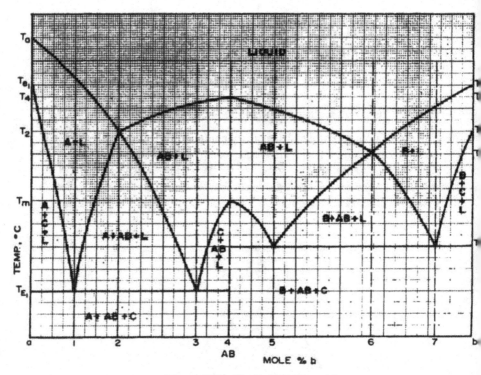

Figure 5.65. Vertical Section a–b

directly to the eutectics, liquids 2 and 6 crystallize two phases as the paths move toward E_1 and E_2, respectively, and liquid 4 is unique inasmuch as crystallization stops at m to yield a mixture of C and AB.

C. Ternary System With a Binary Congruently Melting Compound and a Ternary Peritectic

For a system such as shown in Figure 5.66 the vertical sections at levels of 20% and 50% C are shown in Figures 5.67a and 5.67b, respectively. The section at the 20% level crosses the primary fields of A, AB and B, composition triangles A-AB-C and B-AB-C and involves eight types of starting liquids. The section at the 50% level crosses the primary fields of A, B and C, both composition triangles and six types of starting liquids.

D. Simple Ternary Eutectic System With Immiscible Liquid Region Originating in System A-C

The vertical section A-P through a region of liquid immiscibility is shown in Figure 5.68. This system and one isothermal section have already been shown as

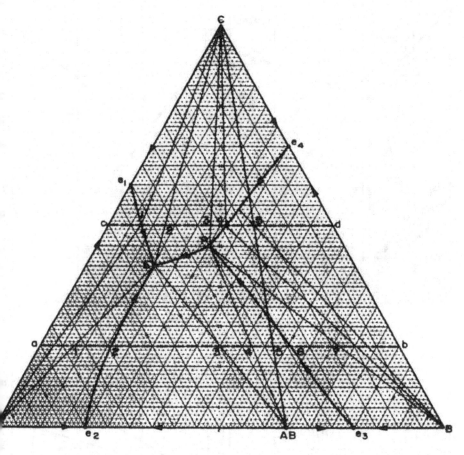

Figure 5.66. Congruently Melting Binary Compound and Ternary Peritectic

Figures 5.38a, 5.38b, 5.39 and 5.40, but the basic configuration is repeated here
as Figure 5.68 in order to more easily follow the details of construction of
Figure 5.69a (section A–p) involving composition points A, 2, 3, 4, 5, p, e_3 and
E. Figure 5.69b (section a–b) involves composition points a, a', a", 2', 3', b",
b' and b. Figure 5.69c (section c–B involves composition points c, K, a'", e_1,
c", c'", and B and Figure 5.69d (section d–B) involves composition points d, d',
d", d'", e_1 and B.

E. Vertical Sections in a Real Oxide System

The system Na_2O–CaO–SiO_2 is shown in Figure 5.70. The vertical sections between $CaSiO_3$ and NC_2S_3 and $Na_2Si_2O_5$ and NC_2S_3 are simple eutectic types,

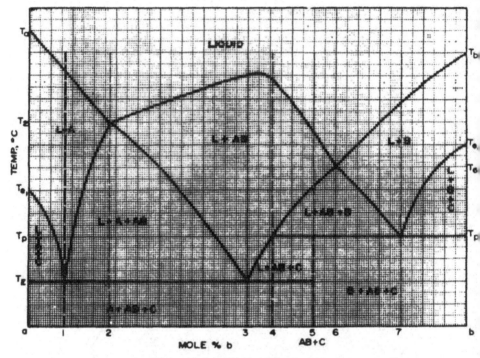

Figure 5.67a. Vertical Section a-b

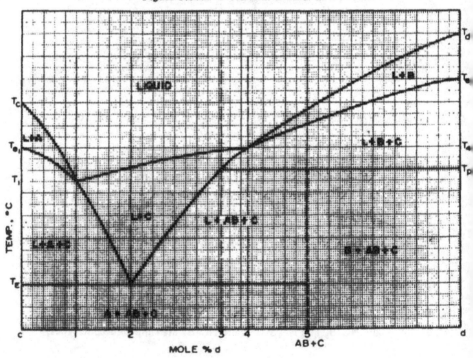

Figure 5.67b. Vertical Section C-d

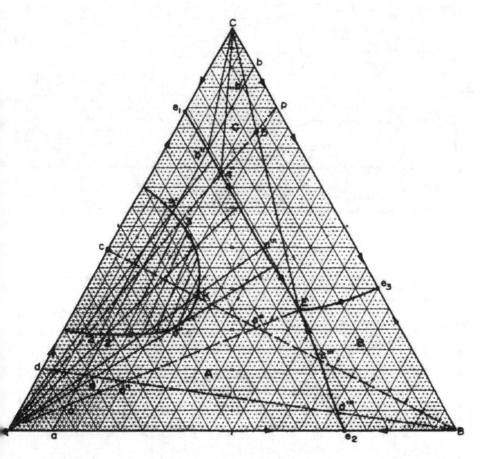

Figure 5.68. Simple Ternary Eutectic System With Immiscible Liquid Region Originating in System A–C

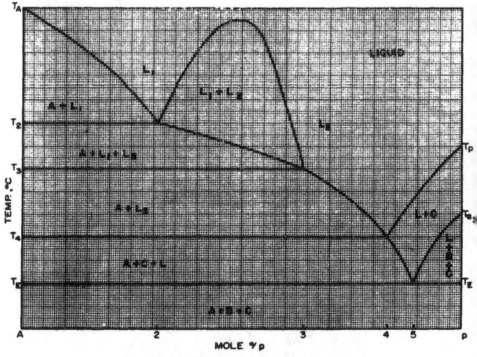

Figure 5.69a. Vertical Section A–p

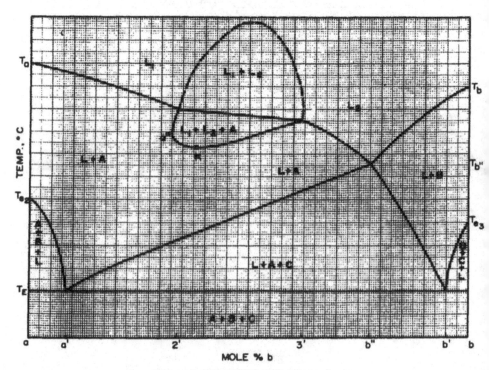

Figure 5.69b. Vertical Section a–b

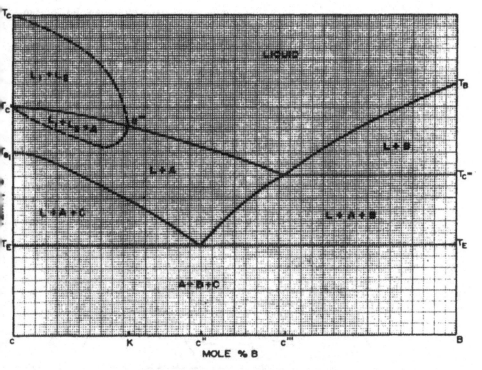

Figure 5.69c. Vertical Section c-B

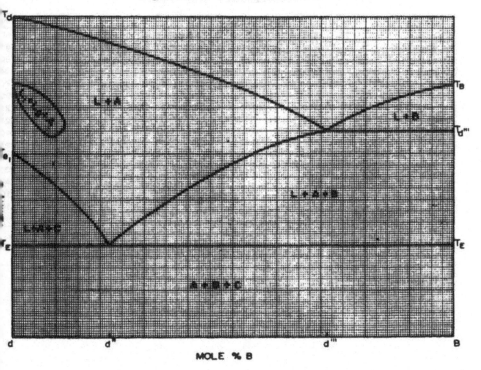

Figure 5.69d. Vertical Section d-B

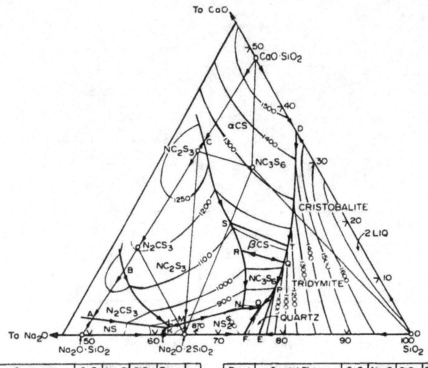

Compounds	CaO	Na₂O	SiO₂	Temp	
SiO₂			100.0	1710	M
αCaO·SiO₂	48.3		51.7	1540	M
βCaO·SiO₂	48.3		51.7	1180	I
Na₂O·SiO₂		50.8	49.2	1088	M
Na₂O·2SiO₂		34.1	65.9	874	M
2Na₂O·CaO·3SiO₂	15.6	34.4	50.0	1141	D
Na₂O·3CaO·6SiO₂	28.5	10.5	61.0	1047	D
Na₂O·2CaO·3SiO₂	31.6	17.5	50.9	1284	M

M = MELTING POINT
D = DECOMPOSITION POINT
I = INVERSION POINT

Point		Crystal Phases	CaO	Na₂O	SiO₂	Temp
1	A	NS·N₂CS₃	3.0			1060
3	B	N₂CS₃·NC₂S₃	11.5			1141
1	C	NC₂S₃·αCS	33.0			1280
1	D	αCS·S	37.0		63.0	1436
5	E	T·QUARTZ		24.3	75.7	870
1	F	QUARTZ·NS₂		26.4	73.6	790
2	K	NS₂·NS·N₂CS₃	1.8	37.5	60.7	821
4	L	N₂CS₃·NC₂S₃·NS₂	2.0	36.6	61.4	827
4	N	NS₂·NC₂S₃·NC₃S₆	5.2	24.1	70.7	740
2	O	NC₃S₆·Q·NS₂	5.2	21.3	73.5	725
5	P	Q·NC₃S₆·T	7.0	18.7	74.3	870
4	Q	T·βCS·NC₃S₆	12.9	13.7	73.4	1035
4	R	NC₃S₆·NC₂S₃·βCS	14.5	19.0	66.5	1030
3	S	βCS·NC₂S₃·αCS	19.5	17.7	62.8	1110
3	T	αCS·S·βCS	15.6	11.4	73.0	1110
1	I	NS·NS₂		38.0		840

1 = BINARY EUTECTIC , 2 = TERNARY EUTECTIC,
3 = DECOMPOSITION POINT , 4 = REACTION POINT,
5 = INVERSION POINT , C = CaO , N = Na₂O , S = SiO₂,
T = TRIDYMITE

Figure 5.70. High SiO₂ Corner of the Ternary System Na₂O–CaO–SiO₂

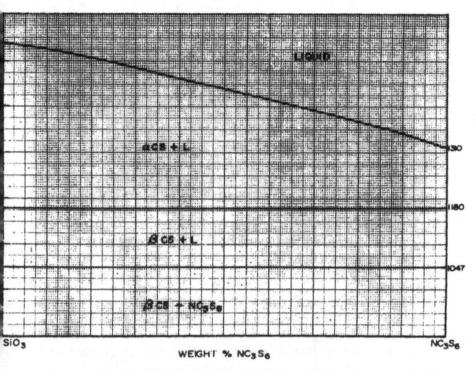

Figure 5.71. Vertical Section $CaSiO_3$–NC_3S_6

true binary systems within the ternary system. However, many other vertical sections between the various composition points in the system are non-binary and
not as simple, as for example that between $CaSiO_3$ and NC_3S_6 (devitrite), as
shown in Figure 5.71.

XI. COMPLEX TERNARY SYSTEMS

Up to this point, the same technique used in Chapter 3 to explain the various
possible geometries in binary systems has been used for ternary systems. That is,
in order to effectively illustrate the isoplethal analysis, crystallization and
melting paths, the Lever Rule, and the influence of binary and ternary compounds and liquid immiscibility on the geometry and interpretation of the
ternary system, these phenomena have been isolated as single events, insofar as
possible.

    In real systems, is it of course, common for several of these geometries to
be combined to yield a complex system containing several binary and ternary

compounds, a variety of primary fields, joins, invariant points and composition triangles and perhaps immiscible liquid regions.

Perhaps the best way to understand a complex ternary system without solid solution is to be required to construct one from experimental data. One would expect to have complete information on the equilibrium diagram of the three binary systems involved. The first step in the construction of the ternary system might be to establish the valid joins at subsolidus temperatures by the use of solid state reactions. One would attempt to acquire the maximum amount of data using a minimum number of compositions and heat treatments. If ternary compounds existed, their composition, polymorphism and thermal behavior would have to be established. It is to be noted that joins will run only between the end members, binary or ternary compounds and that they will never cross each other.

Liquidus temperatures, primary fields, boundary lines, isotherms and invariant points would have to be established by quenching methods or D.T.A. The important point is that all the accumulated data has to be fitted together so that no improper or impossible constructions are made when drawing the completed diagram.

Very few professionals are ever required to construct an equilibrium diagram from experimental data and therefore the next best method of acquiring a complete understanding of the ternary diagram is to do problems on abstract and real systems which illustrate and require the use of all the principles so far discussed. A complete set of problems designed to fulfill this objective is supplied at the end of this chapter.

One illustrative problem on a complex abstract ternary system which contains two incongruently melting binary compounds and one incongruently melting ternary compound as shown in Figure 5.72 will serve to emphasize the kind of analysis needed to completely understand the relative temperature relationships, the boundary lines, the invariant points, the joins and the crystallization paths in this system.

1. First, the primary fields for the compounds AC, BC, ABC and the end members should be identified and labeled as shown in Figure 5.73.

2. Second, the compatibility triangles should be identified and associated with a particular invariant point if possible. For example, the following five invariant points obviously exist:

1) C–AC–ABC   4) B–BC–ABC
2) C–BC–ABC   5) A–B–ABC
3) A–AC–ABC

The joins should be drawn between the composition points which create the proper composition triangles as shown in Figure 5.73.

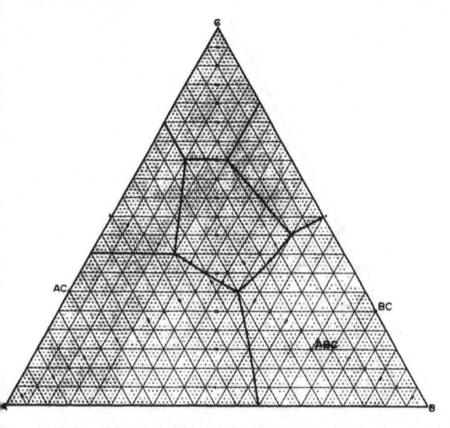

Figure 5.72. Abstract Ternary System

3. Third, ALL invariant points should be labeled with respect to type, that is eutectic, peritectic or reaction, including those in the bounding binary systems. It turns out that only one ternary invariant point is a eutectic lying within its associated compatibility triangle and all others are peritectics, lying outside their associated compatibility triangles.

4. Fourth, arrows should be placed on ALL boundary lines, indicating the direction of falling temperature. It should be noted that the join C-ABC crosses the boundary lines e_3P_1, P_1P_2 and P_2P_4, but does NOT furnish any information relative to the direction of falling temperature on these boundary lines.

It is now possible to trace the crystallization path of liquid X (28% A, 50% B and 22% C), lying in the primary field of B and in the compatibility triangle C-AC-ABC, to its final crystallization at E.

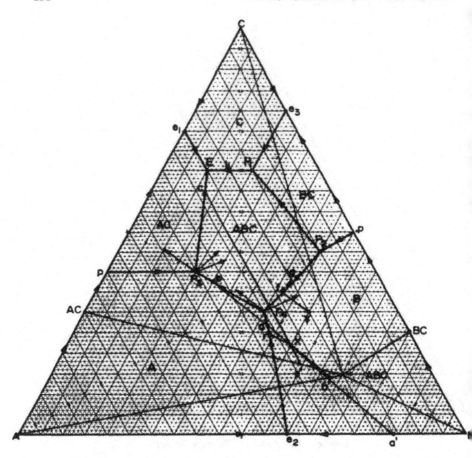

Figure 5.73. Abtract Ternary System With Proper Joins, Composition Triangles, and Invariant Points

B will crystallize as the primary phase to point r, at which temperature both A and B crystallize from the liquid. Quantitative calculations on the amount of A and B in equilibrium with liquid would be referred to the join A-B. For example, tie line a-a'. At P_4, all of B is resorbed during invariant reaction at constant temperature. A is resorbed and ABC crystallizes from the liquid as the path moves toward P_3. (The liquid path does *not* move toward P_2 because B and ABC would then be in equilibrium with liquid, an impossible situation because these three phases are at the apices of a triangle which does not include the original composition point X and projections from any liquid composition on P_4P_2 though point X would not intersect the join B-ABC). While the liquid path moves along P_4P_3, all quantitative calculations are referred to the join A-ABC. For example, tie line b-b'.

At P_3, A is completely resorbed at constant temperature as heat is withdrawn. From P_3 to E, AC and ABC crystallize and quantitative calculation of the amount of solid and liquid and the relative amounts of AC and ABC are referred to the join AC-ABC, as for example tie line c-c'.

Liquid Y, whose composition (20%A, 50% B, 30% C) is not far removed from that of liquid X, will crystallize by an entirely different route under equilibrium conditions as it proceeds to final crystallization at E.

B crystallizes as the primary phase to point f on the boundary line between the primary fields of B and ABC. At this point, B is resorbed and ABC crystallizes. At point g, B is completely resorbed, the liquid path proceeds from g to h across the primary field of ABC. C and ABC crystallize from the liquid from h into E.

Note that liquids lying in the composition triangle A-ABC-B and in the primary fields of either A or B will precipitate A and B on e_2P_4 and then PARTIALLY resorb A and B at P_4 as ABC crystallizes and the invariant reaction takes place at constant temperature. (F=4-4=0)

XII. REAL TERNARY SYSTEMS

A convenient list of real ternary systems taken from the 1964, 1969, 1975 and 1981 collections of "Phase Diagrams for Ceramists" is given in Table 5.5. The systems have been chosen for their value in illustrating the principles discussed in Sections I-X of this chapter. All of these systems are valuable for the purpose of illustrating principles and interpreting the geometry of the various configurations which arise. If one is called upon to select from these systems their practical value in ceramic science and engineering, a very difficult problem arises due to the broad scope of modern "ceramics". The systems $Na_2O-Al_2O_3-SiO_2$ and $K_2O-Al_2O_3-SiO_2$ are of some interest in whiteware, $Na_2O-CaO-SiO_2$ and $Na_2O-B_2O_3-SiO_2$ are of tremendous concern to glass science and technology, the phosphate systems are important in metallurgical slag formulating, the system $CaO-Al_2O_3-SiO_2$ is very important to cement chemistry and the lead systems are of great concern to anyone making low melting fluxes for use with pigments in glass and ceramic decorating processes. Numerous ternary titanate, zirconate and stannate systems are basic for electroceramics and ternary systems with Fe_2O_3 are fundamental for magnetic ceramics. Ternary systems involving rare earth oxides are now of interest in luminescence and laser technology.

In addition to the immiscible liquid divalent metal oxide-silica systems mentioned in Section VII-B, the systems $Li_2O-B_2O_3-SiO_2$, $Na_2O-B_2O_3-SiO_2$, $CaO-B_2O_3-SiO_2$, $BaO-B_2O_3-SiO_2$, $PbO-B_2O_3-SiO_2$, $Na_2O-CaO-SiO_2$, $PbO-GeO_2-V_2O_5$ and $ZnO-SiO_2-Nb_2O_5$ are very important for the understanding

Table 5.5 A Convenient List of Real Ternary Diagrams which Illustrate Congruent Melting, Incongruent Melting, Dissociation, Primary Fields, Boundary Lines, Joins, Compatibility Triangles, Isotherms and Liquid Immiscibility

System	Page	Figure	Remarks
			1964 Collection
$Li_2O-K_2O-SiO_2$	145	377	Fluxing effect of mixed alkalies
$Na_2O-K_2O-SiO_2$	146	381	Fluxing effect of mixed alkalies, polymorphism of SiO_2, low quartz on liquidus
$K_2O-CaO-SiO_2$	149	391	Large number of compounds
$K_2O-MgO-SiO_2$	152	399	
	154	403	
$K_2O-Al_2O_3-SiO_2$	156	407	Feldspar, leucite, 1:1:2 and 1:1:1 compounds, high viscosity glasses
$Li_2O-Na_2O-SiO_2$	162	430	Silica inversions
$Li_2O-MgO-SiO_2$	165	443	
$Li_2O-Al_2O_3-TiO_2$	169–170	460–464	Sequence of compatibility triangles
$Li_2O-B_2O_3-SiO_2$	170	465	Fluxing effect of Li_2O and B_2O_3
$Na_2O-CaO-SiO_2$	174–175	481–482	Devitrite, soda-lime silica glass
$Na_2O-Al_2O_3-SiO_2$	181	501	Feldspar, nepheline and carnegieite–high viscosity glasses
$Na_2O-B_2O_3-SiO_2$	184	515–518	Pyrex, Vycor, metastable liquid immiscibility in all binary systems
$Na_2O-Fe_2O_3-SiO_2$	186	520	
$Na_2O-SiO_2-P_2O_5$	189–191	535–542	Phosphate slag
$BaO-CaO-SiO_2$	193–194	549–551	
$BaO-B_2O_3-SiO_2$	196–197	558–561	Liquid Immiscibility-most complete study of dome
$CaO-ZnO-SiO_2$	217	624	
$CaO-Al_2O_3-SiO_2$	219–221	630–639	Excellent practice system—see problems
$CaO-B_2O_3-SiO_2$	224–225	646–649	Danburite, liquid immiscibility
$CaO-Fe_2O_3-SiO_2$	228–229	656–658	
$CaO-SiO_2-TiO_2$	230–231	660–663	Sphene, liquid immiscibility

System			
$MgO-SiO_2-TiO_2$	249	723	Low melting, immiscibility
$PbO-B_2O_3-SiO_2$	253–254	740–742	Refractory oxides
$Al_2O_3-SiO_2-TiO_2$	261–262	768–771	Refractory oxides
$Al_2O_3-SiO_2-ZrO_2$	262	772	Refractory oxides
$Al_2O_3-TiO_2-ZrO_2$	262	773–774	Refractory oxides
$NaF-KF-MgF_2$	434	1535	Fluoride system
$LiF-NaF-BeF_2$	436	1542	Fluoride system
$NaF-RbF-UF_4$	442	1560	Fluoride system
$NaF-RbF-ZrF_4$	443	1561	Fluoride system
$Na_2O-SiO_2-H_2O$	535–539	1959–1971	Hydrothermal growth of quartz
$K_2CO_3-Na_2CO_3-H_2O$	565	2052	Two Salts and H_2O
			1969 Collection
$Na_2O-MgO-SiO_2$	121	2435	Compare with $K_2O-MgO-SiO_2$
$CaO-Al_2O_3-SiO_2$	138–139	2493–2498	Revisions, p–T relations
$MgO-B_2O_3-SiO_2$	151	2540	
$Al_2O_3-Y_2O_3-SiO_2$	165	2586	Rare earth system
$NaF-KF-CaF_2$	375	3421	Fluoride system
$LiF-NaF-BaF_2$	379	3437	Fluoride system
$NaF-MgF_2-BaF_2$	382	3445	Fluoride system
$NaF-BeF_2-ThF_4$	383	3447	Fluoride system
			1975 Collection
$Na_2O-ZnO-SiO_2$	206	4522	
$Na_2O-B_2O_3-SiO_2$	209	4526	
$BaO-Al_2O_3-SiO_2$	219–220	4543–4544	
$BeO-Al_2O_3-SiO_2$	222	4548	Beryl, Emerald
$CaO-SiO_2-Nb_2O_5$	231	4562	

Table 5.5 (*Continued*)

System	Page	Figure	Remarks
$CaO-SiO_2-Ta_2O_5$	232	4564	
$PbO-B_2O_3-SiO_2$	248	4584–4585	
$LiF-KF-CsF$	365	4830	Fluoride systems
$KF-CsF-MnF_2$	366	4831	Fluoride systems
$LiF-CsF-MnF_2$	366	4832	Fluoride systems
$NaF-CsF-MnF_2$	367	4833	Fluoride systems
$NaF-KF-NbF_5$	367	4834	Fluoride systems
$CaF_2-MgF_2-BaF_2$	369	4837	Fluoride systems
			1981 Collection
$Na_2O-BaO-SiO_2$	168	5319	
$Na_2O-CaO-SiO_2$	170–171	5321, A and B	
$Na_2O-MgO-V_2O_5$	179	5329	
$Na_2O-P_2O_5-MoO_3$	187	5346	
$CaO-MgO-TiO_2$	203	5380	
$CaO-B_2O_3-P_2O_5$	206	5388	
$PbO-GeO_2-V_2O_5$	224	5429	
$ZnO-SiO_2-Nb_2O_5$	227	5436 (A to F)	

of real ternary oxide liquid immiscibility. The metastable immiscibility and practical value of the system $Na_2O-B_2O_3-SiO_2$ will be discussed in Chapter VI.

In order to limit the scope of this text, none of the real systems listed in Table 5.5 will be discussed in detail with respect to their practical value, but a large selection of problems at the end of the chapter will include work with several of them to illustrate basic principles. By doing problems, it is hoped that the student will not only learn to interpret all aspects of a real diagram from a fundamental standpoint, but also become curious with respect to its application in a particular field of specialization in glass or ceramics.

In this connection, it is helpful to have available some of the ten large scale diagrams of E. F. Osborn and Arnulf Muan or the three large scale diagrams of F. A. Hummel, D. W. Johnson and S. C. Samanta which are based on a 500 mm equilateral triangle and easily permit composition to be expressed to 0.5% accuracy. These thirteen large scale diagrams can be purchased from the American Ceramic Society, 65 Ceramic Drive, Columbus, Ohio, 43214.

Problem 29 at the end of Chapter 3 provided some information on the number of binary oxide systems (4,005) which would have to be examined for 90 elements in the Periodic Table (only the most common valence state for each element).

For three-component oxide systems (most common valence state only), the number would be:

$$\frac{90!}{3!(90-3)!} = \frac{90!}{3 \times 2 \times 87!} = \frac{90 \times 89 \times 88}{6}$$
$$= 117,480$$

This indicates that quite a few ternary oxide, silicate, fluoride, sulfide, boride, carbide, nitride and other ceramic systems still need to be examined for phase equilibrium relationships.

PROBLEMS

I. Crystallization paths, heating paths, quantitative calculations and vertical sections for abstract systems:

1. Using Diagram P5.1, draw joins which will create the proper composition triangles for the system. Place arrows on ALL boundary lines in the binary and ternary systems indicating the direction of falling temperature so that temperature maxima on boundary lines may be recognized and so that invariant points may be identified and labeled as eutectics, peritectics, etc.

   a) For a composition containing 80% C, 10% A and 10% B, trace the crystallization path, making sure to describe ALL major events as the liquid cools below the temperature of the invariant point.

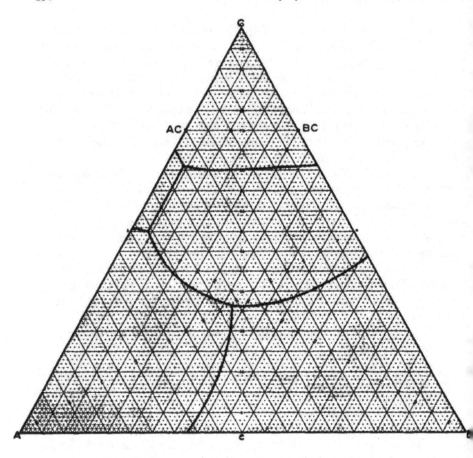

Diagram P5.1. Binary and Ternary System I

b) Show QUANTITATIVELY by calculation that one phase partially dissolves at the invariant point as heat is withdrawn from the system.
c) Show that the amount of A, B, and C in the final crystallized mixture
   is the same as that in the original liquid. (80C, 10A, 10B)
d) Trace the crystallization path of a composition containing 75% C, 15%
   A and 10% B. The composition is on the join AC–BC. Make a quantitative calculation just before and just after final crystallization to show
   exactly what takes place at the invariant point.
e) Trace the crystallization path of a liquid containing 65% C, 20% A and
   15% B on the boundary line between the primary fields of C and BC.

f) How much solid and how much liquid coexist when the liquid composition is 60% C, 30% A and 10% B? What is the composition of the
solid?

g) Make a construction which shows the area in which starting liquids in
the primary field of C and in the composition triangle AC-BC-A will
cross the primary field of BC as they proceed to the eutectic where AC,
BC and A coexist.

h) For the liquid composition given in (e) above, make a quantitative
analysis of the equilibrium just before and just after final crystallization at the eutectic. Show that the amount of A, B, and C in the assemblage of liquid, AC and BC just before final crystallization is the same
as that in the starting liquid (65C, 20A, 15B). Show that the amount of
A, B and C in the final crystallized mixture of A, AC and BC is the
same as that in the starting liquid (65C, 20A, 15B). How much A is
precipitated during final crystallization at the invariant point?

i) Trace the crystallization path of a liquid containing 70% C, 5% A and
25% B. Be sure to describe ALL major events which occur as the
temperature is lowered below the invariant point. Make quantitative
calculations at temperatures where C is in equilibrium with liquid,
where C and BC are in equilibrium with liquid, where only BC is in
equilibrium with liquid, where BC and A are in equilibrium with liquid
and just before and just after final crystallization at the invariant point.
How much B is precipitated during reaction at the invariant point?

j) Draw the vertical section for the Join AC-BC. Assume reasonable
temperature relationships for the system and label each area of the
diagram for phases present. Use a quantitative scale for the horizontal
composition axis.

k) Draw the vertical section between C-c. Assume reasonable temperature
relationships in the system and label each area of the diagram for
phases present. Make sure the horizontal composition axis is reasonably
quantitative between C and c.

2. Using Diagram P5.2, draw joins which will create the proper composition triangles for the system. Place arrows on ALL boundary lines in the binary
and ternary systems indicating the direction of falling temperature, so that
temperature maxima on boundary lines may be recognized and so that invariant points may be identified and labeled as eutectics, peritectics, etc.

a) Trace the crystallization path of a liquid containing 55% C, 15% A and
30% B. This composition is located on the join BC-A. Be sure and
describe thoroughly each major event as the crystallization proceeds to
the invariant point and the reaction at the invariant point. What are the

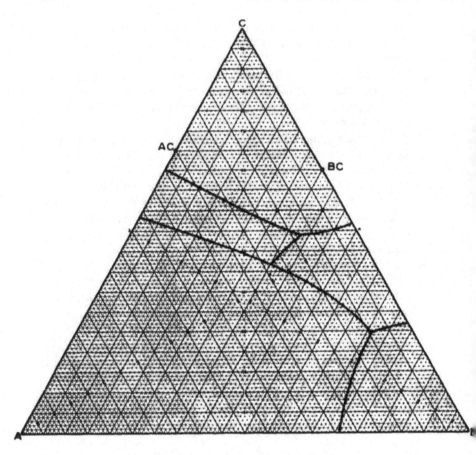

Diagram P5.2. Binary and Ternary System II

percentages of each phase in the mixture after final crystallization?
Give the amount and composition of liquid and solid just before final
crystallization at the invariant point.

b) Several quantitative calculations of the composition and amount of
liquid and solid(s) are possible during intermediate stages of crystallization. Create several problems during crystallization of C, as the liquid
travels the C-AC boundary or the AC-BC boundary. (For the composition given in (a) above.)

c) Trace the crystallization path of a liquid containing 55% C, 5% A, and
40% B, being sure to thoroughly describe ALL major events as the
composition goes from all liquid to all solid. Calculate the amount and

composition of coexisting phases just before and just after final crystallization at the invariant point. Make calculations to show that in both cases the amount of A, B and C in the assemblage is 55C, 5A and 40B.

d) For the liquid composition in (c) above, what is the liquid composition when equal parts of solid and liquid are in equilibrium. Ans. 45% C, 10% A and 45% B; solid is BC.

e) Draw the vertical section AC-BC. Assume reasonable temperature relationships and use a quantitative composition scale on the horizontal axis. Label each area of the diagram for phases present.

f) Draw the vertical section at the 10% C level. Assume reasonable temperature relationships and use a quantitative scale for composition along the horizontal axis. Label each area of the diagram for phases present.

g) Two of the most interesting starting liquids have been used for the problems given in a–d above, but many other problems can be formulated in this relatively simple system by using starting liquids which lie in the primary fields of A, B, C, AC or BC, on the two joins in the system or on any boundary line in the ternary system. Crystallization paths or heating paths may be traced first and then quantitative calculations can be made when only one crystalline phase is in equilibrium with liquid, when two crystalline phases are in equilibrium with liquid and just before and just after final crystallization at invariant points.

3. Using Diagram P5.3, draw joins which will create the proper composition triangles for the system. Place arrows on ALL boundary lines indicating the direction of falling temperature, so that temperature maxima on boundary lines may be recognized and so that invariant points may be identified and labeled as eutectics, peritectics, etc. This includes the three binary systems, A-C, A-B and B-C. Label primary fields.

a) Trace the crystallization path of a liquid containing 5% A, 60% B and 35% C. Detail ALL important events which occur between the stages of "all liquid" and "all crystalline". What is the composition of the final crystallized mixture of phases?

b) What is the composition of the liquid at a point where 50% crystals and 50% liquid are in equilibrium? Show that the amount of A, B and C in the 50–50 mixture is the same as that in the original liquid (5A, 60B, 35C).

c) What is the composition of the liquid at a point where 75% crystals and 25% liquid are in equilibrium? What is the composition of the liquid at a point where 25% crystals and 75% liquid are in equilibrium?

d) Trace the crystallization path of a liquid containing 5% A, 75% B and

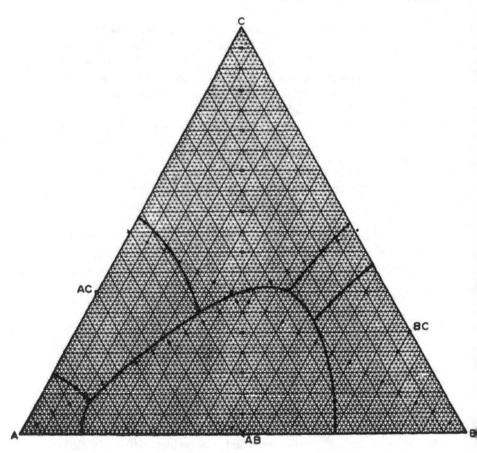

Diagram P5.3. Binary and Ternary System III

20% C. Detail ALL important events as the path goes to the invariant point. Show by a simple vector diagram what happens at the invariant point and then show by quantitative calculation what happens during reaction at the invariant point.

e) Draw the vertical sections AB–BC, C–AB, AC–AB or at 10, 20, 30, 40, 50, 60 and 70% levels of component C. Assume reasonable temperature relationships and be as quantitative as possible with respect to the composition axis. Label phases present in each area of the vertical section.

4. Using Diagram P5.4, which shows a system containing an incongruently melting binary compound AB and a congruently melting ternary compound,

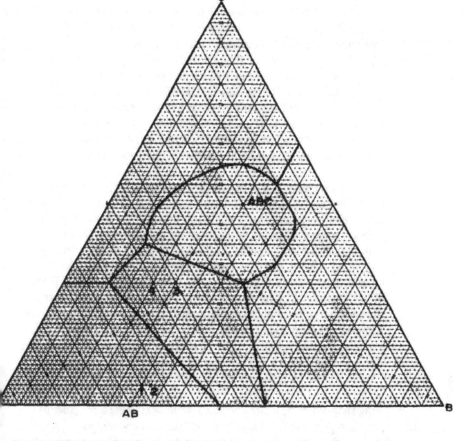

Diagram P5.4. Incongruently Melting Binary Compound and Congruently Melting Ternary Compound

draw joins which create the proper composition triangles. Place arrows on ALL boundary lines in the binary and ternary systems which indicate the direction of falling temperature, so that temperature maxima on boundary lines may be recognized and so that invariant points may be identified and labeled as eutectic, peritectic, reaction, etc.

a) Trace the crystallization path of a liquid containing 65% A, 30% B and 5% C. Describe ALL important events which occur during crystallization to and below the temperature of final crystallization. What is the composition of the liquid at the temperature where one third of the

mixture is still liquid? What is the solid composition at this point?
What is the liquid composition at a temperature where 80% of the mixture is solid? What is the percentage of each crystalline phase in the
final assemblage after all the liquid has disappeared?

b) Trace the crystallization path of a liquid containing 63% A, 32% B, and
5% C, making sure to describe ALL important events during equilibrium crystallization. On the diagram, show at what point (liquid composition) 50% liquid and 50% crystalline solids would coexist and give
the relative proportion of each crystal in the mixture at that temperature. Indicate at what point (liquid composition) 75% solid and 25%
liquid would be in equilibrium. What is the solid composition at this
point?

c) Trace the crystallization path of a melt containing 45% A, 25% B and
30% C. Indicate ALL important events which occur during crystallization. Prove by quantitative calculation that a phase is RESORBED
during final crystallization at the invariant point. Draw a vector diagram at the invariant point to qualitatively confirm that the ternary
compound ABC is partially resorbed during final crystallization.

d) Trace the crystallization path of a melt containing 50% A, 20% B and
30% C. Describe ALL important events which occur until such temperature as the last trace of liquid disappears. Make a quantitative analysis
of the equilibrium assemblage just before and just after the peritectic
reaction, and just before and just after the eutectic reaction. Show that
the percentage of A, B and C in the final crystallized assemblage of A,
C and AB is the same as that in the original liquid.

5. Using Diagram P5.5, which shows a system containing a congruently
melting binary compound AC, a congruently melting ternary compound A′B′C′
and an incongruently melting ternary compound ABC, draw joins which create
the proper composition triangles for the system. Place arrows on ALL boundary
lines in the binary and ternary systems which indicate beyond doubt the directions of falling temperature, so that temperature maxima on boundary lines may
be recognized and so that invariant points may be identified and labeled as
eutectics, peritectics, reaction, etc.

a) Trace the crystallization path of a liquid containing 8% A, 65% B and
27% C (lying ON the join). What is the ratio of liquid to solid just
before final crystallization? What is the percentage of each solid in the
final crystallized mixture?

b) Trace the crystallization path of a liquid containing 20% A, 50% B and
30% C. Make a complete quantitative analysis of the equilibrium at a
point where the ratio of solid to liquid is 1:1. Give composition of

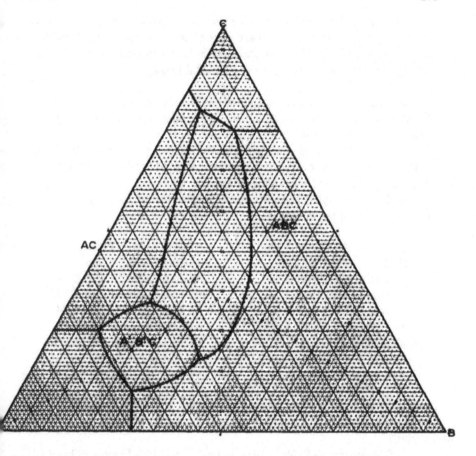

Diagram P5.5. Congruently Melting Binary Compound, Congruently Melting Ternary Compound, and an Incongruently Melting Ternary Compound

liquid and solids and show that the percentage of A, B and C in the assemblage of liquid and two solids is the same as that in the original melt.

c) For a liquid containing 10% A, 30% B and 60% C, make a quantitative calculation of the equilibrium assemblages just before and just after final crystallization at the invariant point and draw a conclusion about the nature of the invariant point and the amount of resorbtion of a phase during final disappearance of the liquid. Show that the distribution of A, B and C in the final crystallized mixture is the same as that in the original liquid.

d) Draw the vertical sections for the joins AC–ABC, C–ABC, and B–ABC, or at a level of 25% C. Make reasonable assumptions about relative temperatures, use a quantitative scale for composition, and label phases present in each area of the completed diagram.

6. Study the vertical sections shown in Figures 5.63a and 5.63d in the text and make sure you can account for the temperatures, compositions, boundary lines and labeling.

7. Study the vertical sections shown in Figures 5.67a, 5.67b, 5.69a–d in the text and make sure you can quantitatively account for the temperatures, compositions, boundary lines and labeling.

8. Draw a complete and quantitative vertical section for the joins shown in the following figures in the text:

a) Join C–AB, Figure 5.15   j) Join B–ABC, Figure 5.33
b) Join C–AB, Figure 5.17   k) Join C–ABC, Figure 5.33
c) Join C–AB, Figure 5.21   l) Join A–ABC, Figure 5.34
d) Join C–AB, Figure 5.22   m) Join B–ABC, Figure 5.34
e) Join C–AB, Figure 5.28   n) Join C–ABC, Figure 5.34
f) Join C–AB, Figure 5.30   o) Join SiO_2–ABC, Figure 5.35
g) Join A–ABC, Figure 5.32   p) Join AC–ABC, Figure 5.35
h) Join AB–ABC, Figure 5.32   q) Join C–AB, Figure 5.43
i) Join A–ABC, Figure 5.33   r) Join C–AB, Figure 5.44

II. Crystallization paths, heating or melting paths, quantitative calculations, isothermal and vertical sections in real systems shown in "Phase Diagrams for Ceramists".

1. The System $K_2O–MgO–SiO_2$, pages 152 and 154, Figures 399 and 403, 1964.

a) Trace the crystallization paths of the following liquid composition, being certain to describe all the major events which occur during cooling:

i) 30% $2MgO \cdot SiO_2$ iii) 30% $2MgO \cdot SiO_2$
    40% SiO_2 20% SiO_2
    30% $K_2O \cdot MgO \cdot 5SiO_2$ 50% $K_2O \cdot MgO \cdot 5SiO_2$
ii) $K_2O \cdot 5MgO \cdot 12SiO_2$ iv) 20% $2MgO \cdot SiO_2$
                                         25% SiO_2
                                         55% $K_2O \cdot MgO \cdot 5SiO_2$

b) Draw the isothermal plane for 1100°C. Identify the phases present in each area of the isothermal plane, draw tie lines in appropriate two phase regions.

2. Study the configurations shown in the systems $Li_2O-K_2O-SiO_2$, $Na_2O-K_2O-SiO_2$, and $Li_2O-Na_2O-SiO_2$, noting especially the primary fields for the silica polymorphs and the isotherms. (See Table 5.5, Section XII)

3. The System $Li_2O-Al_2O_3-TiO_2$, pages 169-170, Figures 460-464, 1964.

Explain the significance of this series of compatibility triangles in terms of the thermal behavior of the binary and ternary compounds which exist, especially $Li_2O\cdot TiO_2$, $2Li_2O\cdot 5TiO_2$, $Li_2O\cdot 3TiO_2$, $Al_2O_3\cdot TiO_2$ and $Li_2O\cdot Al_2O_3\cdot 4TiO_2$. Obviously the subsolidus compatibility triangulation changes for each temperature range due to 1) changes in the system Li_2O-TiO_2 and 2) the appearance of a ternary compound. Using precise, accepted phase equilibrium language regarding the thermal behavior of binary and ternary compounds, explain the reasons for the changes in triangulation.

4. The System $Na_2O-CaO-SiO_2$, pages 174-175, Figures 481-482, 1964. Using Figure 482,

a) List the six compatibility triangles shown. Define composition (or compatibility triangle). Which of the composition triangles shown comprise a true ternary subsystem? Go down the list of six triangles and explain *why* each one *is* or *is not* a true ternary subsystem.

b) Trace the path of crystallization of a melt of the composition NC_3S_6. Just before final crystallization of the liquid, what is its composition in terms of Na_2O, CaO and SiO_2? How much liquid is in equilibrium with the solid and what is the composition of the solid?

c) Locate a composition containing 10% Na_2O, 35% CaO and 55% SiO_2. Trace the crystallization path of this liquid until it disappears at the invariant point. Describe FULLY all the important events which take place during cooling of the system.

Make a quantitative analysis of the equilibrium at $1100°C$. Calculate percent liquid and solid and the percent of each phase in the solid mixture. What is the liquid composition?

Demonstrate by calculation that one crystalline phase partially dissolves as the liquid disappears at the invariant point.

d) Draw the isothermal plane for $1100°C$. Draw tie lines in two phase regions and label the phases present in each area of the isothermal plane. Draw similar isothermal planes for $900°$, $1000°$, $1200°$, $1300°$ and $1400°$.

e) Locate a liquid of the composition 13% Na_2O, 20% CaO and 67% SiO_2. Trace the crystallization path until all the liquid disappears. At what temperature will 50% liquid and 50% crystals exist in equilibrium? Make a quantitative analysis of the equilibrium just before and just after final crystallization at the invariant point (eutectic). Show that the amounts of Na_2O, CaO and SiO_2 in the assemblages before and after final crystallization are the same as those in the original liquid.

f) Draw vertical sections for the following joins in the system:

$SiO_2-NC_3S_6$ N_2CS_3-NS

$CaSiO_3-NC_3S_6$ $N_2CS_3-NS_2$

$Na_2O \cdot 2SiO_2-NC_3S_6$ $NC_2S_3-NS_2$

$CaSiO_3-NC_2S_3$ $NC_2S_3-NC_3S_6$

$NC_2S_3-N_2CS_3$

g) Using Figure 481, draw in the proper joins which will create the proper composition triangles for the system. Be sure and draw as many triangles as are justified by the data in the figure, but do not draw in triangles which are *not* justified by the data.

h) The system $Na_2O-CaO-SiO_2$ is the basis for many commercial glasses. Study references 19, 20, 21, 24, and 50 in Table 1.5, Appendix A as a basis for understanding microphase separation and structure of these commercial glasses.

5. The system $Na_2O-B_2O_3-SiO_2$ is the basis for the commercial production of Pyrex, Vycor and other common borosilicate type glasses. Study references 4, 9, 20, 26, 27, 30, 36, 37, 50, 60, 64, 65, 67 and 70 in Table 1.5, Appendix A as basis for understanding liquid phase separation and its relation to chemical and physical properties of the commercial glasses.

6. Study the system $BaO-B_2O_3-SiO_2$, shown as Figures 558-561 on pages 196-197 of the 1964 collection, and the original papers by Levin and Ugrinic and Levin and Cleek as a basis for the understanding of ternary borosilicate immiscibility, stable and metastable types.

7. The System $Na_2O-Fe_2O_3-SiO_2$, page 186, Figure 520, 1964.

a) Trace the equilibrium heating or crystallization path of a composition containing 20% SiO_2, 75% Fe_2O_3, and 5% $Na_2O \cdot SiO_2$. Include a description of all major events as the composition changes from complete liquid to complete solid or vice versa. Make a quantitative analysis of the equilibria just before and just after final crystallization and demonstrate what happens at the invariant point. Show that the amounts of tridymite and hematite in the equilibrium assemblage just BEFORE final crystallization are the same as they were in the original melt (20% and 75%).

b) A composition containing 45% SiO_2, 30% Fe_2O_3 and 25% $Na_2O \cdot SiO_2$ is very intimately mixed and heated slowly under equilibrium conditions until completely melted. Trace the equilibrium path from the start of heating until complete melting occurs. Mention the temperatures at which major events occur during heating. At what temperature will the mixture be 10% liquid? 50% liquid? Completely melted?

c) During equilibrium heating of a mixture containing 80% Fe_2O_3, 10% SiO_2, and 10% $Na_2O \cdot SiO_2$, at what temperature would liquid first appear? At what temperature would liquid first appear during equilibrium heating of a mixture containing 30% SiO_2, 20% Fe_2O_3 and 50% Na_2SiO_3? There is a difference of 60% in the Fe_2O_3 content of these two mixtures. Could you comment on the relationship of composition to the temperature of initial melting for these two mixtures?

8. The System $CaO-Al_2O_3-SiO_2$, page 219-221, Figures 630-639, 1964; revisions, page 138-139, Figures 2493-2498, 1969.

a) Trace the crystallization paths of the following compositions:

CaO	Al_2O_3	SiO_2	CaO	Al_2O_3	SiO_2
57	8	35	37	45.5	17.5
55	5	40	40	35	25
5	70	25	35	40	25
5	60	35	45	35	20
10	10	80	45	30	25
45	50	5			

For each composition, make a quantitative calculation of the equilibrium assemblage just before and just after final crystallization.

b) If a pyrometric cone consisting of 20% CaO, 30% Al_2O_3 and 50% SiO_2 begins to deform when 33% liquid is present in the composition (assume equilibrium heating), at what temperature would you expect deformation to start in a mixture of $CaSiO_3$, anorthite and silica? How much of each mineral would be needed to make up the pyrometric cone? If the wollastonite ($CaSiO_3$) was omitted from the composition by mistake or oversight, what is the lowest temperature at which you would expect liquid to form?

c) A fusion containing 57% CaO, 8% Al_2O_3 and 35% SiO_2 is cooled slowly, but the Ca_2SiO_4 which precipitates sinks to the bottom of the melting furnace or container and out of contact with the remaining liquid until the liquid reaches the temperature at which gehlenite appears. If equilibrium crystallization is now maintained until all the liquid disappears, what is the percentage of phases in the final completely crystalline mixture, including equilibrium and non-equilibrium phases?

d) Draw the isothermal sections for $1300°$, $1400°$ and $1500°$.
e) Construct the following vertical sections:

  i) Al_2O_3–anorthite
  ii) $CaO \cdot 6Al_2O_3$–anorthite
  iii) $CaO \cdot 6Al_2O_3$–gehlenite
  iv) $CaO \cdot 2Al_2O_3$–gehlenite
  v) $3CaO \cdot SiO_2$–$3CaO \cdot Al_2O_3$
  vi) $2CaO \cdot SiO_2$–$3CaO \cdot Al_2O_3$
  vii) At a level of 20, 30, 40, 50, 60, 70, 80 or 90% SiO_2

  Use real composition and temperature scales and label phases present in each area of the diagram.

9. The System CaO–ZnO–SiO_2, page 217, Figure 624, 1964. Quantitatively analyze the crystallization behavior at the invariant point of a melt which originally contains 10% ZnO, 40% CaO and 50% SiO_2. What happens as crystallization takes place?

10. The System CaO–TiO_2–SiO_2, page 230, Figure 660, 1964.

a) Draw the binary system CaO–TiO_2 from information contained in the ternary diagram. Place important temperatures and compositions on the diagram and label areas for "phases present".
b) For a composition containing 40% CaO, 30% TiO_2 and 30% SiO_2, what phases and how much of each are in equilibrium at $1300°C$?
c) Draw the vertical section between SiO_2 and $CaTiO_3$. (Ans. in Figure 663, page 231).

III. Miscellaneous problems on ternary systems

1. Draw one complex ternary system A-B-C containing each of the following:

a) Two congruently melting binary compounds
b) Two incongruently melting binary compounds
c) A dissociating binary compound with an upper temperature limit of stability with a primary field in the ternary system
d) One congruently melting ternary compound
e) One incongruently melting ternary compound

Show primary fields for each end member and compound, boundary lines, place arrows on all boundary lines to indicate the direction of falling temperature, connect composition points to form the proper joins and composition triangles and label all invariant points which you create to indicate E, P, R, e, p, etc. Label temperature maxima, m, on boundary lines in the ternary system. The diagram must be consistent and contain no improper constructions.

2. Draw ONE complex ternary system A-B-C containing each of the following:

a) Two congruently melting binary compounds, one in the system A-C and the other in the system A-B. (A_2C_8) and $(A_{7.5}B_{2.5})$

b) Three incongruently melting binary compounds, one in the system A-C and two in the system B-C. (A_4C_6), (B_7C_3) and B_4C_6.

c) One congruently melting ternary compound (60% C, 20% A, and 20% B).

d) One incongruently melting ternary compound (50% B, 30% C, 20% A).

Follow instructions given in Problem 1 above.

3. A ternary system is composed of components A, B, and C with congruent melting points of 1600°, 1500° and 1800°, respectively. The binary systems AC and BC are simple eutectic systems with eutectics at 1400° and 80% C and 1300° and 75% C, respectively. The ternary system is bounded by a third binary system which contains an intermediate compound AB (70% A, 30% B) which has a lower limit of stability at 1000°C. AB melts incongruently at 1300° to A and a liquid containing 60% A. The eutectic point in the system AB is at 1150° and 30% A. The ternary eutectic contains 15% A and 70% C at 700°. Using triangular paper, sketch in the boundary lines for the ternary system, indicate direction of falling temperature on ALL boundary lines (including binary systems) using arrows, label primary fields, and draw in a compatible set of isotherms based on the data given. Label temperatures at important invariant points by assumption or as provided by the above data.

Draw the vertical section between AB and C from the data given and your own assumptions about the liquidus surface.

Draw isothermal planes for each of the following temperatures:

a) 600°C   c) 1100°C
b) 900°C   d) 1400°C

4. On page 43 of the 1964 collection of diagrams is shown Figure 27, parts a, b, c and d. This is a join in the ternary system Al-Fe-O. Using precision phase equilibrium language, explain the outstanding feature of this series of diagrams as the oxygen pressure is changed.

What do Figures 27a and 27b infer with respect to the synthesis and stability of the intermediate phase? What favors the formation of the intermediate phase? The compositions of the phases corundum (Al_2O_3) and hematite (Fe_2O_3) are well known to you. What is (or what might be) the composition of the SPINEL solid solution shown in these diagrams? Use standard phase equilibrium language at all times, in a quantitative manner.

5. Draw *one* ternary system containing *all* of the following:

a) Primary fields for compounds A_2B, AB and AB_2 which melt congruently, incongruently and congruently, respectively.

b) Primary fields for the incongruently melting compound BC and the dissociating compound BC_2 (upper limit of stability).

c) Primary fields for the incongruently melting compound A_2C and the compound AC. (The latter compound has a lower limit of stability and melts incongruently.)

d) Primary fields for two incongruently melting ternary compounds and one congruently melting ternary compound.

Draw correct boundary lines, connect composition points to form correct joins and composition triangles and draw arrows on *all* boundary lines indicating the direction of falling temperature. Label all composition points and primary fields clearly.

6. Using the CaO–MgO–SiO_2 system shown on page 210 as Figure 598 of the 1964 collection:

Give the temperature at which liquid would be expected to appear during equilibrium heating of compositions located in the following composition triangles:

a) MgO–$3CaO \cdot SiO_2$–$2CaO \cdot SiO_2$
b) Ca_2SiO_4–MgO—Merwinite
c) Ca_2SiO_4—Merwinite-Akermanite
d) Ca_2SiO_4–$3CaO \cdot 2SiO_2$—Akermanite
e) Monticellite-Merwinite-Akermanite

6

NON-EQUILIBRIUM IN CERAMIC SYSTEMS

I. GENERAL

The definition of equilibrium and the phenomena of non-equilibrium and meta-stability have been set forth and illustrated in portions of Chapters 1, 2 and 3 and in Chapter 5. The background of the student should now be sufficiently advanced so that it is appropriate to further expand and illustrate non-equilibrium and metastability, using examples chosen from abstract or real one, two and three component systems.

From the practical, experimental standpoint, it is often very difficult to know when equilibrium has been established in an oxide, silicate or ceramic system. The following quotation from the definitions given in Levin, Robbins and McMurdie (1) is especially significant:

"Three criteria have been used variously either singly or together:

1. The time criterion, based on the constancy of phase properties with the passage of time;
2. The approach from two directions criterion, yielding under the same conditions phases of identical properties, e.g., from undersaturation or supersaturation, from raising or lowering the temperature, or by raising or lowering the pressure; and
3. The attainment by different procedures criterion, producing phases having identical properties when the same conditions, with respect to the variants, are reached.

None of these criteria are entirely adequate for excluding metastable relationships. In silicate systems, in particular, metastable "equilibrium" is common and may persist for long periods of time and at high temperatures. In the final analysis, interpretation and judgment by the investigator are of prime importance."

With respect to the time criterion, the major difficulty lies in the definition of what is a "long time". The active research lifetime of the average experimentalist may be about thirty years and he is inclined to believe that eight hours is a long time to wait for an equilibrium to establish itself. Yet Schairer and Bowen (2) readily admitted that equilibrium was not established in three years when working on certain portions of the system $K_2O-Al_2O_3-SiO_2$. Glasses on the leucite-silica join did not crystallize after five years at 1250°C and one composition near the join did not crystallize after seven years. Mason (3) reported that lunar tridymite and cristobalite had not converted to quartz in three billion years. The work on the system $K_2O-Al_2O_3-SiO_2$ also shows that it is not always feasible to approach the equilibrium from two directions; that is, glasses will not crystallize after seven years at 1200°C. On the other hand, in some cases, certain low temperature (400°-600°) crystalline phases can only be prepared in a reasonable time by crystallization from glass; if synthesis is attempted by solid state reaction, the reaction does not take place in a reasonable time. In the remainder of this chapter, a few specific cases will be discussed to illustrate the possibilities for non-equilibrium and their influence on the microstructure and properties of the system.

II. ONE COMPONENT SYSTEMS

A. Silica

Reference should be made to the discussion of silica in Chapter 2 and Figures 2.4 and 2.5. Several metastable states have already been treated in Chapter 2, but others can now be discussed in the light of the definition given in I.

Silica glass, α-cristobalite, and α-tridymite are all metastable with respect to α-quartz at room temperature and one atmosphere under the present concept of the stability relationships of these phases. None of these three phases shows any tendency whatever to convert to α-quartz at room temperature and one atmosphere; this is true for silica glass, α-cristobalite or α-tridymite, whether found in nature or made in the laboratory from more or less pure starting materials. However, they should be *tending* to convert to α-quartz. Each of these three phases should be tending to convert to α-quartz at 574°C and one atmosphere. (Tridymite and cristobalite would now be in their high temperature forms as β'' and β respectively.) No one has ever held these phases at 574° for fifty years to observe whether or not they do convert, but it is known that there is no conversion after a few months. Likewise, these three phases (glass, β'' tridymite and β cristobalite) should convert to β-quartz if held long enough at one atmosphere between 573° and 870°. No one has so far held these phases in this temperature interval for fifty years to see if the conversion would actually occur, but it is known that silica glass will yield β-cristobalite (not β-quartz) if held for several months near 870°. Presumably β-cristobalite is "less metastable" than silica glass at 870° and one atmosphere. IF tridymite is the equilibrium silica phase between 870° and 1470°, β-cristobalite should convert to β'' tridymite in this temperature range at one atmosphere, but apparently no one has held β-cristobalite under these conditions for fifty years to determine whether or not the conversion would occur.

The high pressure phases coesite and stishovite seem to be able to exist indefinitely in a metastable state with respect to α-quartz at room temperature and one atmosphere, but it is known (4) that coesite will change to quartz between 1100°-1350° and subsequently to cristobalite at one atmosphere. A treatment of 1160° for forty-two hours completely converted coesite to cristobalite. Stishovite becomes amorphous to x-rays after heating for five minutes between 650°-750°, forms quartz after half an hour to eighteen hours between 1000°-1100°, and eventually transforms to cristobalite if held long enough between 500°-1100°, all at one atmosphere. An extended study of the kinetics and conditions under which stishovite reverts to coesite or quartz has been made (5). Below 225°, stishovite is highly resistant to conversion to any other phase under a variety of persuasions, but at 20k bars and 400°, only a few days are required for complete conversion to quartz.

The lesson to be learned from this discussion and Chapter 2 is that the system SiO_2 is extremely sluggish in many of its reactions and the times necessary to investigate the true stability relationships may have to be measured in decades or perhaps hundreds of years. Many experiments need to be done by holding pure SiO_2 at various temperatures and pressures for very long periods of TIME. The reversibility criterion (#2) needs to be applied and demonstrated and finally, the influence of a variety of starting materials (criterion #3). This, combined with calculations from thermodynamic data, may enable the construction of a true equilibrium diagram for SiO_2, rather than the (weakly) probable or "not-impossible" diagram shown in Figure 2.5, Chapter 2.

B. Zirconia

This system (see Figure 2.9, Chapter II) is an excellent one for demonstrating the unusual techniques which must sometimes be used to generate metastable phases and the extremely unstable nature of metastable phases, once they are produced (in sharp contrast to the extremely persistent nature of the metastable silica phases discussed in (A) above).

So far no one has been able to bring pure cubic ZrO_2 to room temperature using a starting material of ordinary particle size ($>1\mu$). The rapid cubic to tetragonal inversion invariably occurs during cooling. Many additives to ZrO_2 will enable the cubic form to be brought to room temperature in a "stabilized", more or less permanent condition, but the system is then composed of two components.

However, by decomposition of an alkoxide under controlled conditions (6), both cubic and tetragonal pure ZrO_2 can be obtained in a very finely divided 50Å condition at room temperature and one atmosphere. Decomposition of zirconium oxychloride will also yield tetragonal ZrO_2 (pure) at room temperature. If the 50Å alkoxide cubic ZrO_2 is now reheated (7) to 305° for 150 hours it transforms to the less metastable tetragonal form, and after twenty-four hours at 400° the tetragonal form reverts to the stable monoclinic form. If metastable tetragonal ZrO_2 produced by the alkoxide method is heated to 400°-800°, it will also revert in a matter of hours to the monoclinic form which is stable from room temperature to (approximately) 1000°. If the particle size of the metastable cubic ZrO_2 lies between 80-100Å, transformation to metastable tetragonal takes place between 250°-270°, demonstrating the considerable influence of particle size on the rate of a metastable → metastable transition.

It is worthwhile to repeat that:

1. these metastable transitions in ZrO_2 are in sharp contrast to those in the system SiO_2,
2. very unusual techniques are sometimes required to produce metastable phases,
3. the rate of metastable → metastable or metastable → stable transitions is often highly dependent on the particle size of the material, and
4. it is sometimes possible to generate reproducible and persistent metastable states in a system which can easily be mistaken for true equilibrium states. (in the system SiO_2 especially)

III. TWO COMPONENT SYSTEMS

A. General

In two-component systems there are many ways in which metastability or non-equilibrium can develop.

i. Invariant Reactions
When F=3-3=0 at constant temperature and pressure, the following invariant equilibrium reactions can occur:

i. Eutectic, $A + B \rightleftarrows L$

ii. Peritectic, $AB \rightleftarrows A + L$

iii. Monotectic $A \rightleftarrows L_1 + L_2$

iv. Eutectoid $\alpha_{ss} \rightleftarrows \beta SS_A + \beta SS_B$

v. Peritectoid $\beta SS_A + \beta SS_B \rightleftarrows \gamma SS_A$

vi. Dissociation $AB \rightleftarrows A + B$

vii. Two Liquid Formation of Intermediate Compound $AB \rightleftarrows L_1 + L_2$

During heating or cooling at these invariant points, the TIME must be long enough to permit each reaction to completely take place. If heating or cooling is too rapid, many metastable, non-equilibrium situations can occur.

ii. Monovariant Reactions

When $F=3-2=1$, many two phase monovariant equilibrium reactions can occur involving two liquids, liquid and solid, two solids, two solid solutions and polymorphs as indicated by the many two-component diagrams shown in Chapter 3. Again, in all cases, during heating and cooling, the TIME must be long enough to permit each reaction to completely take place. If heating or cooling is too rapid, many metastable, non-equilibrium situations can occur.

B. Non-equilibrium During Heating

i. Non-equilibrium Heating of a Mixture of Two Crystals in a Simple Eutectic System

Consider the reactions which occur during the heating of a mixture of powders of A and B in the proportions shown by X_o on the phase diagram of Figure 6.1.

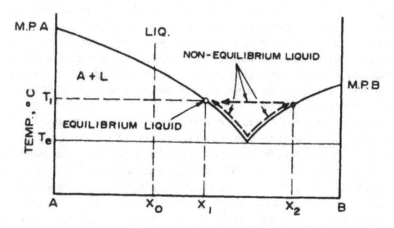

Figure 6.1. Non-equilibrium Liquid Formation During Heating of Composition X_o

Figure 6.2. Microstructure of Composition X_o

After carefully mixing the raw materials, they might be reacted in a crucible or, possibly, pressed into some desired shape prior to being heated. In either case the particles of A and B are randomly distributed and the microstructure before heat treatment would appear approximately as shown schematically in Figure 6.2, the void space accounting for 30-50% of the total volume.

On heating to a temperature just below T_e, no reactions would be expected except for possibly a certain degree of solid state sintering (densification). On heating to temperature T_1, the sample would be expected, when equilibrium is established, to consist of crystals of A in a matrix of melt of the composition given by X_1. The first liquid to form on heating to a temperature above that of the eutectic (T_e) will occur at the contact points between particles of A and B, as shown in Figure 6.3a, and would approximate the composition of the eutectic. The portion of the melt in contact with the crystal of A (Figure 6.3b) will dissolve additional A and thus becomes richer in A causing the melt to approach the equilibrium concentration of A (X_1 in the phase diagram of Figure 6.1) at which concentration the dissolution of A would cease. Similarly, the melt in contact with particle B will become richer in B as more B is dissolved. Dissolution of B will cease when the composition of the melt reaches the equilibrium concentration given by X_2 in Figure 6.1.

The simultaneous dissolution of A at one surface of the melt and that of B at the other surface produces a composition gradient in the melt. This gradient provides the driving force for the counter diffusion of A and B atoms within the melt (Figure 6.3b). The counter diffusion of A and B atoms prevents the A-rich and B-rich melts from reaching their saturation concentrations; consequently, the dissolution process continues until all of the B is consumed and eventually the system consists of a perfect equilibrium of crystals of A in a melt of composition given by point X_1 in Figure 6.1. If the viscosity of the liquids is very high, the solution of A and B and the counter diffusion of A and B in the melt could be very slow. If composition X_o was not heat treated for a period of time

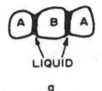

LIQUID

a

Figure 6.3a. Liquid Formation Between A and B Crystals

sufficient to attain equilibrium, the final microstructure could contain some crystals of B, less liquid phase than that predicted by the phase diagram, and a liquid phase which was not homogeneous, containing concentration gradients. The path of the non-equilibrium liquids as they approach $T_1 X_1$ is shown in Figure 6.1.

ii. Compound Formation

To an inexperienced person, the formation of any of the hundreds of binary compounds shown in oxide phase diagrams or otherwise reported in the literature would seem to be a simple matter, merely requiring that fine powders of pure raw materials be intimately mixed and heated to some "appropriate" temperature for perhaps a few hours. Compound formation producing 100% yield is not that simple and many unexpected events occur when the objective is to produce a totally reacted pure compound, free of unreacted starting materials or other impurity phases.

The reaction of several of the alkaline earth oxides with SiO_2 provide especially interesting illustrations. Merely combining BeO and SiO_2 in the ratio to

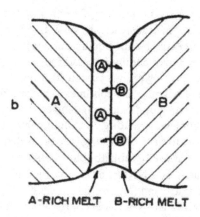

A-RICH MELT   B-RICH MELT

Figure 6.3b. Counter Diffusion of A and B in Liquid Region

form the mineral phenacite does not produce Be_2SiO_4. Heating in a temperature range between 1200–1500° does not result in appreciable reaction in any reasonable time and one must resort to a "mineralizer" such as Zn_2SiO_4 in order to effectively cause the reaction of BeO and SiO_2 to occur. One to two percent of Zn_2SiO_4 promotes the Be_2SiO_4 reaction at the temperatures mentioned above and a high yield can be obtained at one atmosphere.

Orthosilicates prefer to form in a mixture of MgO and SiO_2 or CaO and SiO_2 rather than metasilicates. If one desires to synthesize $MgSiO_3$ or $CaSiO_3$ by combination of a pure source of MgO, CaO and SiO_2, it is found that short term heating of 1:1 mole ratios produce mainly Mg_2SiO_4 or Ca_2SiO_4. Only long term heating at appropriate temperatures combined with repeated grinding, mixing and reheating will cause the orthosilicate to react with SiO_2 to produce the desired metasilicate. It should be mentioned that the high temperature forms of Ca_2SiO_4 form with much greater ease than Mg_2SiO_4 (forsterite). Like Be_2SiO_4, the synthesis of Mg_2SiO_4 is greatly assisted by the addition of "mineralizing" or "fluxing" agents.

The above examples emphasize the importance of the MECHANISM of oxide reactions as well as the KINETICS of these reactions. These fields are extremely important in ceramics and solid state science, but these topics and the manner in which mineralizers function to aid reactions are beyond the scope of this text.

The reaction of Al_2O_3 with SiO_2 to produce a phase-pure mullite is sluggish and requires high temperatures and long times. The minerals kyanite, sillimanite and andalusite cannot be produced by merely combining a 1:1 ratio of Al_2O_3 and SiO_2, no matter how long the mixture is heated at elevated temperatures; only the combined influence of pressure and temperature will permit the synthesis of these minerals.

iii. Incongruently Melting and Dissociating Compounds

In general, the incongruent melting reaction, $AB \rightleftarrows L + A$ and the dissociation reaction, $AB \rightleftarrows A + B$, may easily be non-equilibrium reactions if the heating is too rapid. The compound AB may continue to exist at temperatures above the equilibrium incongruent melting or dissociation temperature. Likewise, compositions near the compound AB may also continue to contain the compound AB above the equilibrium melting or dissociating temperature if the heating is too rapid. In the case of incongruent melting, the liquids which form may also be of a non-equilibrium composition when heating is so rapid that the binary compound continues to exist above the equilibrium reaction temperature.

It was once thought that $Na_2O \cdot 2SiO_2$ was the only compound intermediate between $Na_2O \cdot SiO_2$ and SiO_2 in the system $Na_2O–SiO_2$ (see 1964 collection, page 94, Figure 192 and 1981 collection, page 87, Figure 5122). However, by quenching compositions in a narrow compositional range between 72 and 77.5 mole percent SiO_2, Williamson and Glasser (8) showed that a new compound existed, $3Na_2O \cdot 8SiO_2$, melted incongruently at $808° \pm 2°$ and had a lower limit of stability at $700° \pm 10°$, yielding $\beta Na_2 Si_2 O_5$ and quartz. Single crystal cell volume, density data and calculation of the molar refractivity were used to supplement the quenching, optical and powder diffraction data and fix the stoichiometry of the compound. The compound could be prepared using $\alpha Na_2 Si_2 O_5$ and quartz, $Na_2 CO_3$ and quartz, or glass as the starting material, but it should be noted that times of 69-144 hours were used to prepare the 3:8 compound from crystalline materials and 200 hours when crystallized from glass. The compound could be made phase-pure in as short a time as twenty-four hours at 796°. The dissociation to quartz and liquid at 808° was admittedly sluggish and it is inferred that the 3:8 compound might be easily superheated above 808° without decomposition. On the other hand, it should be noted that hydrothermal runs were used to assist the determination of the lower limit of stability at 700°. The 3:8 compound is metastable at room temperature and at all temperatures up to 700°. Finally, the authors showed that a previously claimed stoichiometry of 1:3 was not correct. These results would of course affect the study of all three component systems involving $Na_2 O$ and SiO_2. (See 1969 collection, page 81, Figure 2286.)

A similar study by West and Glasser (9) showed that $Li_2 Si_3 O_7$ can be crystallized in five to six minutes from glasses in the system $Li_2 O-SiO_2$ as a totally metastable phase. It will persist indefinitely at room temperature and above, but it is decomposed by heating for twelve to twenty-four hours at 650°, even though it was originally crystallized at 890° from glass. This occurrence is especially interesting because it is superimposed on the well-known metastable two-liquid immiscibility which develops in this system below 850°. If phase separation of the liquids is allowed to occur, the formation of $Li_2 Si_2 O_5$ and SiO_2 is favored. In order to encourage the formation of the metastable trisilicate, it is necessary to by-pass metastable two liquid formation by holding a homogeneous glass between 850-920°C for several minutes. If the metastable trisilicate is held above 650° to form disilicate and SiO_2, the disilicate itself is a metastable solid solution containing up to 72 mole percent SiO_2. Finally, it was pointed out that the T-X limits of formation of the metastable trisilicate lie just outside the metastable liquid immiscibility boundary and it is likely that the liquids are already "structured", perhaps assisting the nucleation and growth of the trisilicate in preference to stable phases.

The case of zircon ($ZrSiO_4$) in the system $ZrO_2 -SiO_2$ is ideal for illustrating the effect of non-equilibrium superheating. For many years, zircon was thought to melt congruently at 2550° or 2430°. Rapid heating of the natural mineral or a synthetic mixture to what is now known to be the temperature of complete liquifaction, combined with rapid cooling would lead to this conclusion, especially if the product was amorphous or poorly crystallized at room temperature. In 1949, a claim for incongruent melting at $1775°$ was made; in 1953 a claim for solid state decomposition (dissociation) at $1540°$; and in 1957, another claim for incongruent melting at $1720°$. Butterman and Foster (10) have summarized the history of work on the system, and finally concluded that $ZrSiO_4$ dissociated to ZrO_2 and SiO_2 at $1676°$, very close to the eutectic temperature at $1687°$. (See 1969 collection, page 110, Figure 2400.) The proximity of the dissociation and eutectic temperatures explain the claims for incongruent melting. The ease with which natural zircon or even a synthetic product can be superheated and the ease of formation of a siliceous liquid which can be quenched to room temperature as a glass explain the claims for congruent melting. If $ZrSiO_4$ is not heated too far above its dissociation temperature, the ZrO_2 resulting from the decomposition would not have an opportunity to grow in the highly viscous siliceous liquid which forms, especially at short holding times. At room temperature, the product might appear to be totally amorphous.

iv. Polymorphic Inversions

There are many cases in which a low temperature polymorph of a binary compound fails to invert promptly to its high temperature form when heated, similar to the case of single oxides. This behavior leads directly to the study of the kinetics of polymorphic inversions, a vast field. A few of the low temperature binary phases which can be substantially superheated into the stability range of the high temperature form are: enstatite ($MgSiO_3$), wollastonite ($CaSiO_3$), and $MnSiO_3$. Such behavior is expected in the more structurally complex chain and double chain silicates in which silicon-oxygen tetrahedra are directly linked.

C. Non-equilibrium During Cooling

i. Glass Formation

The preparation of a glass involves cooling a melt from above the liquidus temperature of the mixture to room temperature without the formation of a crystalline phase. Commercial glass compositions are formulated in such a way as to increase the ease with which siliceous melts supercool to a metastable liquid which subsequently exists at room temperature as the solid material called "glass". Nearly all common silicate, borate and phosphate glasses as well as special glasses exist indefinitely at room temperature in a metastable condition, and also when reheated to moderate temperatures. (up to $700°$). Silica glass resists crystal formation to well above $800°C$.

In order for a crystal to form within a supercooled melt, a nucleus composed of atoms in crystalline array must be formed and must be of sufficient size to be thermodynamically stable. The probability of a sufficient number of atoms of suitable energy and orientation coming together within the melt to form a stable nucleus is very low at small undercoolings. However, foreign surfaces within the melt may serve as nucleation sites and will reduce the number of atoms

required for a stable nucleus. Surfaces such as container walls, bubbles, the inter-
face between an immiscible liquid phase and its liquid matrix, or foreign material
within the melt frequently serve as nucleation sites for crystal growth.

In many oxide melts and especially in siliceous melts the rate at which
stable nuclei are formed, even in the presence of foreign surfaces, is very low
owing in large measure to the low mobility of the atoms and the complex atomic
groups within the melt.

Consider a simple melt of composition X_0 in Figure 6.4. When this melt
is cooled to temperature T_1 it becomes super-saturated in A and a small quantity
of crystalline A precipitates from the melt under equilibrium conditions. If there
are no stable nuclei present, crystallization of A will not occur and the melt is
said to be in a metastable condition. With further cooling, such as to T_2 or T_3 in
Figure 6.4, the degree of supersaturation becomes greater, the tendency (thermodynamic driving force) for crystallization becomes greater and the probability of
forming stable nuclei is enhanced.

As the temperature decreases, the viscosity of the melt increases rapidly
and the mobility of the atomic species within the melt is decreased thereby re-
ducing the rate at which stable nuclei can be formed. These two opposing forces,
i.e. the greater thermodynamic driving force for nucleation and the decreasing
atomic mobility, give rise to a relatively narrow temperature range over which
crystallization is likely to occur on cooling a melt. Fairly rapid cooling through
this critical temperature range precludes crystallization and a glassy or vitreous
phase is formed.

It should be noted that the binary compounds Li_2SiO_3 and $CaSiO_3$ are
used as temperature standards, especially for "quenching" studies. In addition to

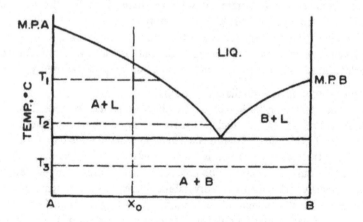

Figure 6.4. Supercooling of Glass-Forming Composition X_0 to Temperatures
T_1, T_2 and T_3. Metastable Liquid Formation

being compositionally stable at the melting point (they melt congruently and do not change composition when held for periods up to an hour), they also form glasses when quenched rapidly to room temperature from one to ten degrees above the M.P., yet crystallize within seconds when held one or two degrees below their melting temperature. These are the requirements which a good quenching standard must meet. Many other materials could be considered as temperature standards, providing they have glass-forming and crystallization characteristics like Li_2SiO_3 or $CaSiO_3$, and providing they remain compositionally constant at the M.P.

ii. Metastable Liquid Immiscibility

Metastable two-liquid separation is now known to be very common in borate and silicate systems and is easily generated by:

1. Undercooling of separated liquids which were generated in a temperature range where the phases were stable, as for example, $RO-B_2O_3$ and $RO-SiO_2$ systems,
2. Slow or rapid cooling of an originally homogeneous liquid to a temperature range where metastable separation takes place, followed by prolonged heat treatment in this temperature range, or
3. Quenching of an originally homogeneous liquid to room temperature, followed by prolonged heating in a temperature range where metastable separation develops.

The latter two methods are used for the alkali borate and alkali silicate systems. Levin (11) gives a summary of the work which has been done on binary metastable immiscibility in borate and silicate systems, including the work of Shaw and Uhlmann (12) on borates and a variety of authors on the silicates.

The work on the binary systems is important as it relates to many physical properties such as light scattering, small angle x-ray scattering, electrical properties, nucleation and growth theory, and spinodal decomposition theory.

iii. Suppression of an Intermediate Compound

The ease with which a crystalline phase may be nucleated is dependent upon a number of interrelated factors such as the free energy difference between the crystal and melt, the liquid-solid interfacial energy, the complexity of the crystal structure, and the degree of association of the melt. These factors may be different for each crystal-melt combination considered and can result in marked differences in nucleation behavior among crystalline phases within the same system. Consider melt X in the diagram of Figure 6.5. On cooling to point K crystals of AB would precipitate under equilibrium conditions. If, however, the crystalline phase AB is not readily nucleated, the melt may supercool to point H (the intersection of isopleth X with the metastable extension of the liquidus line for

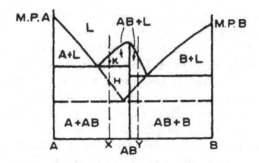

Figure 6.5. Metastability in the Area of an Intermediate Binary Compound AB. Crystallization of A or B, Rather Than AB

crystalline A) at which temperature crystals of A may nucleate and grow. A similar situation may occur for composition Y in which B crystals may form rather than crystals of AB. The metastable equilibrium diagram might thus indicate a simple eutectic system as shown by the dashed lines of Figure 6.5. It is said that the compound $K_2O \cdot 4SiO_2$ is easily bypassed in favor of such a metastable eutectic between $K_2O \cdot 2SiO_2$ and SiO_2 in the system K_2O-SiO_2 (page 87, Figure 167, 1964 collection). Non-equilibrium behavior of this kind may be expected in other glass-forming binary systems involving K_2O due to the relatively high viscosity of the liquids. (B_2O_3, GeO_2)

iv. Polymorphic Inversions

Very often, a high temperature form of a single oxide may have some highly desirable chemical or physical property which is unattainable at room temperature due to the great speed with which inversion to a lower temperature form occurs on cooling. Examples of such oxides and their properties are 1) the low thermal expansion of the β-cristobalite form of silica (metastable below 1470°C), 2) the characteristics of the β-quartz form of GeO_2 (metastable with respect to α-quartz and rutile type GeO_2 below 1049°C), and 3) the superb chemical, thermal and dimensional stability of cubic ZrO_2 (metastable below 2372° ± 50°). In the latter case, additions of a second component such as MgO, CaO or Y_2O_3 will permit the existence of cubic ZrO_2 at room temperature. It should be quickly pointed out, however, that in the case of MgO stabilization (13) the cubic phase is really metastable below about 1400°, yet it exists indefinitely at room temperature and also at temperatures up to 800°-900°C. If held for a long period above 1240°C, it will dissociate to MgO and tetragonal ZrO_2 solid solution, more rapidly as the temperature increases toward 1400°. CaO and Y_2O_3 seem to provide complete stabilization from room temperature to the liquidus. In these cases either 1) the cubic phase is thermodynamically stable in the range room temperature to the liquidus, or 2) the kinetics of dissociation to other phases are so slow as to be negligible in a time span measured in years.

The system $Li_2O-B_2O_3$ can be used to illustrate sluggish versus rapid reactions and metastable relationships. The portion of the system between Li_2O and $Li_2O \cdot B_2O_3$ (14) is generally non-glass forming and the melts crystallize spontaneously during cooling. The system was studied mainly by solid state reaction or reheating of crystallized melts, and by D.T.A. The compound $2Li_2O \cdot B_2O_3$ exists in two polymorphic forms in the narrow interval between $600°$ and $650°$ and both can be quenched to room temperature as metastable phases. The high temperature β form of 2:1 could persist below $618°$ in the stability field of α 2:1.

In the region of the system between $Li_2O \cdot B_2O_3$ and B_2O_3 (15), glasses were easily obtained at room temperature, as well as the compounds 2:5, 1:3 and 1:4, yet only the 1:4 is really stable at room temperature. The 2:5 compound is metastable below $696°$, the 1:3 below $595°$ and the glasses at all temperatures below the liquidus. Full crystallization of glasses containing 84.5 to 87.5 B_2O_3 took over seven days in the temperature interval between $500°$-$600°C$.

If one considers the case of true binary joins within ternary systems, many cases of stabilization of a high temperature polymorph to room temperature by means of the addition of a second component can be cited. For example, so-called $\beta-Zn_3(PO_4)_2$ can be stabilized by additions of $Mg_3(PO_4)_2$ or $Mn_3(PO_4)_2$, producing solid solutions which appear to be stable between room temperature and liquidus temperatures. Pure $\beta-Zn_3(PO_4)_2$ is only stable above $942°$.

$\alpha Sr_3(PO_4)_2$ is the stable form to $1320°$, when it inverts rapidly and reversibly to the β form which has the whitlockite $[Ca_3(PO_4)_2]$ structure. Additions of $Mg_3(PO_4)_2$, $Zn_3(PO_4)_2$, or $Ca_3(PO_4)_2$ will permit the existence of the β form at room temperature as a solid solution with the whitlockite structure. As always, it is difficult to decide whether the solid solution has produced 1) a truly stable β solid solution phase at room temperature or up to $1400°$ or 2) merely slowed the rate of the transition to the point where it is not perceptible in a "reasonable" period of time.

v. Solid Solutions

Cooling a sample through a two-phase region consisting of a solid solution and a liquid under equilibrium conditions involves a continuous change in the compositions of both the solid and the liquid phases. Diffusion in the solid state is very slow and thus extremely slow cooling may be required to maintain equilibrium.

Consider a melt of composition X in the binary system A–B (Figure 6.6) in which the components are completely miscible in both the liquid and solid states. Under equilibrium cooling conditions, the first crystals to precipitate from the melt will have the composition α_1 and the liquid will be of composition

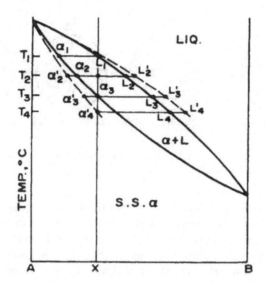

Figure 6.6. Non-equilibrium Cooling of a Melt of Composition X in a Solid Solution System

L_1. On cooling to temperature T_2, the overall composition of the crystalline phase will change from α_1 to α_2 (B atoms will diffuse into the existing crystals and A atoms will diffuse outward in order to accomplish this compositional change); similarly, the liquid composition will change from L_1 to L_2. Final solidification will take place at temperature T_3; the last melt to solidify will have the composition L_3 and the composition of the crystals will be that of α_3.

If the melt is cooled too quickly for equilibrium conditions to be maintained, the crystalline and liquid phases will develop compositional gradients, but owing to the faster diffusion of atoms in the liquid than in the solid, the gradient in the liquid would be considerably less than that in the solid. A schematic drawing of the microstructural and compositional changes which occur during non-equilibrium cooling is shown in Figure 6.7a, b. As the sample cools from temperature T_1 to T_2, the additional crystalline material either precipitates on the existing crystals of α_1 or forms new (equilibrium) crystals if conditions for nucleation are favorable. It is more probable that most of the growth will occur on the existing crystals. The outside layers of the crystal become progressively richer in B as the system attempts to reach the equilibrium compositions of the solid and liquid corresponding to temperature T_2. Concurrently, there is a diffusion of B atoms into the interior and a diffusion of A atoms toward the exterior owing to the compositional gradient in the crystal. Because equilibrium is not attained, a compositional gradient remains and the average composition of

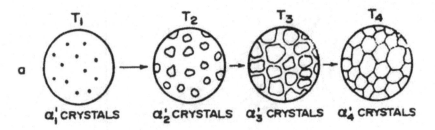

Figure 6.7a. Changes in Gross Microstructure During Non-equilibrium Cooling

the crystalline phase is deficient in B as shown by α'_2. The average composition of the liquid phase will be correspondingly slightly richer in B than predicted by the equilibrium phase diagram, as shown by L'_2. (If the system were held at this temperature for a length of time sufficient to allow the composition gradients to be eliminated, the composition of the crystalline phase would reach α_2 and that of the liquid would reach L_2).

On cooling to temperature T_3 the average composition of the solid and liquid phases will deviate from the equilibrium compositions as shown by the points α'_3 and L'_3. Under equilibrium conditions T_3 is the temperature of final solidification but as seen in Figure 6.6, under non-equilibrium cooling, a small quantity of melt remains until the sample temperature reaches T_4. Thus, when cooling is too rapid to permit attainment of equilibrium the temperature of final disappearance of the liquid is decreased and the crystals are of non-uniform composition (Figure 6.7a). The metallurgist refers to this phenomenon as "coring". To simplify the above discussion, it could be assumed that the diffusion in the liquid state was so fast that equilibrium could be maintained and that liquids L_2, L_3 and L_4 would coexist with non-equilibrium solids α'_2, α'_3 and α'_4. The vis-

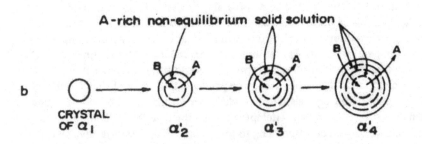

Figure 6.7b. Crystals of Average Composition During Non-equilibrium Cooling of Melt X. Outside Layers Become Richer in B than Inside Layers

cosity of some liquid metals may be low enough to permit equilibrium diffusion, but it is not likely to occur in highly viscous borate or silicate melts, especially the latter.

In solid solution systems undercooling normally precedes the start of freezing; that is, the liquid cools below L_1 before freezing starts and the first solid solution to appear may have a gross composition nearer to L_1 than α_1. However, the crystallization is usually accompanied by a release of the latent heat (recalescence) which serves to raise the temperature and counteract the original non-equilibrium effect. If a cored structure is found to be objectionable in a metal or ceramic process, it can be modified or eliminated by a homogenizing treatment at some lower temperature where diffusion of atoms or ions is still reasonably rapid. The concept of homogenizing or annealing treatments for the control of grain size and microstructure of solid solutions was highly developed in metallurgy (metallography) before it was much used in ceramics. However, since 1945, a large field of ceramography has developed and it is common practice to control microstructure and grain size in oxide and silicate products. Time, temperature and rates of transition of polymorphs are controlled to yield products with tailor made properties.

Stubican (16) has given some excellent examples of the development of metastable phases in the binary systems $CaO-Yb_2O_3$, $MgO-Al_2O_3$, TiO_2-SnO_2 and $Al_2O_3-Cr_2O_3$.

IV. THREE COMPONENT SYSTEMS

A. General

Chapter 5 shows that the number of possible equilibrium reactions in ternary systems is far greater than for binary systems and hence, the number of possible non-equilibrium reactions is correspondingly greater. Both types of reactions are also more complex in ternary systems with respect to kinetics and mechanisms.

i. Invariant Reactions

In the Rhines textbook (17), after detailed discussions of ternary systems with various types of solid solution, a list of ternary four-phase equilibrium is given on page 207, Table 4. The list contains a total of fifty-seven invariant reactions for three classes of four phase equilibria (F=4-4=0).

In ternary systems without solid solution, some of the more common invariant reactions are:

1. Eutectic $A + B + C \rightleftarrows L$
            $AB + B + C \rightleftarrows L$
            $A + AB + ABC \rightleftarrows L$
            $ABC + B + C \rightleftarrows L$

2. Peritectic

$$AB + A + B \rightleftarrows L$$
$$AC + BC + C \rightleftarrows L$$
$$AB + ABC + C \rightleftarrows L$$
$$ABC + A'B'C' + C \rightleftarrows L$$

3. Three Liquid Formation

$$A \rightleftarrows L_1 + L_2 + L_3$$
$$ABC \rightleftarrows L_1 + L_2 + L_3$$

4. Dissociation

$$ABC \rightleftarrows A + B + C$$

The above list is very limited and does not include solid solution systems, polymorphism of end-members, binary or ternary compounds or many of the fundamental types of invariant reactions. However, they are a few examples that metastable, non-equilibrium situations can occur if heating or cooling is too rapid.

ii. Monovariant Reactions

When $F=4-3=1$, many, many three phase monovariant equilibrium reactions can occur in systems which do not include solid solutions, but do contain phases such as three liquids, two liquids and a solid, one liquid and two solids, three solids and polymorphs. Once again, if heating or cooling is too rapid, many, many metastable, non-equilibrium situations can occur.

iii. Divariant Reactions

When $F=4-2=2$, many two phase equilibrium reactions can occur in ternary systems with no solid solution. Two liquids, a liquid and solid, two solid phases and polymorphs may be involved in equilibrium or non-equilibrium reactions.

Some specific examples of ternary non-equilibrium will be given in the following sections.

B. Non-equilibrium During Heating

i. Non-equilibrium Heating of a Mixture of Three Crystals
   in a Simple Eutectic System

Particles of three oxides are mixed as intimately as possible and heated slowly in an attempt to reach an equilibrium condition. However, even with extremely fine raw materials, the reaction between A and B, A and C and B and C starts at interfaces as described for two component systems. A-rich, B-rich and C-rich liquids will develop at interfaces or contact points according to the respective binary eutectic reactions. These binary eutectic liquids would then diffuse and interact to produce the ternary eutectic liquid, and, if the original composition was in the primary field of A, at some temperature all of the B and C crystals would dissolve to yield only A in equilibrium with eutectic liquid. It is easily seen that a non-equilibrium reaction could develop if insufficient time is allowed for solution of B and C and the diffusion in the liquid state. One final non-equilibrium assemblage would consist of crystals of A, B and C and a variety of non-equilibrium liquids. If crystals of C dissolve rapidly and with great ease, another non-equilibrium assemblage would consist of A, B and a different variety of non-equilibrium liquids. Finally, even if B and C rapidly and easily dissolved and lost their identity as crystals, A might well coexist with a variety of non-equilibrium liquids due to high viscosity and slow diffusion.

ii. Compound Formation

Ternary compound synthesis may be carried out by several methods. The simplest method might seem to be to intimately mix three extremely finely divided oxides and heat to a prescribed temperature. No matter how finely divided, the reaction will start at contact points or interfaces and unless sufficient time is allowed for SOLID STATE diffusion, the reaction to form a ternary compound may be incomplete. For example, β-eucryptite ($Li_2O \cdot Al_2O_3 \cdot 2SiO_2$) and β-spodumene ($Li_2O \cdot Al_2O_3 \cdot 4SiO_2$) are very easy to synthesize by direct reaction of Li_2CO_3, Al_2O_3 and SiO_2, due to the rapid diffusion and mineralizing action of lithia. The direct reaction of Na_2CO_3 or K_2CO_3 to form ternary sodium aluminosilicates or potassium aluminosilicates is much more difficult due to slow diffusion, especially in the case of the potassium compounds. In general, in the case of alkaline earth aluminosilicates, beryl ($BeO \cdot Al_2O_3 \cdot 6SiO_2$) cannot be made by dry reaction of oxides and cordierite is very difficult to synthesize and completely react, even when using natural raw materials such as clay and talc which introduce mineralizing impurities. Foster (18) has made a detailed analysis of the reactions which occur in the system $MgO-Al_2O_3-SiO_2$ which is instructive with respect to solid state reaction and liquid formation. Calcium, strontium and barium aluminosilicates form more easily than cordierite when using pure carbonates, Al_2O_3 and SiO_2 for the synthesis.

The preferential formation of binary compounds in a ternary oxide mixture is very common.

At times, it is advantageous to deliberately synthesize a binary compound as an intermediate in order to facilitate the formation of a ternary compound. For example, in making various apatites by dry reaction, the combination of three moles of orthophosphate with a fluoride or chloride at an appropriate temperature is a very efficient method of synthesis (19) of $3[Ca_3(PO_4)_2] \cdot CaF_2$, $3[Sr_3(PO_4)_2] \cdot SrF_2$, $3[Cd_3(PO_4)_2] \cdot CaCl_2$, etc. The two components react rapidly to form the apatite and little or no loss of the fluoride by oxidation or volatilization occurs.

Equilibrium reaction of three component oxide, fluoride or sulfide compounds to give essentially 100% yield can be facilitated by reaction of gaseous species, oxidation of metal powders, the use of gelatinous starting materials, confining vessels such as silica glass, gold or platinum alloys, and the use of pressure.

iii. Incongruently Melting and Dissociating Compounds

Like binary compounds, many incongruently melting and dissociating ternary oxide compounds can be superheated beyond their equilibrium decomposition point at one atmosphere pressure. Potash feldspar is a most notable case among silicate ternary compounds. $K_2O \cdot Al_2O_3 \cdot 6SiO_2$ melts at $1150 \pm 20°$ to leucite $(K_2O \cdot Al_2O_3 \cdot 4SiO_2)$ and liquid, but it is well known in the whiteware field that feldspathic bodies of various kinds (dental porcelain, hotel china, etc.) which have been heated for short times in the range 1200-$1400°C$ contain crystals of microcline or orthoclase when examined at room temperature.

iv. Polymorphic Inversions

Like binary compounds, many room temperature polymorphs of ternary compounds can easily be superheated beyond the equilibrium inversion temperature to the higher temperature polymorph. Like the binary compound zircon, $ZrSiO_4$, many ternary compounds in the form of 1) relatively pure natural minerals OR 2) extremely pure synthetic compounds can be superheated beyond their inversion temperature to their high temperature polymorphic structure.

C. Non-equilibrium During Cooling

i. Glass Formation

The number of commercial glasses made from one or two component oxide systems is very small. Silica glass is virtually the only one component glass which is useful and Vycor (96% SiO_2 -4% B_2O_3) is outstanding among the two component glasses. It is likely that several other two component glass systems are useful commercially for special purposes, but most of the large tonnage commercial glasses contain at least three components. Two of the most basic commercial glass systems are based on Na_2O-CaO-SiO_2 and Na_2O-B_2O_3-SiO_2. Increasing the number of components in a silicate, borate or phosphate system increases the limits of the glass-forming region and many exploratory studies of the "glass forming region" in ternary systems have been made. Addition of the third component not only increases the probability of glass forming capability, but also lowers the melting temperature and improves workability, durability, and other important chemical and physical properties. At this point the student is referred to several excellent general texts on glass such as Morey (20), Stanworth (21), Rawson (22), Doremus (23) and others for a systematic survey of composition, constitution and properties of glass-forming systems.

ii. Metastable Liquid Immiscibility

a. The Na_2O-CaO-SiO_2 System: The solid-lime-silica based glasses are the most historical type of glasses known around the world. The earliest glassmakers were Egyptians, Mesopotamians and Romans, dating back hundreds upon hundreds of years and until 1965, the most common glass compositions in the world were thought to be the most homogeneous compositions ever known. In 1965, Ohlberg, Golob, Hammel and Lewchuk (24), showed that the historical, homogeneous soda-lime-silica glass could be transformed to a metastable immiscible liquid structure if held for seventeen hours or more at temperatures below 701°C. Although it was known in 1927 that stable liquid immiscibility was present in the binary system $CaO-SiO_2$, it was not until about 1965 that the systems Na_2O-SiO_2 and $Na_2O-CaO-SiO_2$ were known to develop metastable immiscible liquid structures at appropriate temperatures and time.

b. *The $Na_2O-B_2O_3-SiO_2$ System*: The famous Pyrex glass was developed by W. C. Taylor of Corning Glass Works in the period 1915-1920 (approximately 82% SiO_2, 12% B_2O_3, 4% Na_2O, etc.). Vycor glass (approximately 96% SiO_2, 4% B_2O_3) was developed by Harrison Hood and Martin Nordberg (25,26) and patented in 1938.

Hood and Nordberg were researching the chemical durability of $Na_2O-B_2O_3-SiO_2$ glasses and the great solubility to acid attack of compositions in the area of 62% SiO_2, 28% B_2O_3 and 10% Na_2O led to the idea of making a specialized high silica glass. A glass of the approximate composition listed above was fabricated by several conventional glass-forming methods, heat treated around 580°C for substantial periods of time, treated with acid to remove essentially all of the Na_2O and all but approximately 4% B_2O_3. After acid leaching the surface area was so tremendous that reconstruction of a high density product could take place by reheating around 1250°C. After leaching, the product was partially opaque, but after reheating at 1250°C or there about, the product became transparent, like conventional glasses.

During the early years of production of Vycor, certain scientific people such as Dr. George W. Morey questioned the immiscibility of glasses in the system $Na_2O-B_2O_3-SiO_2$ and felt that extremely finely divided crystalline material was being leached out of the system to produce the final 96% SiO_2, 4% B_2O_3 glass composition.

However, after the many studies of metastable liquid immiscibility in the systems $Na_2O-B_2O_3$, Na_2O-SiO_2 and $Na_2O-B_2O_3-SiO_2$, it is now certain that the heat treatment in the neighborhood of 580°C creates liquid phase separation and the generation of a stable silica-rich phase and a very acid soluble Na_2O, B_2O_3-rich phase. (Chapter 4, Appendix A, Table 4.5, especially references 3, 4, 9, 21, 23, 24, 26, 27, 30, 36, 37, 50, 60, 67 and 70).

Although it is now known that Pyrex type glasses could become immiscible if held sufficiently long at an appropriate temperature, it is the Vycor glass which is most significant commercially and technologically because it is necessary to heat treat to produce immiscibility in order to create commercial products of 96% SiO_2, 4% B_2O_3 which in many cases are more suitable than fused SiO_2 due to the crystallization characteristics in the 900-1250°C region.

iii. Non-equilibrium Crystallization Paths

Chapter 5 was almost entirely devoted to a discussion of equilibrium crystallization paths in some of the more common types of ternary systems which contained primary fields of end members, binary and ternary compounds and regions of liquid immiscibility. Some of the more common cases of non-equilibrium crystallization will now be discussed, using the same examples given in Chapter 5.

Non-equilibrium crystallization of a liquid may arise even in a simple eutectic type system. Referring to Figures 5.6 and 5.8, equilibrium crystallization of C and then A and C occurs as the liquid composition moves into the eutectic E. In the primary of C, it is possible that cooling would be so rapid that not all of crystalline C would precipitate as liquid X moved to 6'. As the liquid moved from 6' to E, it is again possible that non-equilibrium crystallization of C and A would occur. It is even possible that no C, no A or no C and A would crystallize from 6' to E. Let us now assume that the rate of cooling is too rapid for the perfect equilibrium crystallization of A, B, and C at the eutectic. If cooling is so fast that no B crystallizes, the final non-equilibrium assemblage will consist of crystals of C and A and a liquid (perhaps a glass) of the eutectic composition. The amounts of A and C in the assemblage will be less than those present after a perfect equilibrium crystallization.

Figure 5.17 can be used to illustrate several important non-equilibrium situations which may arise during cooling of liquid Z. As the liquid composition arrives at the boundary line pP, crystals of B should begin to dissolve or resorb as the temperature decreases along the boundary line. Let us assume that a very refractory crystal such as Al_2O_3 or ZrO_2 had been crystallizing in the primary field labeled B. At Z', these crystals would be reluctant to dissolve as the temperature decreases from Z' to Z". If none of B dissolved as the liquid moved from Z' to Z", and if equilibrium crystallization occurred from Z" to E, the final assemblage would contain crystals of A, C, AB and B. During perfect equilibrium crystallization, only A, C, and AB would be contained in the final assemblage.

A similar situation may arise during crystallization of liquid W in Figure 5.21. Due to the fact that liquid W is on the join C–AB, the final equilibrium assemblage should be 40% C and 60% AB. During perfect equilibrium crystallization at P, all of the B which has crystallized should disappear. If B is a very refractory crystal such as Al_2O_3 or ZrO_2, perhaps only little or no B will dissolve or resorb as the peritectic reaction goes on. The final assemblage would then contain three phases, C, AB and B. It might even be possible to have an assemblage of four phases in the final mixture. Assume a non-equilibrium reaction at P which would mean that crystals of B, AB and C would coexist with liquid as it moved toward E. If cooling was so rapid at E that no A was permitted to crystallize, one would be left with crystals of B, AB and C and a liquid (perhaps glass) in the final mixture. If A did crystallize at E, a crystalline four phase assemblage would coexist, A, B, C and AB.

One final variation might be possible. Suppose that some A crystallized from the liquid at E, but not the amount that would ordinarily appear during perfect equilibrium crystallization. The final mixture would then contain five phases, crystals of B, C, AB, A and liquid (perhaps glass).

As an exercise, this kind of analysis should be applied to other systems where crystallization, solution or resorbtion takes place as the temperature decreases, as for example in Figures 5.28, 5.30, 5.33, 5.34, 5.41 and 5.73. The resorbing process may take place along a boundary line or during an invariant reaction at a peritectic or reaction point.

iv. Polymorphic Inversions and the Influence of Pressure

Numerous examples of the metastable existence of a high temperature polymorph of a ternary compound at room temperature can be cited.

Starting with the alkali aluminosilicates, it is well known that β-eucryptite ($Li_2O \cdot Al2O_3 \cdot 2SiO_2$) and β-spodumene ($Li_2O \cdot Al_2O_3 \cdot 2SiO_2$) can exist indefinitely at room temperature and one atmosphere pressure. However, at the appropriate temperature and pressure each of these minerals can be converted to the stable low temperature (α) forms, coinciding with the natural minerals which occur in the earth.

Munoz (27) has determined the stability relations for the bulk composition $Li_2O \cdot Al_2O_3 \cdot 4SiO_2$ as shown in the p-T diagram of Figure 6.8. Note the large stability region of low temperature (α) spodumene, the smaller stability region of β-quartz solid solutions, the still smaller stability region for β-spodumene, and the coexistence of β-spodumene solid solutions and β-quartz solid solutions. A detailed discussion of these relationships is postponed until solid solution in ternary systems has been treated in Chapter 7.

Nepheline, $Na_2O \cdot Al_2O_3 \cdot 2SiO_2$, is a naturally occurring mineral which inverts at $1248°$ to a form known as carnegieite, which melts at $1526°$ (at one atmosphere). Once carnegieite is formed above $1248°$, it does not revert to nepheline during cooling at one atmosphere. Instead, it inverts to a low temperature form around $705°$, which is then metastable with respect to nepheline at room temperature and one atmosphere.

It should be noted that the mineral jadeite ($Na_2O \cdot Al_2O_3 \cdot 4SiO_2$) does not appear in the stability diagram for the system $Na_2O-Al_2O_3-SiO_2$ at one atmosphere (see 1964 collection of diagrams, pages 181-183). The p-T diagrams (28,29) for $NaAlSi_2O_6$ and $NaAlSi_3O_8$ (albite) shown in Figures 6.9 and 6.10 show the relationship between these two phases as well as their relationship to nepheline and coesite.

The high temperature form of cordierite ($2MgO \cdot 2Al_2O_3 \cdot 5SiO_2$) which is developed during firing of ceramic bodies is metastable with respect to a low

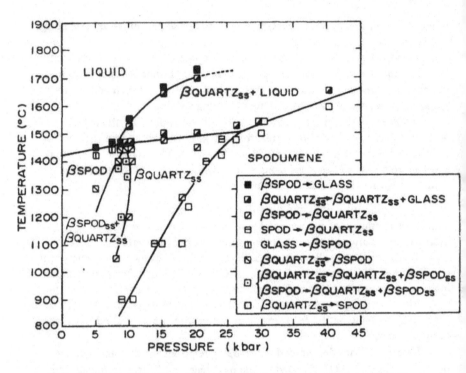

Figure 6.8. P–T Stability Relations for Bulk Composition LiAℓSi$_2$O$_6$. (Li$_2$O·Aℓ$_2$O$_3$·4SiO$_2$)

temperature form (β) at room temperature and one atmosphere. However, the high temperature (α) form is so persistent that its metastable existence has virtually no effect on its use as a semi-refractory or special purpose ceramic body.

The above discussion is intimately related to the mechanisms involved in polymorphic inversions of one, two and three component oxide compounds. This subject is beyond the scope of this text, but many original papers and reviews have been written on the nature of transitions and transformations. One of the most recent is by Wang and Gupta (30) in which many references to classical papers are given.

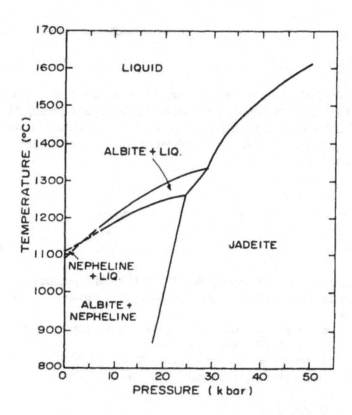

Figure 6.9. P–T Plane for Jadeite, Albite and Nepheline

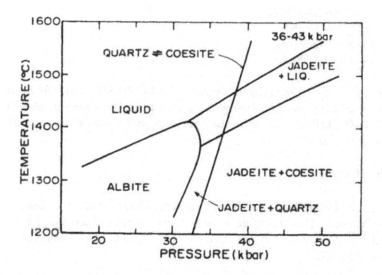

Figure 6.10. P–T Plane for Albite Between 36–43 kbars; Quartz-Coesite Equilibrium; Albite Melting Curve

PROBLEMS

Refer to the problems at the end of Chapter 5 (Sections I, II and III) and:
1. Use Figures 1-5 in Section I on "abstract systems" and develop NON-
   EQUILIBRIUM paths for each composition. For example, using Figure 1, it
   was requested that equilibrium crystallization paths be traced for the following compositions:

1	2	3	4
10% A	15% A	20% A	5% A
10% B	10% B	15% B	25% B
80% C	75% C	65% C	70% C

For each Figure (1-5), develop NON-EQUILIBRIUM crystallization events
for the compositions listed for EQUILIBRIUM crystallization and isoplethal
analyses.
   Choose many other UNLISTED compositions in Figures I-V and develop NON-EQUILIBRIUM situations.
2. Use the following "real" systems in Section II and develop NON-EQUILIB-
   RIUM paths, isoplethal analyses and quantitative calculations:

   a. $K_2O-MgO-SiO_2$
   b. $Na_2O-CaO-SiO_2$
   c. $BaO-B_2O_3-SiO_2$
   d. $Na_2O-Fe_2O_3-SiO_2$
   e. $CaO-A\ell_2O_3-SiO_2$
   f. $CaO-ZnO-SiO_2$
   g. $CaO-TiO_2-SiO_2$
   h. $CaO-MgO-SiO_2$

3. After constructing the complex ternary EQUILIBRIUM diagrams for Section
   III, select any number of ternary compositions and develop any number of
   NON-EQUILIBRIUM crystallization paths, isoplethal analyses and quantita-
   tive calculations.

REFERENCES

1. Ernest M. Levin, Carl R. Robbins and Howard F. McMurdie, Phase Diagrams
   for Ceramists, 1964, compiled at the National Bureau of Standards, Edited
   and Published by the American Ceramic Society, 65 Ceramic Drive, Columbus, Ohio, 43214.

2. J. F. Schairer and N. L. Bowen, The System $K_2O-Al_2O_3-SiO_2$, Am. J. Science, 253, 681-746, 1955.

3. Brian Mason, Lunar Tridymite and Cristobalite, Amer. Min., 57, Nos. 9 and 10, page 1530, Sept.-Oct., 1972.

4. Frank Dachille, Robert J. Zeto and Rustum Roy, Coesite and Stishovite: Stepwise Reversible Transformation, Science 140, 3570, pages 991-993, May, 1963.

5. Paul D. Gigl and Frank Dachille, Effect of Pressure and Temperature on the Reversal Transitions of Stishovite, Meteoritics, 4 (2), pages 123-135, October, 1968.

6. K. S. Masdiyasni, C. T. Lynch and J. S. Smith, Preparation of Ultra-High Purity Submicron Refractory Oxides, J. Am. Ceramic Soc., 48 (7), 372-375, 1965.

7. K. S. Masdiyasni, C. T. Lynch and J. S. Smith, Metastable Transitions of Zirconium Oxide Obtained from Decomposition of Alkoxides, J. Am. Ceramic Soc., 49 (5), 286-287, 1966.

8. J. Williamson and F. P. Glasser, Phase Relations in the System $Na_2Si_2O_5-SiO_2$, Science 148, [3677], pages 1589-1591, June, 1965.

9. A. R. West and F. P. Glasser, Crystallization of Lithium Trisilicate, $Li_2Si_3O_7$, from Li_2O-SiO_2 Melts, Mat. Res. Bull, 5 (9), pages 827-842, 1970.

10. William C. Butterman and Wilfrid R. Foster, Zircon Stability and the ZrO_2-SiO_2 Phase Diagram, Amer. Min., 52, Volume 5-6, page 880, May-June, 1967.

11. Ernest M. Levin, Liquid Immiscibility in Oxide Systems, Chapter 5 of Volume 3, Phase Diagrams, Materials Science and Technology, Academic Press, 1970.

12. R. R. Shaw and D. R. Uhlmann, Subliquidus Immiscibility in Binary Alkali Borates, J. Amer. Cer. Soc., 51 (7), pages 377-381, 1968.

13. D. Viechnicki and V. S. Stubican, Mechanism of Decomposition of the Cubic Solid Solutions in the System ZrO_2-MgO, J. Amer. Cer. Soc., 48 (6), page 292, 1965.

14. B. S. R. Sastry and F. A. Hummel, Studies in Lithium Oxide Systems: V, $Li_2O-Li_2O\cdot B_2O_3$, J. Amer. Cer. Soc., 42 (5), pages 216-218, May, 1959.

15. B. S. R. Sastry and F. A. Hummel, Studies in Lithium Oxide Systems: I, $Li_2O\cdot B_2O_3-B_2O_3$, J. Amer. Cer. Soc., 41 (1), pages 7-17, January, 1958.

16. V. S. Stubican, Metastable Phases in Some Oxide Systems, Colloques Internationaux, C.N.R.S., No. 205, Etude des Transformations Crystallines a Haute Temperature, pages 447-451, 1972.

17. Frederick N. Rhines, Phase Diagrams in Metallurgy, McGraw-Hill Book Co., 1956.

18. W. R. Foster, Contribution to the Interpretation of Phase Diagrams by Ceramists, J. Amer. Cer. Soc., 34 (5), 151-160, May, 1951.

19. Eric R. Kreidler and F. A. Hummel, The Crystal Chemistry of Apatite; Structure Fields of Fluor- and Chlorapatite, Amer. Min., Volume 55, pages 170-184, Jan.-Feb., 1970.

20. G. W. Morey, The Properties of Glass, 2nd Ed., 591 pp., Reinhold, New York, 1954.

21. J. E. Stanworth, Physical Properties of Glass, 224 pp., Clarendon Press, Oxford, England, 1950.

22. H. Rawson, Inorganic Glass Forming Systems, 317 pp., Academic Press, 1967.

23. R. H. Doremus, Glass Science, Wiley-Interscience, N.Y., 1973.

24. S. M. Ohlberg, H. R. Golob, J. J. Hammel and R. R. Lewchuk, Noncrystalline Microphase Separation in Soda-Lime-Silica Glass, J. Amer. Cer. Soc., 48 (6), 331-332, 1965.

25. H. P. Hood and M. E. Nordberg, Treated Borosilicate Glass, U.S. Patent 2,106,744, 1938.

26. H. P. Hood and M. E. Nordberg, Borosilicate Glass, U.S. Patent 2,221,709, 1940.

27. James L. Munoz, Stability Relations of $LiAlSi_2O_6$ at High Pressures, Special Publications No. 2, Pyroxenes and Amphiboles: Crystal Chemistry and Phase Petrology, Mineralogical Society of America, 1969.

28. P. M. Bell, Carnegie Inst. Washington, Yearbook, 63, 172, 1963-64.

29. P. M. Bell and E. H. Roseboom, Jr., Carnegie Inst. Washington, Yearbook, 64, 140, 1964-65.

30. F. Y. Wang and Kedar P. Gupta, Transformation in the Oxides, Metallurgical Transactions 4, 2767-2779, December, 1973.

7

SOLID SOLUTION IN TERNARY SYSTEMS

I. INTRODUCTION

Historically, solid solution was so common in commercial metallurgical systems that early phase literature in metallurgy almost disregarded the existence of the intermetallic compound. Of course, compound formation and solid solution are equally important in both metal and ceramic systems as far as basic phase relations are concerned.

At this point it should be emphasized that due to the importance of solid solution in metallurgical systems (binary, ternary, quaternary, etc.), the excellent books of two outstanding metallurgists ("Phase Diagrams in Metallurgy" by Frederick N. Rhines and "Alloy Phase Equilibria" by A. Prince) contain the most basic, the greatest amount, and the most essential information on solid solution systems. Although this chapter will attempt to provide sufficient information for ceramists on ternary solid solution systems, it will not be qualitatively or quantitatively as significant as the books by Rhines and Prince.

In Chapter 3 it was shown that solid solution in binary systems could be complete or partial and that considerable complexity could result if the end members or intermediate solid solution phases existed in two or more polymorphic forms. More complexity could result from solid solution unmixing or "exsolution".

Ternary systems can also be composed entirely of solid solution phases; that is, a ternary system may be bounded by binary systems in which solid solution is either complete or partial. One must refer to Chapter 3 for 1) the liquidus relationships among the three types of complete solid solution systems and the two types of partial solid solution, 2) the effect of polymorphic inversion at subsolidus temperatures which leads to eutectoid and peritectoid configurations, and 3) unmixing or exsolution relationships.

On the other hand, ternary systems which contain both solid solution and intermediate compound formation are common. They would combine the phenomena and configurations to be discussed in this chapter with those which have already been discussed in Chapter 5.

II. THE COMPLETE SOLID SOLUTION SYSTEM;
THE EQUILIBRIUM CRYSTALLIZATION PATH

The space model for the complete solid solution or ternary isomorphous system is shown in Figure 7.1. It consists of a homogeneous liquid region, a homogeneous solid solution region and a lens-shaped region of coexistence of liquid and solid solution phases. It would be possible to have a ternary maximum or ternary minimum point on the liquidus surface, but this is not shown in Figure 7.1. Two general isothermal sections which might be drawn for this system are shown in Figure 7.2 (a and b).

When discussing the equilibrium crystallization path of any liquid in this system, it is important to realize that ternary liquids will be in equilibrium with ternary solid solutions as the path proceeds through the two phase region. That is, the solid solutions as well as the liquids contain all three components and therefore, in general, the path of the liquid and the path of the solid solution will be some kind of CURVED path. In Chapter 5 it was obvious that the initial path of the liquid was a straight line path as crystallization began and continued in the primary field of one of the components or the primary field of a binary or ternary compound. The composition of the crystallizing solids remained constant in these cases.

In ternary solid solution systems, as in binary solid solution systems, the composition of the crystallizing solid changes as the composition of the liquid changes and in general, the liquid path is then forced into a curvature dictated by the exact compositions involved.

The fundamental diagram for crystallization of any liquid selected from Figure 7.1 is shown in Figure 7.3. The first solid solution to crystallize from liquid X will be γ_1 in an infinitesimal amount. As the temperature is lowered, and as the liquid proceeds to L_2, under perfect equilibrium conditions, it will co-exist with solid solution γ_2. The combination $L_3\gamma_3$ represents another intermediate stage and at L_4, the last trace of liquid will disappear as heat is withdrawn from the system and a solid solution of composition X (γ_4) will remain. Note that all of the tie lines run through X, the original liquid composition, when conditions for equilibrium are perfect.

If conditions for perfect equilibrium are not met, that is, if cooling is too rapid to permit complete equilibrium diffusion in both liquid and solid, some

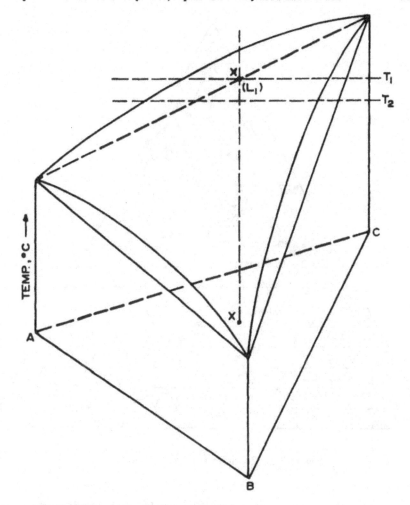

Figure 7.1. Solid Model for Complete Solid Solution (No Maximum or Minimum in Liquidus Surfaces)

non-equilibrium paths will result as shown by the dashed lines in Figure 7.3. This figure assumes that the final trace of liquid will have the composition L_4, but that the final solid solution composition to precipitate will be different from $X (\gamma_4)$, say γ_5.

Variations on the configuration shown in Figure 7.1 occur when any or all of the bounding binary systems have maxima or minima in their liquidus curves.

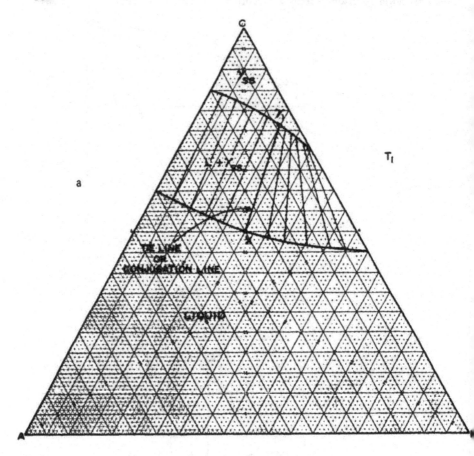

Figure 7.2a. Isothermal Section, Temp., T_1

The number of types of ternary systems with complete solution relationships would be as follows with respect to bounding binary relationships:

1. Three bounding binary systems with no maximum or minimum in the liquidus (Figure 7.1).
2. Three bounding binary systems, each with a maximum in the liquidus.
3. Three bounding binary systems, each with a minimum in the liquidus.
4. Two bounding binary systems with no maximum or minimum and one system with a maximum.
5. Two bounding binary system with no maximum or minimum and one system with a minimum.

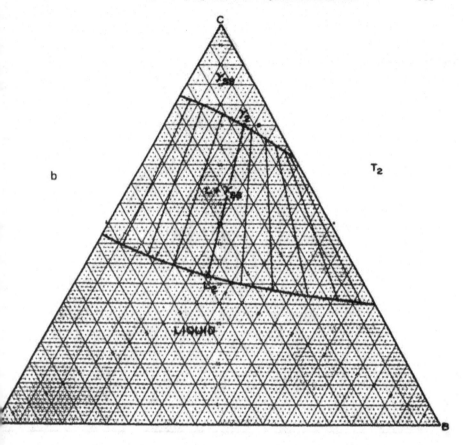

Figure 7.2b. Isothermal Section, Temp., T_2

6. One bounding binary system with no maximum or minimum and two systems with a maximum.
7. One bounding binary system with no maximum or minimum and two systems with a minimum.
8. One bounding binary system with no maximum or minimum, one system with a maximum and one system with a minimum.
9. Two bounding binary systems with a maximum and one with a minimum.
10. Two bounding binary systems with a minimum and one with a maximum in the liquidus surface.

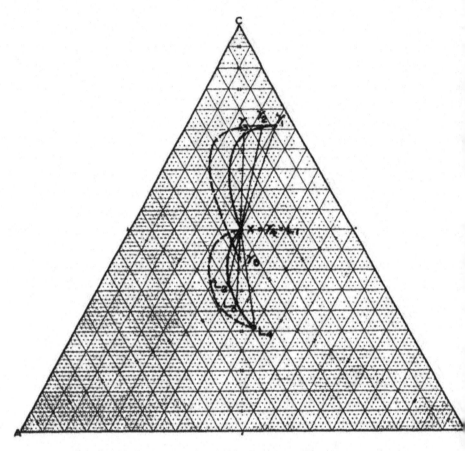

Figure 7.3. Equilibrium and Non-equilibrium Crystallization Path for Liquid X

These events would not change the fundamental concepts involved in the crystallization path, as described for Figure 7.3, but systems of type 9 or 10 may often create a specialized liquidus surface with a "saddle" point as shown in Figure 7.4. As mentioned previously, it is always possible for a ternary maximum or minimum point to exist in the liquidus surface, as well as a saddle point.

The student should try to construct various vertical slices through the solid model shown in Figure 7.1.

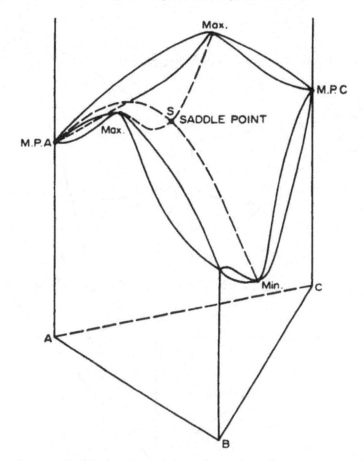

Figure 7.4. Saddle Point on Ternary Liquidus Surface of Complete Solid Solution System; Systems A–B and A–C with Maximum in Liquidus and System B–C with Minimum in Liquidus

III. COMPLETE SOLID SOLUTION IN ONE BINARY SYSTEM COMBINED WITH TWO SIMPLE EUTECTIC SYSTEMS; ISOTHERMAL PLANES; THREE PHASE BOUNDARY LINES; QUANTITATIVE ANALYSIS OF THE CRYSTALLIZATION PATH (ISOPLETHAL ANALYSIS); THREE PHASE TRIANGLES

The solid model for this system is shown in Figure 7.5 and two associated isothermal sections are shown in Figures 7.6 and 7.7.

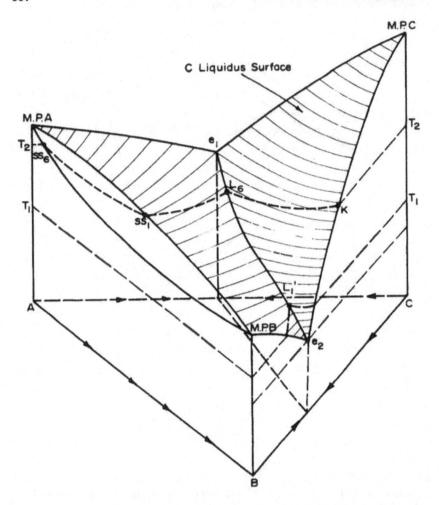

Figure 7.5. Solid Model for Binary Solid Solution System (no maximum or minimum in liquid surface) Combined with Two Simple Eutectic Binary Systems

Note that the isothermal section at T_2 (Figure 7.6) contains one-three phase region, three two phase regions and a region of homogeneous liquid. The isothermal plane has intersected the boundary line $e_1 e_2$ at L_6 which is known from experimental data to be in equilibrium with SS_6 at T_2. This creates a three phase triangle involving C, SS_6 and L_6.

The tie lines in the region bounded by C, A, and SS_6 show that all compositions in this region at T_2 will be mixtures of C and solid solutions ranging from

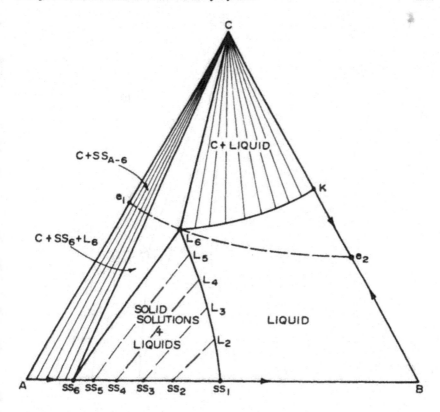

Figure 7.6. Isothermal Section at T_2

A to SS_6. The region C + Liquid is a type which was much discussed in Chapter 5 where tie lines radiate from C (whose composition is constant) to various points on the saturation curve, KL_6. In contrast to this region is the one marked solid solutions plus liquids where a series of variable crystalline compositions (solid solutions) are in equilibrium with a variety of liquids ranging from L_2 to L_6. The remainder of the system is liquid at T_2.

The isothermal section shown in Figure 7.7 is very similar, except that at the lower temperature T_1, the two phase region bounded by C, A and SS1′ is now much larger, the three phase region is composed of C, SS1′ and L1′ and the liquid region is much smaller. The characteristic region of solid solutions plus liquids now involves solid solutions whose compositions run from SS1′ to B and liquids L1′, L3′, L5′, etc.

It is now possible to trace the equilibrium crystallization path of a liquid X which lies in the primary field of the solid solutions. It must be emphasized

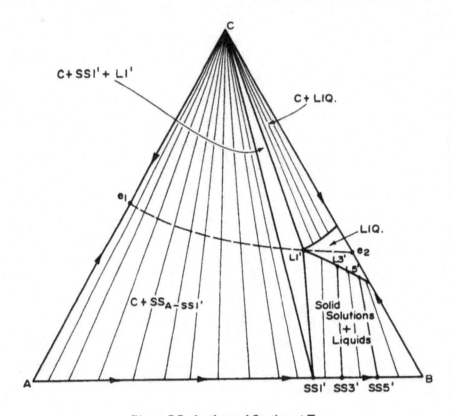

Figure 7.7. Isothermal Section at T_1

that the liquid is a ternary liquid containing a certain amount of A, B and C, but that (in this special case) the solid solutions are binary and contain only A and B. Figure 7.8 shows liquid composition X, its crystallization path which ends at L_d, boundary line $e_1 e_2$, various tie lines and various three phase boundary lines. A three phase boundary line is a special kind of tie line whose one extremity lies on the boundary line $e_1 e_2$.

It is now imperative to explain how the liquid X gets from its starting point to L_d on the boundary line $e_1 e_2$, where all liquid disappears and leaves a final crystallized mixture of C and the solid solution labeled $SS_{FINAL}(SS_d)$.

As the temperature of liquid X drops and encounters the liquidus surface (Figure 7.9), an infinitesimal amount of solid solution SS_x is precipitated. The composition of SS_x cannot be predicted, but must be known purely from experimental data which has been previously accumulated on the binary system

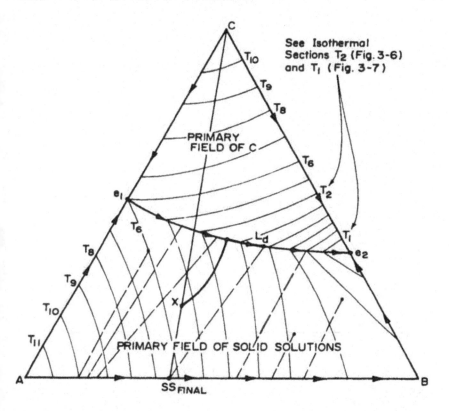

Figure 7.8. Projection of Liquid Surface on a Two-Dimensional Plane Showing Isotherms, Boundary Line $e_1 e_2$, Three Phase Boundary Lines, Tie Lines and a Crystallization Path for Liquid X

A-B. SS_x can be identified and characterized based on its optical and x-ray properties and perhaps other properties.

As the temperature is lowered, the path of the liquid moves from X to L_a as shown and the solid solution SS_a is now in equilibrium with L_a. The path is curved because the composition of the solid which precipitates from the liquid is VARIABLE (not constant as in all cases discussed in Chapter 5). Solid solutions of composition SS_x to SS_a crystallize from liquids of composition X to L_A as the temperature decreases. The percentage of solid solution SS_a at L_a is given by $\frac{XL_a}{SS_a L_a} \times 100$ and the percentage of liquid by $\frac{XSS_a}{SS_a L_a} \times 100$. The compositions of each are read directly from a ruled diagram.

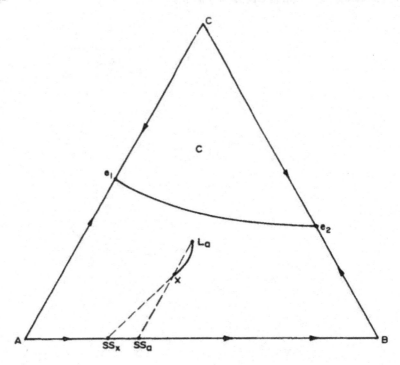

Figure 7.9. Crystallization Path for Liquid X From X to L_a. Quantitative Analysis at L_a

    This process continues as the liquid path proceeds toward the boundary line $e_1 e_2$ and at the moment the path reaches the boundary line, the liquid L_b and the solid solution SS_b are in equilibrium with an infinitesimal amount of C. (Figure 7.10)

    As heat is withdrawn, the liquid path now moves down boundary line $e_1 e_2$ (Figure 7.11) and at a temperature corresponding to liquid composition L_c, a three phase triangle may be constructed with apices L_c, SS_c and C. The percentages of liquid and solid in the mixture at this temperature are given by $\frac{KX}{KL_c} \times 100$ and $\frac{L_c X}{KL_c} \times 100$, respectively. Furthermore, the amounts of C and SS_c in the solid mixture are given by $\frac{KSS_c}{CSS_c} \times 100$ and $\frac{KC}{CSS_c} \times 100$, respectively.

    The final "drying up" point or the point at which all liquid disappears occurs when the solid solution composition SS_d, X and C all fall on a straight line, which is one leg of the three phase triangle whose apices are L_d, SS_d and C.

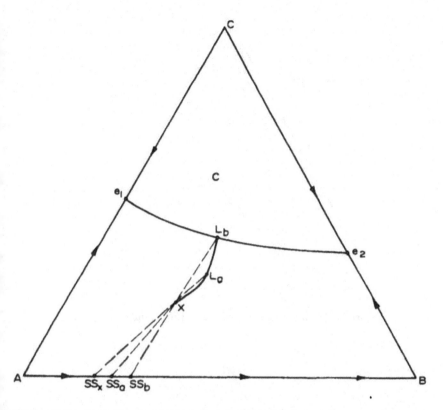

Figure 7.10. Crystallization Path at the Moment it Encounters Boundary Line $e_1 e_2$. (L_b)

(Figure 7.12) After all the liquid has disappeared, the composition of the solid mixture is given by $\dfrac{XSS_d}{CSS_d} \times 100$ (amount of C) and $\dfrac{XC}{CSS_d}$ (amount of SS_d).

Once again it should be emphasized that this process of equilibrium crystallization assumes perfect diffusion in both the solid and liquid by infinitesimal increments as the temperature is lowered to the point where all liquid has disappeared.

For an extended discussion of a ternary solid solution system containing a binary continuous solid solution with a minimum in the liquidus, the classical paper of Osborn and Schairer (1) should be consulted. Fractionation curves and the general principles involved in the interpretation of this type of system are discussed as well as the course of equilibrium crystallization curves.

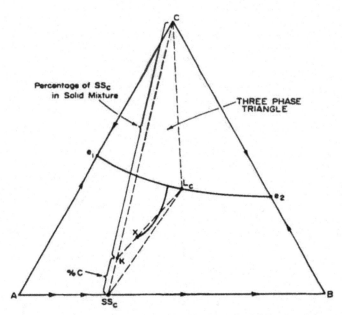

Figure 7.11. Quantitative Analysis of the Equilibrium at L_c

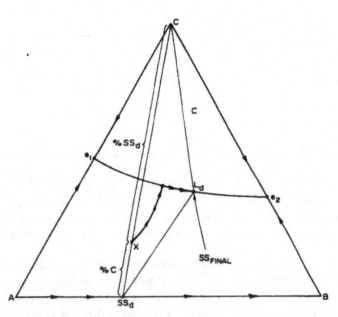

Figure 7.12. Final "Drying Up" Point (L_d); Equilibrium Crystallization Path of Liquid X; Final Three Phase Triangle

IV. TERNARY SOLID SOLUTIONS CRYSTALLIZING FROM TERNARY LIQUIDS; ONE COMPLETE SOLID SOLUTION SYSTEM, TWO PARTIAL SOLID SOLUTION SYSTEMS; CRYSTALLIZATION PATH; ISOTHERMAL AND VERTICAL SECTIONS

A variation of the above system is now presented. The only fundamental difference between the two systems lies in the fact that, in the system to be discussed, the crystallizing solid solutions are TERNARY; that is, like the liquids from which they are crystallizing, they contain all three end members, A, B and C.

In the ternary system shown in Figure 7.13, there is complete solid solubility between components A and B to form the solid solution phase α. In the binary system A-C, there is partial solubility of C in A but no solution of A in C. Similarly, in the binary side B-C, C is partially soluble in B but B is not soluble in C. Thus, there is only one solid solution (Phase α) in the ternary system, and its composition contains A, B, and C (except, of course, those compositions which fall in the binary system A-B).

The binary systems A-C and B-C have eutectics at e_1 and e_2, respectively, which are joined by a three-phase boundary line. Along e_1-e_2, liquid is in equilibrium with crystals of C and α solid solutions. A plane projection of the liquidus surface is shown in Figure 7.14.

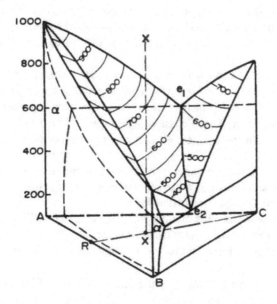

Figure 7.13. Ternary Solid Solutions Crystallizing From Ternary Liquids

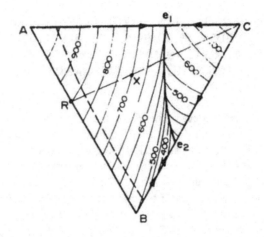

Figure 7.14. Plane Projection of Liquid Surface of System Shown in Figure 7.13

The system is composed of six state spaces: α, Liquid, α + Liquid, C + Liquid, α + C + Liquid, and α + C. An isothermal plane passed through the figure at 550° intersects all of the six spaces, as shown in Figure 7.15 and Figure 7.16. In the two-phase regions, C + L and C + α, tie lines can be drawn from C to various points on the curves representing the compositions of the liquid and solid solutions, respectively. However, in the two-phase region, α + L, where both phases have variable composition, the tie lines must be determined experimentally.

The three-phase region, α + C + L, is called a *tie triangle*; it is composed of three tie lines. All composition points which fall within the triangle are composed of various proportions of the three phases represented by the apices of the triangle. The proportion of each phase is determined by the position of the composition point within the triangle, as discussed for composition triangles in Chapter 5.

Within the two single-phase regions, α and L, compositions are determined in terms of A, B, and C; that is, these compositions can be read directly from a properly ruled ternary equilateral triangle.

In order to determine the cooling path for the melt or solid portion of a given sample, it is necessary to have isothermal sections in which the conjugate phases are shown by tie lines. The following isoplethal study will illustrate the use of such sections.

The composition represented by the point X shown in Figures 7.13 and 7.14 has been selected for analysis. Note that on cooling, this isopleth will intersect the liquidus surface at 700°. At all temperatures above 700°, the sample will

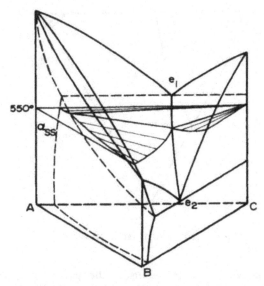

Figure 7.15. Intersection of Isothermal Plane at 550°C

consist of liquid only. For example, in Figure 7.17 an isothermal section at 750°
shows the point X to be located in the area of liquid only.

At 700° (Figure 7.18) the point X intersects the boundary of the α + L
area. Thus, the sample is composed of liquid of composition X and an infinitesimal quantity of crystals of α solid solution of a composition given by the end of
the tie line designated 1′ in the diagram.

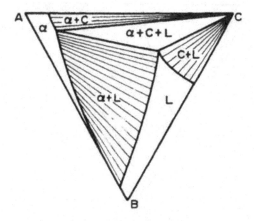

Figure 7.16. Isothermal Section at 550°C, Tie Lines in Two Phase Regions

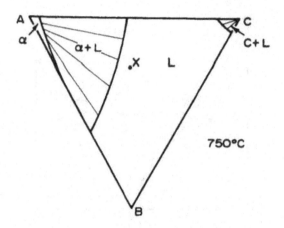

Figure 7.17. 750°C Isothermal Section

At 650° (Figure 7.19) point X intersects the space of α + L and falls on the tie line 2'-2 which gives the composition of the conjugate phases. The proportions of α and liquid are determined by applying the lever rule. For example, the percentage of liquid is given by $\frac{X2'}{2'2} \times 100$, and the percentage of α is given by $\frac{X2}{2'2} \times 100$.

At 600° (Figure 7.20) the point X falls on the tie line 3'-3. Note that the compositions as well as the relative proportions of α and liquid are changing as the sample cools.

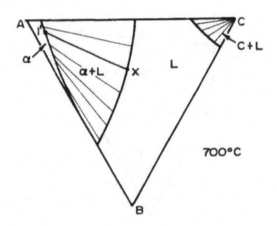

Figure 7.18. 700°C Isothermal Section

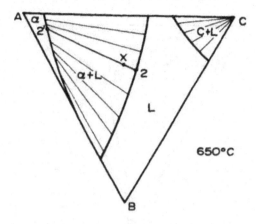

Figure 7.19. 650°C Isothermal Section

At 500° (Figure 7.21) the point X falls on the tie line 4–4′. The composition of the melt lies on the three-phase boundary line e_1 –e_2 ; thus, an infinitesimal quantity of crystals of C will be present at this temperature in addition to α and liquid.

At 475° (Figure 7.22) the point X falls within the tie triangle 5′–C–5; therefore, α + C + L are the conjugate phases. Their proportions are determined by the position of point X within the triangle. For example, the tie line 5–P–5″ can be constructed as shown. The percentage of melt of composition 5 is $\dfrac{5''X}{5''5}$ X

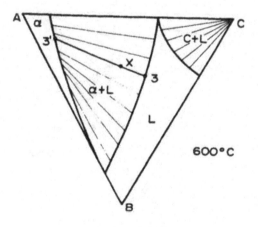

Figure 7.20. 600°C Isothermal Section

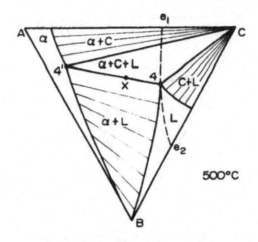

Figure 7.21. 500°C Isothermal Section

100 and the proportion of the sample which is solid is $\frac{5X}{5''5} \times 100$. The solid portion, however, is composed of α solid solution and C in the following proportions: % $\alpha = \frac{5''C}{5'C} \times 100$, and the % of C $= \frac{5'5''}{5'C} \times 100$.

At 450° (Figure 7.23) the point X falls on the join 6'–C which is a side of
the tie triangle 6'–C–6. The sample is composed of α solid solution and C. The

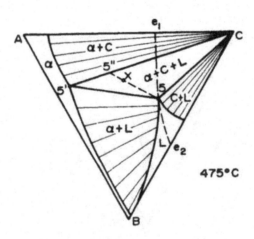

Figure 7.22. 475°C Isothermal Section

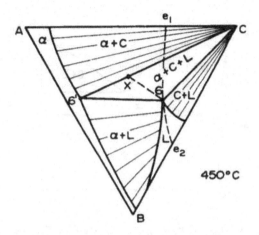

Figure 7.23. 450°C Isothermal Section

composition of the last trace of liquid to solidify is given by point 6. The proportion of α in the sample is $\frac{XC}{6'C} \times 100$ and the percentage of C is $\frac{6'X}{6'C} \times 100$. The sample is now completely solid.

These changes in composition of the solid and liquid portions of the sample during cooling are summarized in Figure 7.24. The liquid followed the path X–2–3–4–5–6 and the solid followed the path 1'–2'–3'–4'–5'''–X.

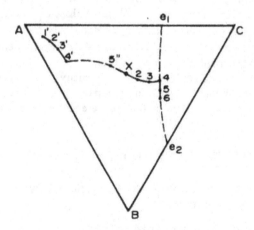

Figure 7.24. Cooling Paths of Liquid and α Solid Solutions for Composition X

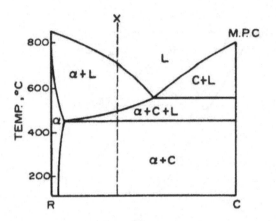

Figure 7.25. Vertical Section C–X–R (see Figs. 7.13 and 7.14)

A vertical section through the system of Figures 7.13 and 7.14 is shown in Figure 7.25. The section was taken on a line from C through the point X to its intersection along the side A-B. The position of the isopleth X is shown by the dashed line.

In order to demonstrate isoplethal analysis and crystallization paths in ternary systems in Sections II, III and IV, binary systems with complete solid solution (section II), binary systems with one complete solid solution and two simple eutectic systems (section III) and binary systems with one complete solid solution and two partial solid solutions (section IV) were used (Figures 7.1, 7.5 and 7.13). In section II, two phase equilibria in a ternary solid solution system was characteristic and in sections III and IV, three phase equilibria in ternary solid solution systems was characteristic. In section II, the number of possible types of ternary systems which could be produced by combinations of binary systems with complete solid solution was listed.

It is now appropriate to demonstrate that many more types of ternary solid solution systems bounded by either complete or partial binary solid solution systems and containing three or four phase equilibrium situations are possible.

V. SOLID SOLUTION SYSTEMS WITH THREE PHASE EQUILIBRIA

A. Three Phase Equilibrium Involving One Complete and Two Partial Binary Solid Solution Systems

i. A System With Eutectics in the Partial Solid Solution Systems
Three phase equilibria has already been discussed in sections III and IV. Figure 7.26 shows another possible variation which involves complete solid between

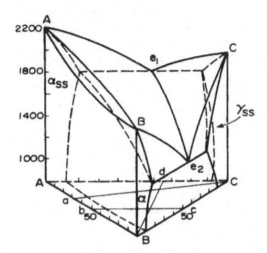

Figure 7.26. Ternary System with Eutectics in the Two Binary Partial Solid Solution Systems

A and B, but also ternary solid solution in the neighborhood of component C.

A typical 1600°C isothermal section (between e_1 and e_2) for this type of system is shown in Figure 7.27. After the liquid has disappeared below Te_2, only solid solutions of the α and γ type coexist. (Figure 7.28) Figures 7.29a-c show vertical sections C-a, b-c and B-d of Figure 7.26.

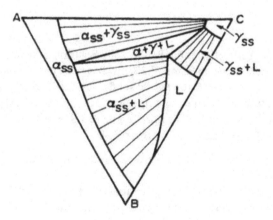

Figure 7.27. Isothermal Section at 1600°C

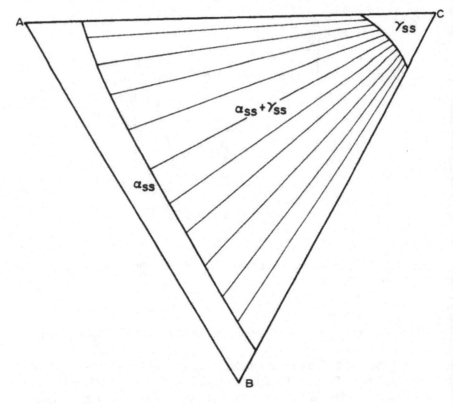

Figure 7.28. Isothermal Section at 1000°C

ii. A System with Peritectics in the Partial Solid Solution Systems
Another variation is shown in Figure 7.30, involving one continuous binary solid solution and two binary peritectic partial solid solutions. Typical isothermal sections at temperatures between p_1 and p_2 are shown in Figures 7.31a–d and an isothermal section below M.P.B is shown in Figure 7.31e. Figures 7.32a–c show vertical sections C–a, b–c and B–d of Figure 7.30.

iii. A System with a Eutectic and a Peritectic in the Partial Solid Solution Systems
Figure 7.33 shows three tie triangles at temperatures T_1, T_2 and T_3 in this type of system where the liquidus surface boundary line runs between p and e.

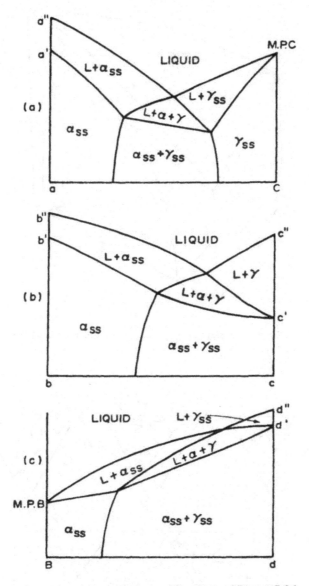

Figure 7.29a–c. Typical Vertical Sections of Figure 7.26

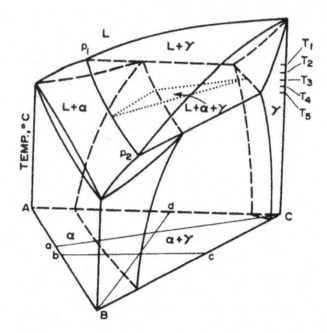

Figure 7.30. Ternary System with Peritectics in the Two Binary Partial Solid Solution System

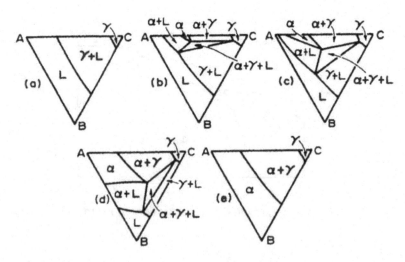

Figure 7.31a–e. Typical Isothermal Sections for Figure 7.30

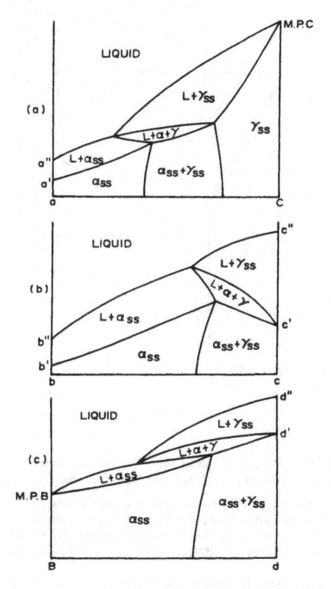

Figure 7.32a–c. Typical Vertical Sections of Figure 7.30

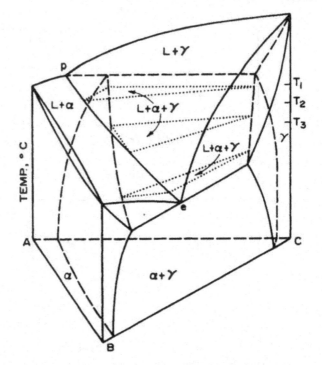

Figure 7.33. Tie Triangles in a System with a Eutectic and Peritectic in the Binary Systems

B. Three Phase Equilibria Involving Two Complete Solid Solution Systems and One Eutectic Type Binary Partial Solid Solution System

Figures 7.34a and 7.34b illustrate systems in which the ternary boundary line which starts from the eutectic point in the binary systems terminates at a particular point on the ternary liquidus. Three phase equilibria exists in the ternary system, starting from either binary system A-C or A-B until the boundary line terminates.

Figures 7.35a-d illustrate vertical sections a-b, a'-b', c'-c", and A-b' of Figure 7.34a.

As previously stated, the basic systems shown in sections C, D and E become far more complex if liquid immiscibility, polymorphism or exsolution exist in certain areas of the three-dimensional solid models.

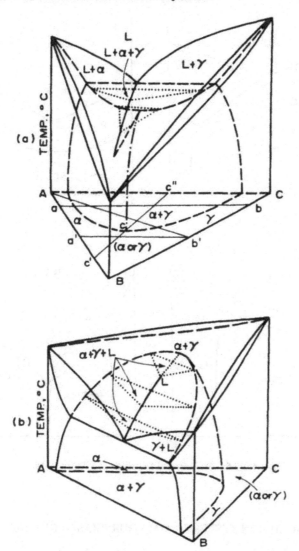

Figure 7.34a,b. Ternary Systems in which the Ternary Boundary Line Terminates on the Liquidus Surface

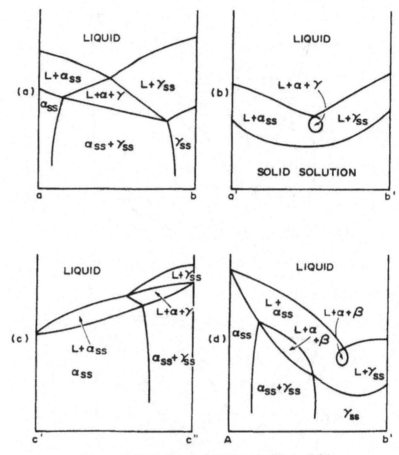

Figure 7.35a–d. Vertical Sections in Figure 7.34a

VI. SOLID SOLUTION SYSTEMS WITH FOUR PHASE EQUILIBRIA

A. Four Phase Equilibria Involving Three Binary Systems with Eutectic Partial Solid Solution

The characteristic equilibrium for this type of system is $L \rightleftarrows \alpha_{SS} + \beta_{SS} + \gamma_{SS}$. The solid model is shown in Figure 7.36 and typical isothermal sections are shown in Figure 7.37. After final disappearance of the liquid, three solid solutions will coexist, α_{SS}, β_{SS}, and γ_{SS}. The generation of a EUCTECTOID system of this type is shown in Figure 7.38. Three complete solid solution systems with

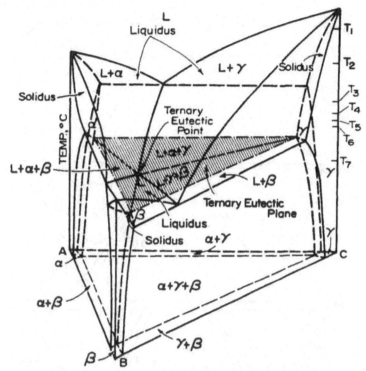

Figure 7.36. Ternary System with Eutectics in the Three Binary Partial Solid Solution Systems

no maximum or minimum in the liquidus are in equilibrium with a homogeneous liquid.

B. Four Phase Equilibria Involving One Binary Peritectic and Two Binary Eutectic Partials

The characteristic equilibrium for this type of system is $L + \alpha \rightleftarrows \beta + \gamma$. The solid model is shown in Figure 7.39 and typical isothermal sections are shown in Figure 7.40.

C. Four Phase Equilibria Involving Two Binary Peritectic and One Binary Eutectic Partial

The characteristic equilibrium for this type of system is $L + \alpha + \beta \rightleftarrows \gamma$. The solid model is shown in Figure 7.41 and typical isothermal sections are shown in Figure 7.42.

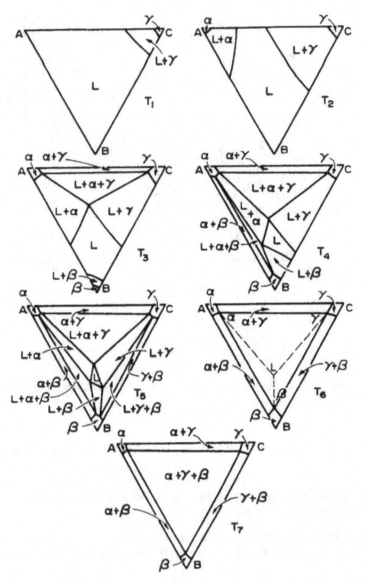

Figure 7.37. Isothermal Sections of Figure 7.36 (T_1 to T_7)

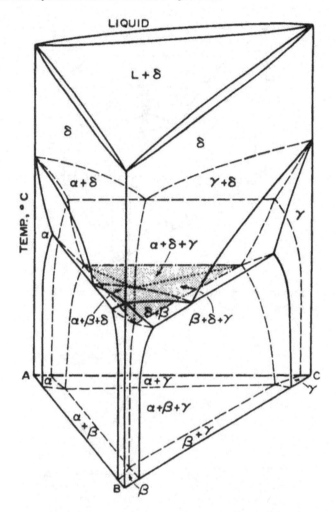

Figure 7.38. Ternary Eutectoid System with Eutectoids in each Binary Partial Solid Solution System

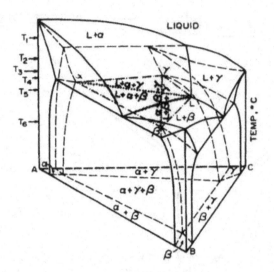

Figure 7.39. Ternary System with Two Binary Eutectic and One Binary Peritectic Partial Solid Solution Type Systems

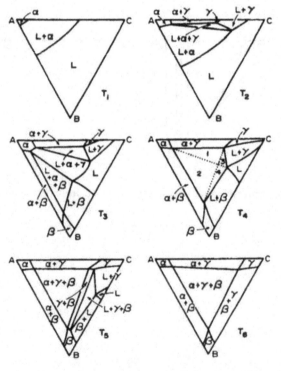

Figure 7.40. Isothermal Sections of Figure 7.39 (T_1 to T_6)

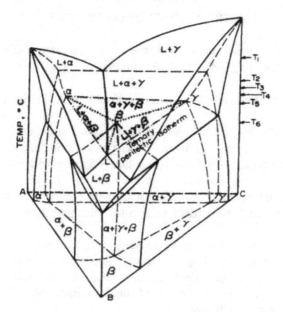

Figure 7.41. Ternary System with Two Binary Peritectic and One Binary
Eutectic Partial Solid Solution Type Systems

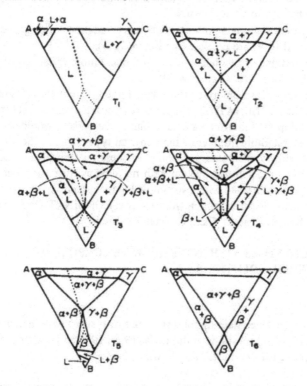

Figure 7.42. Isothermal Sections of Figure 7.41 (T_1 to T_6)

VII. ADDITIONAL BASIC SOLID SOLUTION SYSTEMS

The several basic diagrams shown in Sections II to VI are only a limited number of very fundamental basic diagrams. For example,

1. Only ONE of the ten basic diagrams of Section II (Figure 7.1) was shown.
2. Not all of the basic diagrams of the types used in Sections IV, V and VI were shown. To demonstrate just once, in Section VI, the four phase equilibria involving "Three Binary Systems with Peritectic Partial Solid Solution" was not shown.
3. Many ternary solid solution systems with combinations of complete (three types) and partial solid solution were not shown.

In addition to the many, many diagrams mentioned above which were not presented, the several basic diagrams shown in Sections II to VI are only a very limited number of diagrams which represent solid solution relationships in ternary systems. An enormous number of additional basic diagrams are possible under the following conditions:

1. If liquid immiscibility is present, many more variations of diagrams in Sections II to VI are possible,
2. If polymorphism is present, many more variations of diagrams in Sections II to VI are possible (see Figure 7.38),
3. If unmixing or exsolution is present, many more variations of diagrams in Sections II to VI are possible
4. If intermediate binary or ternary compounds are present (as in Chapter 5), the COMBINATION of complete and/or partial solid solution relationships and intermediate compound relationships would create an almost indefinite number of fundamental type diagrams. Section III is a very simple, elementary example of these relationships (A complete binary solid solution system with no maximum or minimum in the liquidus combined with two simple binary eutectic type systems). Consider the situations which would arise if COMBINATIONS of ternary systems in Chapter 5 and Chapter 7 existed.

VIII. REAL SYSTEMS WITH INTERMEDIATE COMPOUNDS AND SOLID SOLUTIONS

A. General

A list of real systems containing solid solutions or systems with intermediate compounds and solid solutions in the 1964, 1969, 1975 and 1981 collections of "Phase Diagrams for Ceramists" are shown in Table 1.

Table 7.1 Solid Solution in Ternary Systems

	Page	Figure
	1964 Edition	
$Li_2O-A\ell_2O_3-SiO_2$	166–168	449–456
$CaO-FeO-SiO_2$	204–208	586–593
$CaO-MgO-SiO_2$	210–213	598–610
$FeO-MgO-SiO_2$	236–238	682–687
$MgO-ZnO-SiO_2$	244	705
$MgO-ZnO-P_2O_5$	244–245	707–709
$MgO-A\ell_2O_3-SiO_2$	246	712
$MgO-SiO_2-GeO_2$	248	717
$LiF-NaF-ZrF_4$	438	1548
	1969 Edition	
$BaO-CaO-A\ell_2O_3$	125	2451, 2452
$BaO-CaO-SiO_2$	125, 126	2453, 2454
$CaO-MgO-SiO_2$	130–134	2471–2483
$CaO-A\ell_2O_3-Fe_2O_3$	137	2491–2492
$CdO-ZnO-P_2O_5$	140, 141	2504–2507
$MgO-A\ell_2O_3-SiO_2$	146–151	2526–2539
	1975 Edition	
$BaO-CaO-SiO_2$	217	4538
$CoO-ZnO-TiO_2$	235	4567
$NiO-ZnO-TiO_2$	247	4583
$SrO-A\ell_2O_3-Fe_2O_3$	252	4591, 4592
$ThO_2-ZrO_2-P_2O_5$	259, 260	4608
	1981 Edition	
$BaO-SrO-A\ell_2O_3$	193	5362
$BaO-SrO-SiO_2$	195, 196	5365, A–D
$CaO-MgO-SiO_2$	200–203	5375–5379
$CaO-CeO_2-ZrO_2$	207, 208	5390, A–D
$FeO-MgO-TiO_2$	213, 214	5406, A–C
$FeO-A\ell_2O_3-Cr_2O_3$	215	5408, A–D
$Y_2O_3-ThO_2-WO_3$	231	5442, A, B

Each of these systems can be used to illustrate the interpretation of ternary solid solution diagrams and several of them are extremely important in industrial applications of phase data.

It is not possible to discuss all of these systems, but a few of them will be selected on the basis of their combined value in illustrating principles and their technological importance (use and application in industry).

B. The System $Li_2O-Al_2O_3-SiO_2$

The pertinent diagrams are shown in Figures 7.43 (Figure 449, page 166) and 7.44 (Figure 456, page 168). See Table 1, 1964 Edition.

Note first (Figure 7.43) that it is conventional to show solid solubility in ternary systems by the use of cross-hatching, especially on the joins between composition points.

In this system the join $Li_2O \cdot Al_2O_3 - SiO_2$ is critical as shown in Figure 7.44, because the solid solubility involves a wide range of β-spodumene and β-eucryptite compositions ranging from less than two moles of SiO_2 to at least eight moles of SiO_2. The phases which occur in nature are α-eucryptite (Be_2SiO_4-type structure) (2), α-spodumene ($Li_2O \cdot Al_2O_3 \cdot 4SiO_2$), and petalite (3), which has its own structure, different from that of the 1:1:8 β-spodumene structure which results from the heating of petalite.

Skinner and Evans (4) refined the solid solution limit of β-eucryptite in β-spodumene and it is now felt that the spodumene solid solution extends to

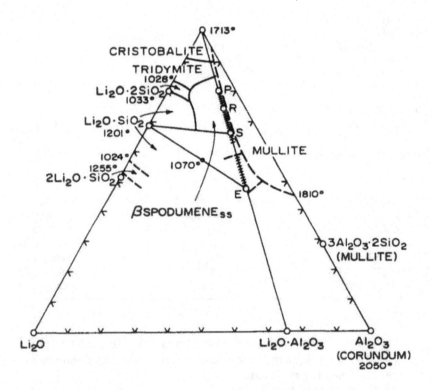

Figure 7.43. The $Li_2O-Al_2O_3-SiO_2$ System

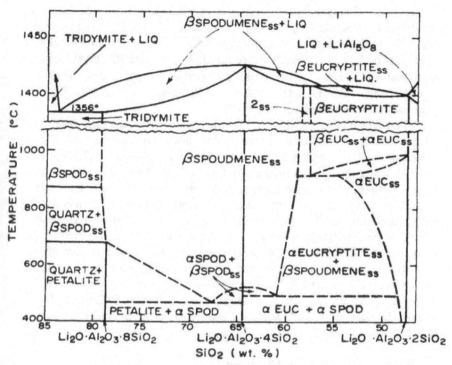

Figure 7.44. The $LiAlO_2$–SiO_2 System

about $Li_2O \cdot Al_2O_3 \cdot 3.8SiO_2$. Munoz (5) has presented the most recent picture of the p–T relationships of the composition $Li_2O \cdot Al_2O_3 \cdot 4SiO_2$. Much detailed work has been done recently on the crystal structures of β-spodumene and β-eucryptite which is beyond the scope of this text and cannot be repeated here.

The two synthetic minerals β-spodumene and β-eucryptite became important in ceramics and glass-ceramics due to the discovery of the ultra low expansion of (β) $Li_2O \cdot Al_2O_3 \cdot 4SiO_2$ and its solid solutions and the negative expansion of (β) $Li_2O \cdot Al_2O_3 \cdot 2SiO_2$ and its solid solutions by Hummel in 1946–1950, as published in 1948 (6), 1951 (7), and 1952 (8) and patented in 1957 (9).

In preparing sintered β-spodumene (solid solution) bodies, it is important to avoid the development of eutectic glasses whose high thermal expansions (10) are incompatible with the extremely low expansion β phases. In making glass-ceramics, other oxides must be added to insure meltability, workability and control of the nucleation and crystallization of the β phases. The axial thermal expansions of β-eucryptite and β-spodumene solid solutions are now known due to the work of Gillery and Bush (11), Ostertag, Fisher and Williams (12), Pillars and Peacor (13), Schulz (14) and Moya, Verduch and Hortal (15). The technology of the ceramics or glass-ceramics can now be understood, starting with the phase relationships and proceeding through the crystal chemistry, crystal structures, thermal expansions and mechanical properties of these phases. The phases have attracted world-wide attention from the standpoints of basic science and (especially) their practical significance (8).

Two recent books (16,17) contain many references to the early development of lithium aluminosilicate commercial products, based on the solid solution phase equilibria and the thermal expansion data of Hummel (6,7,8,9,10).

The six pages of the introductory Chapter 1 of the book on "Glass-Ceramics" by P. W. McMillan contains highly distorted statements about the origin of glass-ceramics. The origin of the very significant lithium aluminosilicate glass-ceramic was based on the phase equilibrium, crystal chemistry and thermal expansion data in references 2-11 above. Ceramists who are involved in phase equilibrium data or research know that crystalline or partially crystalline commercial ceramics products are produced by sintering, fusion, or by crystallizing glasses (totally or partially). The real basis for the product depends upon the chemical and physical properties of the crystal(s) or glass(es). In the case of the lithium aluminosilicate glass-ceramics it was based on the thermal expansion characteristics (6,7,8,9,10).

C. "Metastable" β-quartz Solid Solutions

Beginning with the reinvestigation of the system $Li_2O-Al_2O_3-SiO_2$ in 1948 by Hummel (6) and in 1949 by Roy and Osborn (18), reports of either an unidentified phase or a quartz-like phase (19) obtained by crystallization of glasses appeared in the literature. Around 1961, after the discovery of a wide range of metastable quartz solid solutions in the system $MgO-Al_2O_3-SiO_2$ by Schreyer and Shairer (20), it was realized that an extremely wide region of glasses in the system $Li_2O-Al_2O_3-SiO_2$ would yield β-quartz structures if given the proper heat treatment. Subsequent work by Eppler (21), Beall, Karstetter and Rittler (22), Badger (23), Ray and Muchow (24), and Ray (25,26) has provided much data on the influence of bulk composition and heat treatment on the appearance of the β-quartz solid solutions. Transparent glass-ceramics can be made from selected compositions by minimizing 1) the difference in index of refraction of the glass matrix and the quartz solid solutions, 2) the birefringence of the quartz solid solution, 3) microfractures, and 4) the size of the quartz solid solution crystals.

With respect to the $LiAlO_2-SiO_2$ join in the ternary system, it is now known that β-eucryptite (β-quartz structure) solid solutions range from 31.3-59.8 weight per cent SiO_2 (those between 47.7-59.8 being STABLE solid solutions), while the β-quartz solid solutions exist from 59.8 wt.% SiO_2 to (presumably) 100% SiO_2. The change from optically negative to optically positive quartz solid solutions occurs around 78 mole percent SiO_2 (23).

It has been known (27,28) for some time that Mg^{++}, Ca^{++}, Sr^{++}, Ba^{++} or Mn^{++} added as carbonates to silicic acid will promote the temporary formation of quartz when heated in the range 1200-1300°C, more than 300°C above the upper end of the β-quartz stability range. Segnit and Gelb (29) have recently produced (probably low) quartz solid solutions by the addition of MgO, FeO, ZnO or CuO to kaolinite and heat treatment between 700-1000°C for 16-24 hours. The change in spacing of the $10\bar{1}1$ peak indicated that the oxides had entered the "free" silica which was produced during metakaolinite formation. Prolonged heating above 1000° caused the quartz phase to disappear in favor of expected equilibrium phases. Much earlier, Bradley and Grim (30) had produced β-quartz solid solution from montmorillonite at 1000°C. A review of quartz solid solutions has been made by Ganguli (31). However, the conclusion by Hummel is that it is surprising that so little crystal chemistry, phase equilibria and structure analysis of quartz solid solutions have been done in thirty years after the stimulating papers of Buerger (32,33).

Recent publications by W. F. Horn and F. A. Hummel (34,35,36,37,38,39) have provided data on solid solution in silica related (cristobalite) systems but not on the quartz phase.

D. The System $MgO-A\ell_2O_3-SiO_2$

The phase relationships are shown in Figures 7.45, 7.46, 7.47 and 7.48.

Figure 7.45 is Figure 712 on page 246, 1964
Figure 7.46 is Figure 2530 on page 147, 1969
Figure 7.47 is Figure 2535 on page 149, 1969
Figure 7.48 is Figure 2538 on page 150, 1969

Figure 7.45 (40) shows the presence of the two important eutectics at 1355° and 1365°, the incongruent melting of cordierite to mullite and liquid, the presence of the solid solution regions around cordierite, mullite and spinel and compound sapphirine, whose composition is thought to be $4MgO \cdot 5A\ell_2O_3 \cdot 2SiO_2$.

Figure 7.46 shows the tie lines more clearly at a temperature just below 1355°, especially those involving the cordierite solid solution in the direction of "Mg-beryl". Figure 7.47 is an isothermal section at 1460° which shows the liquid region and the various two and three phase regions (41).

All of the diagrams shown so far have been for one atmosphere pressure and do not show the important garnet phase, pyrope ($3MgO \cdot A\ell_2O_3 \cdot 3SiO_2$).

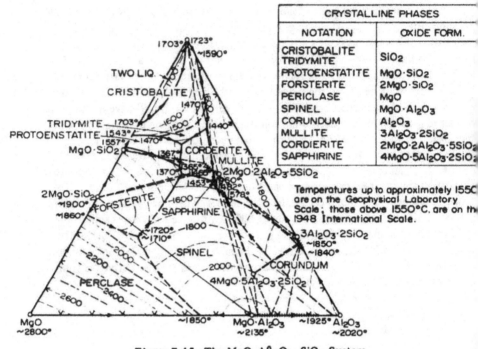

CRYSTALLINE PHASES	
NOTATION	OXIDE FORM.
CRISTOBALITE TRIDYMITE	SiO_2
PROTOENSTATITE	$MgO \cdot SiO_2$
FORSTERITE	$2MgO \cdot SiO_2$
PERICLASE	MgO
SPINEL	$MgO \cdot Al_2O_3$
CORUNDUM	Al_2O_3
MULLITE	$3Al_2O_3 \cdot 2SiO_2$
CORDIERITE	$2MgO \cdot 2Al_2O_3 \cdot 5SiO_2$
SAPPHIRINE	$4MgO \cdot 5Al_2O_3 \cdot 2SiO_2$

Temperatures up to approximately 1550
are on the Geophysical Laboratory
Scale; those above 1550°C. are on th
1948 International Scale.

Figure 7.45. The $MgO-Al_2O_3-SiO_2$ System

Pyrope can only be prepared under pressure and its relationships with other minerals in the system at various pressures have been determined by Boyd and associates (42) as shown for $MgSiO_3$ at 30 kbs. in Figure 7.48.

The interest in cordierite as a ceramic body was generated by the discovery of its low thermal expansion by Singer and Cohn (43) in 1929. The history has been detailed by Singer (44) in 1946 and over the last 55 years, various types of porous, vitrified and glass-ceramic cordierite-based bodies have been made from talc, clay and alumina or from highly purified raw materials. Its use has progressed from saggar bodies and kiln furniture to the most recent application in very sophisticated shapes for exhaust emission control substrates. A recent review paper (45) can be consulted for data on the natural mineral but corresponding data on synthetic ceramics must be sought in individual papers. The papers by Stone (46) and Foster (47) give detailed analyses of the ternary system with respect to phase analyses and application to ceramic firing processes. Its crystal chemical similarity (48) to the mineral beryl can be seen by writing their formulas as follows:

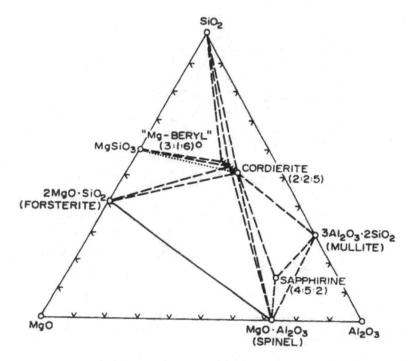

Figure 7.46. Tie Lines in the System $MgO-Al_2O_3-SiO_2$

$Be_3Al_2Si_6O_{18}$
$Al_3Mg_2AlSi_5O_{18}$

which shows the kind of atoms which occupy the four-coordinated, six-coordinated and four-coordinated sites, respectively. The thermal expansion characteristics of beryl and cordierite have been found to be similar as well. Phase relationships in the system $BeO-Al_2O_3-SiO_2$ have been investigated by Ganguli (49). There has been considerable recent literature on the refinement of the crystal structure of beryl and cordierite and on the growth and characteristics of beryl crystals which will not be cited at this time, but this literature and associated literature on the related minerals indialite and osumulite should be consulted by ceramists whose goal it is to develop beryl or cordierite-indialite ceramics. It should be noted that $2FeO\cdot2Al_2O_3\cdot5SiO_2$ and $2MnO\cdot Al_2O_3\cdot 5SiO_2$ cordierite analogues do exist and that petalite-type phases (50,51) have been found to exist in the $MgO-Al_2O_3-SiO_2$ system. The most recent axial

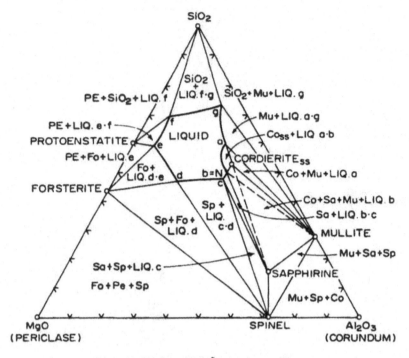

Figure 7.47. The 1460°C Isothermal Section

thermal expansion data on cordierite and chemical modifications is by Evans, Fisher, Geiger and Martin (52).

As stated above, definitive work by Schreyer and Shairer (20) had proved the existence of β-quartz solid solutions in the system $MgO-A\ell_2O_3-SiO_2$. Phases which had been previously labeled "μ-cordierite" were shown to be β-quartz solid solutions and much detailed data was presented on the effect of bulk composition and heat treatment on the appearance of the β-quartz phase. It was shown that the change from optically negative quartz solid solutions to optically positive solid solutions occurred at approximately 70 wt.% SiO_2 on the $MgA\ell_2O_4-SiO_2$ join. The "quartz-like" phases of Prokopowicz and Hummel (53) observed in 1953 in the four-component system $Li_2O-MgO-A\ell_2O_3-SiO_2$ could then be definitely characterized as β-quartz solid solutions.

In the period 1941-1944, when the author worked for Corning Glass Works, he had the idea of crystallizing cordierite from the relatively refractory heat resistant "Top-of-Stove" glass whose molar composition was:

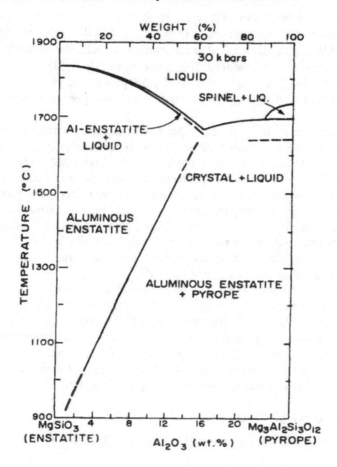

Figure 7.48. The System $MgSiO_3$–$Mg_3Al_2Si_3O_{12}$ at 30 kbars

0.037 Na_2O	0.643 Al_2O_3	2.67 SiO_2
0.208 CaO	0.199 B_2O_3	
0.755 MgO		

With the cooperation of Martin Nordberg (coinventor of Vycor and inventor of the low expansion silica-titania glasses), the author crystallized and measured the expansion (from $0°$-$300°C$) of several magnesium aluminosilicate-based glasses (free of alkali). The data were never published, but this research was really the original basis for the development of glass-ceramics which was announced by Corning in 1957. However, the thermal expansion data on the lithium aluminosilicate compositions listed in the section VIII B references (mainly 1948-1952) is what enabled the less refractory, practical and very highly commercial materials to be developed. The 1953 (54) and 1956 (53) phase equilibrium data on the system $Li_2O-MgO-Al_2O_3-SiO_2$ are also related to the origin and development of ternary, quaternary and multi-component glass-ceramic products.

The system is also extremely important for the interpretation of the heat treatment of "steatite" bodies which after firing are mainly composed of an $MgSiO_3$ phase (clinoenstatite, protoenstatite) and a glassy bond. As for many cordierite bodies, the starting raw materials are talc, clay and alumina (talc dominant). The selection of raw materials is made with great care, partly to adapt properly to processing techniques, but mainly to guard against the formation of low viscosity liquids which form during the firing process. Formation of too much fluid liquid would result in great losses in the kiln due to deformation. For example, high lime talcs cannot be used for high grade steatite, due to the fluidity of the liquid which is induced by CaO.

Previous references (46,47) on the heat treatment of $MgO-Al_2O_3-SiO_2$ compositions are especially helpful for an understanding of steatite technology.

Fosterite (Mg_2SiO_4) refractories and electronic ceramics are of considerable importance and the phase relations in the ternary systems have a bearing on their processing (forming) although the influence of liquifaction is usually less important than for steatite and cordierite. Some ultra pure forsterites are prepared by solid state sintering and little or no glass phase is present.

E. The System $MgO-"FeO"-SiO_2$

The phase relationships are presented in Figures 7.49, 7.50 and 7.51.

> Figure 7.49 is Figure 682 on p. 236, 1964 Edition of "Phase Diagrams for Ceramists"
> Figure 7.50 is Figure 2070 on p. 5, 1969 Edition of "Phase Diagrams for Ceramists"
> Figure 7.51 is Figure 684, on p. 237, 1964 (four diagrams)

Except for the complication due to the presence of a small amount of Fe or Fe_2O_3 in this system even under the most carefully controlled conditions, it is ideal for the demonstration of solid solution relationships. This is due to the similar crystal chemical behavior of Fe^{++} and Mg^{++}, leading to complete solubility of the oxides, complete solubility between the orthosilicates Fe_2SiO_4 and Mg_2SiO_4, and extensive solubility of $FeSiO_3$ in $MgSiO_3$ as shown in Figure 7.49.

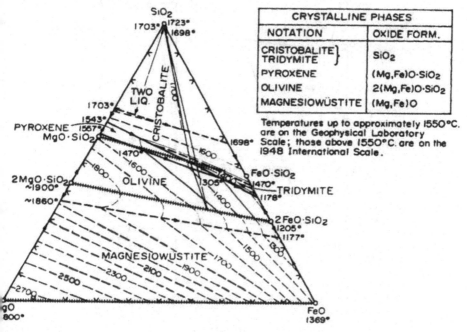

CRYSTALLINE PHASES	
NOTATION	OXIDE FORM.
CRISTOBALITE ⎫ TRIDYMITE ⎭	SiO₂
PYROXENE	(Mg,Fe)O·SiO₂
OLIVINE	2(Mg,Fe)O·SiO₂
MAGNESIOWÜSTITE	(Mg,Fe)O

Temperatures up to approximately 1550°C. are on the Geophysical Laboratory Scale; those above 1550°C. are on the 1948 International Scale.

Figure 7.49. The System MgO–"FeO"–SiO₂

The system is spoken of as MgO–"FeO"–SiO₂ due to the fact that FeO is a non-stoichiometric compound, giving rise to a range of wüstite solid solutions as shown in Figure 7.50. This diagram shows the temperature and oxygen pressure under which wüstite solid solutions and Fe_3O_4 exist in the system Fe-O, and makes clear why the diagram show in Figure 7.49 is described as one in which "oxide phases are in equilibrium with metallic iron." The remaining features of the diagram are best understood by the use of isothermal sections as shown in Figure 7.51 and the crystallization path shown in Figure 7.52.

Consider a melt of composition X (Figure 7.52) in the primary phase field of olivine. The isopleth X intersects the liquidus surface at 1600°, as indicated by its junction with the 1600° isotherm. From the slopes of the tie lines in the olivine + liquid areas of the 1550° and 1527° isothermal sections, it can be estimated that at 1600° the slope of the appropriate tie line might be as shown by the line 1–X in Figure 7.52. The slopes for suitable tie lines at 1550° and at 1527° can be taken from the appropriate sections and are shown as the lines 2–2′, 3–3′, and 4–4′. At 4′, the cooling path for the melt intersects the boundary line between olivine and pyroxene; on further cooling pyroxene will co-precipitate with olivine. The appropriate tie-triangle (estimated from the 1450°

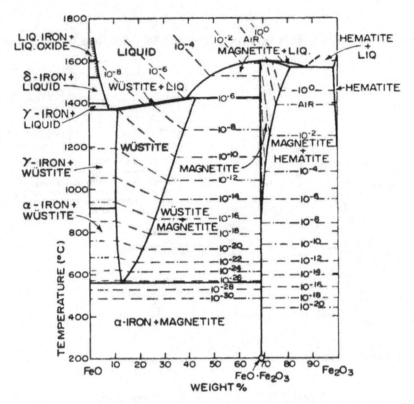

Figure 7.50. The System FeO–Fe$_2$O$_3$

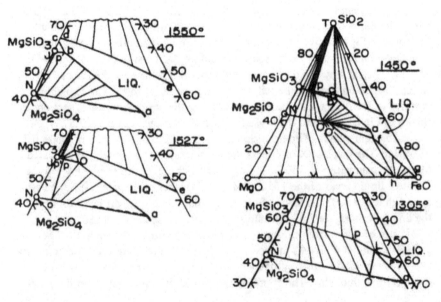

Figure 7.51. Isothermal Sections in the System MgO–"FeO"–SiO$_2$ (1305°, 1450°, 1527°, 1550°)

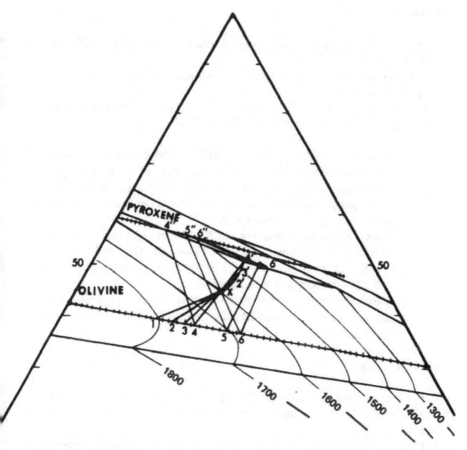

Figure 7.52. Crystallization Path of Composition X Estimated From Slopes of Tie Lines in Isothermal Sections

isothermal section) is given by 4-4'-4". With further cooling the tie line 6-6" passes through the point X. The intersection of the tie line 6-6' with the boundary line indicates that the temperature of final solidification is approximately 1420°. The solidified sample at that temperature will be composed of a mixture of olivine solid solution and pyroxene solid solution of the composition given by the extremities of the tie line 6-6". The proportions of each may be estimated by application of the lever rule.

PROBLEMS

1. Draw vertical sections of any kind of Figures 7.4, 7.13, 7.26, 7.30, 7.36, 7.38, 7.39, or 7.41.
2. Study in detail any of the systems listed in Table I. Analyze the original research papers.
3. Using Figure 915 on page 299 of the 1964 collection of diagrams (the system $CaO \cdot SiO_2 - 2CaO \cdot Al_2O_3 \cdot SiO_2 - 2CaO \cdot MgO \cdot 2SiO_2$)

    a) Draw the isothermal plane for 1400°C. Label each area for phases present and draw tie lines in appropriate areas. In certain areas, you have to assume that you have a knowledge of how the tie lines will run, on the basis of experimental data. The system Gehlenite-Akermanite is shown on page 300 as Figure 918.
    b) For a composition containing 57% $CaSiO_3$ and 33% Gehlenite, what phases are in equilibrium at 1310°C? This point falls on the boundary line in the system. Give the composition of the phases in terms of the end-member components of the system.
    c) When would such a composition (b, above) finish its crystallization and become 100% solid? Describe fully the criterion for the disappearance of the liquid.

REFERENCES

1. E. F. Osborn and J. F. Schairer, The Ternary System Pseudowollastonite-Akermanite-Gehlenite, Amer. J. Sci., Vol. 239, pp. 715–763, October, 1941.
2. H. G. F. Winkler, Polymorphie: II Struktur and Polymorphie des Eucryptit (Tief $LiAlSiO_4$), Beitr. Mineral Petrog 4, 233–42, 1954.
3. Anna Zemann-Hedlik and J. Zemann, The Crystal Structure of Petalite, $LiAlSi_4O_{10}$, Acta Cryst. 8, 781–7, 1955.
4. B. J. Skinner and H. T. Evans, Jr., β-spodumene Solid Solutions on the Join $Li_2O \cdot Al_2O_3 - SiO_2$, Amer. J. Sci. 258–A, 312–324, 1960.
5. James L. Munoz, Stability Relations of $LiAlSi_2O_6$ at High Pressures, Special Publication #2, Pyroxenes and Amphiboles: Crystal Chemistry and Phase Petrology, 1969, Mineralogical Society of America.
6. F. A. Hummel, Thermal Expansion Properties of Natural Lithia Minerals, Foote Prints 20, 3–11, 1948.
7. F. A. Hummel, Thermal Expansion Properties of Some Synthetic Lithia Minerals, J. Amer. Cer. Soc. 34, 235–239, 1951.
8. F. A. Hummel, Significant Aspects of Certain Ternary Compounds and Solid Solutions, J. Amer. Cer. Soc. 35, 64–66, 1952.

9. F. A. Hummel, U.S. Patent 2,785,080, Thermal Shock Resistant Ceramic Body, March 12, 1957; Reissue 24, 795, March 15, 1960.
10. C. E. Brachbill, H. A. McKinstry and F. A. Hummel, Thermal Expansion of Some Glasses in the System $Li_2O-Al_2O_3-SiO_2$, J. Amer. Cer. Soc. 34, 107-109, 1951.
11. F. H. Gillery and E. A. Bush, Thermal Contraction of β-eucryptite ($Li_2O \cdot Al_2O_3 \cdot 2SiO_2$) by X-ray and Dilatometric Methods, J. Amer. Cer. Soc. 42(4), 175-177, 1959.
12. W. Ostertag, G. R. Fisher and J. P. Williams, Thermal Expansion of Synthetic β-Spodumene and β-Spodumene-Silica Solid Solutions, J. Amer. Cer. Soc. 51(11), 651, 1968.
13. William W. Pillars and Donald R. Peacor, The Crystal Structure of Beta Eucryptite as a Function of Temperature, Special Reprint from the American Mineralogist, 58(5-6), p. 681, July-August, 1973.
14. H. Schulz, Thermal Expansion of Beta Eucryptite, J. Amer. Cer. Soc. 57(7), p. 313-318, July 1974.
15. J. S. Moya, A. G. Verduch and M. Hortal, Thermal Expansion of β-Eucryptite Solid Solutions, Trans. and Jour. of the British Ceramic Society, 73(6), 177, 1974.
16. J. H. Fishwick, Applications of Lithia in Ceramics, (Chapters 3, 6 and 7), Cahners Publishing Co., 1974.
17. Phase Diagrams; Materials Science and Technology, Volume 6-V, Chapter VI, R. N. Kleiner and S. T. Buljan, The Use of Phase Diagrams in Development of Silicates for Thermal Shock Resistant Applications, 287-320.
18. R. Roy and E. F. Osborn, The System Lithium Metasilicate-Spodumene-Silica, J. Amer. Chem. Soc. 71, 2086-2095, 1949.
19. E. Henglein, Zur Kenntnis der Hoch temperature-Modifikationen von Lithium-Aluminium-Silikaten, Fortschr Min 34, 40-43, 1956.
20. W. Schreyer and J. F. Shairer, Metastable Solid Solutions with Quartz-type Structures on the join $SiO_2-MgAl_2O_4$, Zeitschrift für Kristallographie, 116(1-2), 60-82, 1961, Geophysical Laboratory Paper No. 1357.
21. R. A. Eppler, Glass Formation and Recrystallization in the Lithium Metasilicate Region of the System $Li_2O-Al_2O_3-SiO_2$, J. Amer. Cer. Soc. 46(2), 97-101, 1963.
22. G. H. Beall, B. R. Karstetter and H. L. Rittler, Crystallization and Chemical Strengthening of Stuffed β-Quartz Glass-Ceramics, J. Amer. Cer. Soc. 50(4), 181-190, 1967.
23. W. B. Badger, Phase Relations in the System $Li_2O-Na_2O-Al_2O_3-SiO_2$ with Special Emphasis on the Silica Polymorphs, PhD thesis in Geochemistry, The Pennsylvania State University, June 1968.
24. Satyabrata Ray and G. M. Muchow, Quartz Solid Solution Phases from Thermally Crystallized Glasses of Compositions $(Li_2O, MgO) \cdot Al_2O_3 \cdot nSiO_2$, J. Amer. Cer. Soc. 51(12), 678-682, 1968.

25. Satyabrata Ray, Solutions in the Keatite Crystal Lattice, J. Amer. Cer. Soc. 54(4), 213-215, 1971.

26. Satyabrata Ray, Study of Ordering in High-Quartz Solid Solutions by Substitutions Affecting Superlattice Reflections, J. Amer. Cer. Soc. 56(1), 42-45, 1973.

27. James H. Schulman, Esther W. Claffy and Robert J. Ginther, Some Observations on the Crystallization of Silicic Acid, Amer. Min. 34(1-2) 68, Jan.-Feb., 1949.

28. L. S. Birks and J. H. Schulman, The Effect of Various Impurities on the Crystallization of Amorphous Silicic Acid, Amer. Min. 35(11-12), 1035, Nov.-Dec., 1950.

29. E. R. Segnit and T. Gelb, Metastable Quartz-type Structures Formed from Kaolinite by Solid State Reaction, Amer. Min. 57(9-10) 1505, Sept.-Oct., 1972.

30. W. R. Bradley and R. E. Grim, High-Temperature Thermal Effects of Clay and Related Minerals, Amer. Min. 36, 182-201, 1951.

31. Dibyendu Ganguli, Crystal Chemistry, Synthesis and Characterization of Quartz Solid Solutions: A Review, Transactions of the Indian Ceramic Society, Volume XXXIV (2), March-April, 1975.

32. M. J. Buerger, Crystals Based on the Silica Structures, Amer. Min. (33), Nos. 11 and 12, pp. 751-52, 1948.

33. M. J. Buerger, Stuffed Derivatives of the Silica Structures, Amer. Min. (39), Nos. 7 and 8, pp. 600-15, 1954.

34. W. F. Horn and F. A. Hummel, The System BPO_4-$A\ell PO_4$, Trans. and Jour. of the British Ceramic Society, 77(5), 158-162, 1978.

35. W. F. Horn and F. A. Hummel, Progress Report on the System BPO_4-SiO_2, Trans. and Jour. of the British Ceramic Society, 78(4), 77-80, 1979.

36. W. F. Horn and F. A. Hummel, The System $A\ell PO_4$-SiO_2, Central Glass and Ceramic Bulletin (India), Volume 26, Nos. 1-4, 47-59, 1979.

37. W. F. Horn and F. A. Hummel, Axial Thermal Expansion of Cristobalite $A\ell PO_4$ and Related Cristobalite Solid Solution Phases, J. Amer. Cer. Soc. 63, Nos. 5-6, May-June, 1980.

38. W. F. Horn and F. A. Hummel, The System BPO_4-$A\ell PO_4$-SiO_2 at 1200°C, Trans. and Jour. of the British Ceramic Society, 79, 109-111, 1980.

39. F. A. Hummel and W. F. Horn, The Quaternary System B_2O_3-$A\ell_2O_3$-SiO_2-P_2O_5; Part I, Literature Review and Exploratory Data on the Ternary Subsystems, Jour. of the Australian Ceramic Society, 17, No. 1, 25-32, 1981.

40. See list of principal references on Page 246 of 1964 collection of Phase Diagrams.

41. W. Schreyer and J. F. Schairer, Compositions and Structural States of

Anhydrous Mg-Cordierites: A reinvestigation of the Central Part of the System $MgO-Al_2O_3-SiO_2$, J. Petrol 2(3), 324–406, 1961.

42. F. R. Boyd and J. L. England, Carnegie Inst. Washington Yearbook, 63, 158, 1963–64.

43. Felix Singer and W. M. Cohn, New Ceramic Bodies: Composition and Expansion Behavior, I–II, Ber Deut. Keram. Ges, 10(6), 269–84, 1929.

44. Felix Singer, Ceramic Cordierite Bodies, J. Can. Cer. Soc., 15, 60–71, 1946.

45. M. Strunz, C. L. Tenneyson and P. J. Vebel, Cordierite Morphology, Physical Properties, Structure, Inclusions and Oriented Intergrowth, Mineral Science and Engineering 3(2), April, 1971.

46. R. L. Stone, Physical Chemistry of Firing Steatite Ceramics, J. Am. Cer. Soc., 26(10), 333–36, 1943.

47. W. R. Foster, Contribution to the Interpretation of Phase Diagrams by Ceramists, J. Amer. Cer. Soc., 34(5), 151–161, May 1951.

48. W. L. Bragg, Atomic Structure of Minerals, Cornell University Press, 1937 (p. 183).

49. Dibyendu Ganguli and Prasenjit Saha, Preliminary Investigations in the High Silica Region of the System $BeO-Al_2O_3-SiO_2$, Trans. Indian Cer. Soc., 24(4), 134–146, 1965.

50. W. Schreyer and J. F. Schairer, Metastable Osumulite and Petalite-Type Phases in the System $MgO-Al_2O_3-SiO_2$, Amer. Min. 47, pp. 90–104, 1962.

51. S. B. Holmquist, A Note on a Mg petalite Phase, Zeit für Kristallographie, 118(5–6), 1963.

52. D. L. Evans, G. R. Fisher, J. E. Geiger and F. W. Martin, Thermal Expansions and Chemical Modifications of Cordierite, J. Amer. Cer. Soc. 63, 11–12, 629–634, 1980.

53. T. I. Prokopowicz and F. A. Hummel, Reactions in the System $Li_2O-MgO-Al_2O_3-SiO_2$: II, Phase Equilibria in the High Silica Region, J. Amer. Cer. Soc. 39(8), 266–78, 1956.

54. M. D. Karkhanavala and F. A. Hummel, Reactions in the System $Li_2O-MgO-Al_2O_3-SiO_2$: I, The Cordierite-Spodumene Join, J. Amer. Cer. Soc. 36(12), 393–397, 1953.

8

QUATERNARY AND MULTICOMPONENT SYSTEMS

I. INTRODUCTION; QUATERNARY SYSTEMS WITHOUT SOLID SOLUTION

In the most general case, when both temperature and pressure are considered, the Phase Rule for a four component system is F=6-P. Once again, at constant pressure, the condensed form of the Phase Rule is F=5-P.

Following the geometry of the equilateral triangle for the representation of ternary systems, the quaternary system is represented as an equilateral tetrahedron as shown in Figure 8.1. This particular system is perhaps the simplest of all four component configurations since each of the six bounding binary and each of the four bounding ternary systems are simple eutectic types, as well as the quaternary system itself. From the standpoint of a metallurgist or from another viewpoint, a quaternary system with 1) no inversions in the four end members, 2) continuous solid solution in the binary and ternary systems which extended to the quaternary system, and with 3) continuous and complete intersolubility in the liquid state, might be considered as the simplest geometry of all possible geometries.

However, the simple eutectic type of system shown in Figure 8.1 is most suitable for demonstrating the important features of a quaternary system without solid solution. It should first be noted that the following kinds of equilibria are characteristic of the system:

1. Five phase eutectic melting point (F=5-5=0) invariant
   $L \rightleftarrows A + B + C + D$ (point) (E_s)

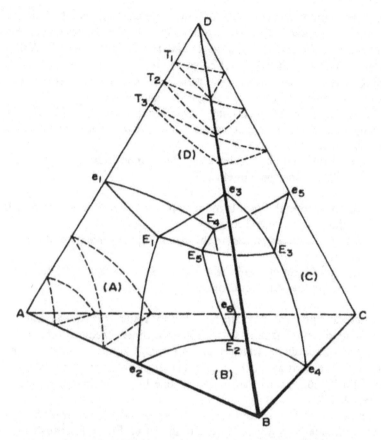

Figure 8.1. Solid Model of Quaternary System Composed of Four Components A, B, C and D. Isothermal Surfaces at T_1, T_2 and T_3. Quaternary Eutectic E_5, Ternary Eutectics E_1, E_2, E_3, E_4 and Binary Eutectics e_1, e_2, e_3, e_4, e_5 and e_6

2. Four phase space boundaries (lines) (F=5-4=1) univariant
   $L \rightleftharpoons A + B + C$ (line E_2-E_5)
3. Three phase space surfaces (F=5-3=2) divariant
   $L \rightleftharpoons A + B$ (e_2-E_2-E_5-E_1)
4. Two phase paths (volume) (F=5-2=3) trivariant
   $L \rightleftharpoons A$ (six sides)
5. One phase volumes (primary phase volumes) (F=5-1=4)

All of these features can be identified in Figure 8.1. The method of specifying composition within the tetrahedron on a weight or molar basis should be obvious after learning the technique for ternary systems. Any point within the tetrahedron is composed of a total of 100% of A + B + C + D. Planes could be passed parallel to the base A–B–C at levels of 0–100% of component D. A similar series of isocompositional planes could be created parallel to the other bounding ternary systems which would then form a convenient three-dimensional grid which could be used for locating a specific composition.

In general, temperature cannot be conveniently represented WITHIN the quaternary tetrahedron, except at invariant points whose temperature could be specified by labeling. (such as E_5)

The equilateral tetrahedron has some special geometrical properties, similar to the equilateral triangle. For example,

1) on a line from the apex D to the opposite face ABC, the ratio of A:B:C in the compositions is constant.
2) on a plane passed parallel to the base ABC, the percentage of D is constant.
3) on a plane passed parallel to the edge AD to the face BCD, the ratio of B:C in the composition is constant.

What was a primary field (area) in a ternary system becomes a primary phase volume or crystallization space, isotherms (lines) become isothermal surfaces and compatibility triangles become compatibility tetrahedra. The ternary invariant point at which, in a condensed system, three solid phases coexist with liquid, becomes an univariant line in a quaternary system ($F=4-4+1=1$).

It is not common practice to show primary phase volumes or isothermal surfaces in a quaternary system, especially the latter. Figure 8.1 shows four primary phase volumes and a series of isothermal surfaces at temperatures T_1, T_2 and T_3. It is sometimes common practice to show compatibility tetrahedra in the event that the geometry is simple enough. An "exploded" view of the quaternary tetrahedra is especially helpful in gaining a picture of equilibrium compatibility relationships between sets of four solid phases.

There are four possible types of invariant points as shown in Figure 8.2.

In the case of simple eutectic melting (case a, Figure 8.2), the invariant composition will fall within its appropriate compatibility tetrahedron. In all other cases (b, c and d), the invariant point will be located outside of its corresponding compatibility tetrahedron.

It is obvious that composition points (end members, binary compounds, ternary compounds, quaternary compounds) may be joined in a very large number of ways in a quaternary system. However, only one set of joins is the PROPER set, as determined by the total number of sets of four compatible solid phases, forming the correct set of compatibility tetrahedra. A join in a quaternary

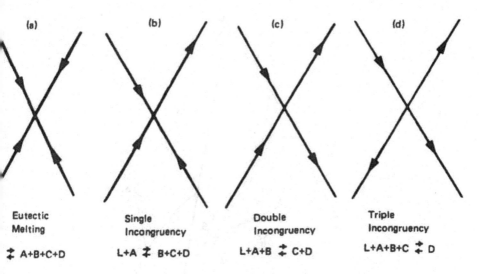

Figure 8.2. Types of Quaternary Invariant Points

system may occasionally be a true binary join within the quaternary system; that is, the composition of all solid and liquid phases appearing on the join may be expressed in terms of the compositions at the extremities of the join. It is also possible to have a true ternary system existing within the quaternary system, provided that the composition of all solid and liquid phases appearing in the ternary system can be expressed in terms of the end members of the ternary system. In the case of eutectic melting in a complex quaternary system involving binary and ternary compounds, the particular quaternary compatibility tetrahedron involved will be a true quaternary tetrahedron within the main quaternary system.

In general, many joins and many ternary triangles within the quaternary system are likely to be non-binary or non-ternary similar to the non-binary join in a ternary system. That is, phases will appear across the joins or within the triangles whose compositions cannot be expressed in terms of the end members of the join or triangle, as the case may be.

II. CRYSTALLIZATION PATH IN A SIMPLE QUATERNARY EUTECTIC TYPE SYSTEM; ISOPLETHAL ANALYSIS

Figure 8.3 shows the system A–B–C–D and the appropriate primary phase volumes, three phase surfaces, four phase boundary lines and the invariant point E_5 (quaternary eutectic). It is obvious that the composition point X, consisting

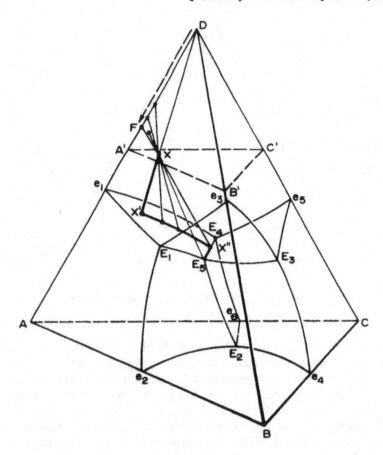

Figure 8.3. Cooling Path of Melt of Composition X (A = 29%, B = 6%, C = 4%, D = 61%)

of 61% D, 29% A, 6% B and 4% C is located in the compatibility tetrahedron A–B–C–D (the only one in the system) and that final crystallization will produce a mixture of crystals of A, B, C and D of the same proportions as those in the original liquid (composition X).

Composition X is located in the primary phase volume of D, and as the temperature of the liquid is lowered to the known liquidus temperature, an infinitesimal amount of D will precipitate from the liquid. As the temperature is lowered below the liquidus, the amount of D precipitating increases and the composition of the liquid is driven on a straight line away from D (through point X). The liquid composition will eventually intersect the surface between the primary phase volumes of A and D at X', thus causing the precipitation of an infinitesimal amount of A.

As heat is withdrawn and temperature decreases, A and D will crystallize together and the liquid composition will be driven along the surface e_1-E_1-E_5-E_4, moving from X' to X''. Calculations on the amount of liquid and solid(s) could be made (quantitatively) by the use of the Lever Rule as the liquid composition moved from X to X' to X''. The relative amounts of A and D will move from D to F as the liquid moves from X' to X''.

At point X'', an infinitesimal amount of crystals of C will appear, since E_4-E_5 is the boundary separating the primary phase volumes of D, A and C. The liquid composition proceeds from X'' to E_5 as the temperature is lowered. As the liquid moves from X'' to E_5, the relative amounts of liquid and solids D, A and C are determined by the tie line which runs from the liquid composition, through X, to the composition triangle D-A-C. The relative amounts of D, A and C are determined by the position of the extremity of the tie line in composition triangle D-A-C. At E_5, crystals of A, B, C and D all precipitate from the liquid as heat is withdrawn at constant temperature ($F=C-P+1=4-5+1=0$). After the liquid disappears, the crystalline assemblage consists of a mixture of A, B, C and D in the same proportions in which they were present in the starting liquid.

It should be obvious that this crystallization path is the simplest or least complicated of all possible paths in a quaternary system. It corresponds to the first crystallization path which was discussed for ternary systems with no solid solution in Chapter 5.

If the system has even one intermediate binary compound, the crystallization path and isoplethal analyses could be much more complex if incongruent invariant points were present as shown in Figure 8.2. During crystallization, it is entirely possible to have phases resorbing (or dissolving) on surfaces, boundary lines or at invariant points as heat is withdrawn from the system, similar to the cases of resorbtion discussed in Chapter 5. The general complex quaternary system will, of course, contain several binary compounds, a number of ternary compounds and possibly even quaternary compounds, even though the probability of occurrence of the latter is relatively low. If a reasonable number of these intermediate compounds occur, the possible number of geometries increases enormously. That is, the number of primary phase volumes and compatibility tetrahedra would increase to the point where a separate textbook would be needed to treat the various specific geometries, crystallization paths and isoplethal analyses which would be involved.

In quaternary systems without solid solution, if polymorphism of one, two, or three or four component compounds exists, if dissociation (upper or lower temperature of stability) of one, two, three or four component compounds exists, and if two, three or four phase liquid immiscibility exists, (each

Table 8.1 Quaternary Systems

System	Page	Figure
1964 Edition		
$K_2O-MgO-A\ell_2O_3-SiO_2$	270–274	802–811
$Li_2O-MgO-A\ell_2O_3-SiO_2$	275, 276	813–819
$Li_2O-A\ell_2O_3-B_2O_3-SiO_2$	276, 277	820–825
$Na_2O-CaO-A\ell_2O_3-SiO_2$	278–284	828–851
$CaO-FeO-A\ell_2O_3-SiO_2$	287–289	866–874
$CaO-MgO-A\ell_2O_3-SiO_2$	291–301	880–923
$CaO-MgO-Fe_2O_3-SiO_2$	301, 302	927, 928
$CaO-MgO-SiO_2-TiO_2$	302, 303	930–934
$CaO-MgO-SnO_2-TiO_2$	304	939, 940
$CaO-A\ell_2O_3-Fe_2O_3-SiO_2$	305–309	943–953
$MgO-A\ell_2O_3-SiO_2-ZrO_2$	312	966, 967
1969 Edition		
$Na_2O-CaO-A\ell_2O_3-SiO_2$	173, 174	2619–2622
$Na_2O-MgO-A\ell_2O_3-SiO_2$	175–178	2623–2631
$Na_2O-A\ell_2O_3-Fe_2O_3-SiO_2$	178–181	2632–2639
$CaO-FeO-MgO-SiO_2$	182–191	2641–2664
$CaO-ZnO-A\ell_2O_3-SiO_2$	193–195	2669–2674
1975 Edition		
$K_2O-MgO-A\ell_2O_3-SiO_2$	261	4611
$CaO-MgO-A\ell_2O_3-Cr_2O_3$	266	4623
$CaO-MgO-A\ell_2O_3-SiO_2$	267–277	4624–4638
$CaO-MgO-Cr_2O_3-SiO_2$	279	4640
$CaO-A\ell_2O_3-Cr_2O_3-SiO_2$	280	4641
$CaO-SiO_2-TiO_2-Nb_2O_5$	280–285	4642–4648
1981 Edition		
$Li_2O-MgO-ZnO-SiO_2$	233–236	5449, 5450
$Na_2O-CaO-MgO-SiO_2$	238–240	5457, A–E
$CaO-MgO-A\ell_2O_3-SiO_2$	246–261	5465–5480
$CaO-MgO-SiO_2-V_2O_5$	264, 265	5485, 5486
$FeO-MnO-SiO_2-ZrO_2$	268, 269	5491, A–C

one in addition to congruent or incongruent melting of one, two, three or four component compounds), the geometry of four component systems is almost beyond the realm of human comprehension. That is, if four component systems contained liquid immiscibility, polymorphism of several compounds, upper or lower temperature of dissociation of several compounds, in addition to congruent and incongruent melting of several compounds, at least one additional textbook would be required to present some of the more common solid models.

Following research on real quaternary oxide systems, it is very rare that data are available on 1) quantitative isoplethal analyses, 2) isothermal analyses, or 3) isothermal sections, in the manner that they can be determined in binary or ternary systems. Instead of the above data, in quaternary systems, it is more common to 1) represent relationships between quaternary invariant points, 2) show primary phase volumes, and 3) show compatibility tetrahedra.

A convenient list of quaternary systems in the 1964, 1969, 1975 and 1981 editions of "Phase Diagrams for Ceramists" is shown in Table 1.

III. METHOD OF REPRESENTING RELATIONSHIPS BETWEEN QUATERNARY INVARIANT POINTS

It has been common practice for persons working experimentally in quaternary systems to represent the relationships between quaternary invariant points in the manner shown in Figures 8.4 and 8.5.

Figure 8.4 is Figure 808, p. 273, 1964 collection.
Figure 8.5 is Figure 825, p. 277, 1964 collection.

The relationships shown in Figure 8.4 is characteristic of the extreme thoroughness of the experimental work done on multicomponent systems of geochemical and petrological interest by J. F. Shairer (1) and associates at the Geophysical Laboratory. Note that the phases involved are all of interest in ceramic science as well as geochemistry (cordierite, forsterite, leucite, silica, mullite, potash feldspar) and that the paper was most appropriately published in the Journal of the American Ceramic Society. Other diagrams of this type may be seen in the 1964 ($CaO-FeO-A\ell_2O_3-SiO_2$), 1969 ($Na_2-MgO-A\ell_2O_3-SiO_2$, $Na_2O-A\ell_2O_3-Fe_2O_3-SiO_2$, $CaO-ZnO-A\ell_2O_3-SiO_2$), 1975 ($CaO-MgO-A\ell_2O_3-SiO_2$) and 1981 ($Na_2O-CaO-MgO-SiO_2$ and $CaO-MgO-A\ell_2O_3-SiO_2$) editions of Phase Diagrams for Ceramists.

The relationships shown in Figure 8.5 are related to an example of a quaternary system (2) which is of extraordinary importance to ceramics and glass-ceramics based on β-spodumene or β-eucryptite phases. The quaternary systems $Li_2O-MgO-A\ell_2O_3-SiO_2$ (3,4) and $Li_2O-Na_2O-A\ell_2O_3-SiO_2$ (5) are also of great significance to the commercial production of glass-ceramics, especially. Much of the data were generated several years before the crystallization of glasses to make commercial "glass-ceramic" products became common practice.

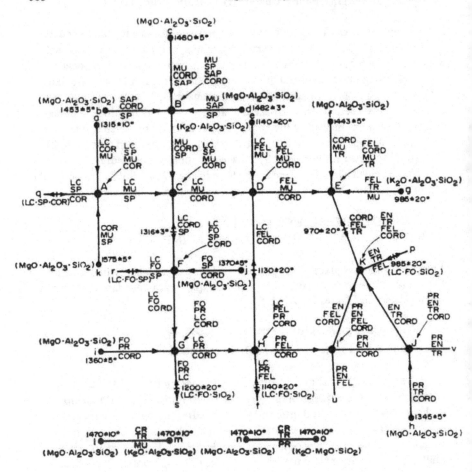

Figure 8.4. Invariant Point Relationships in the System K_2O–MgO–Al_2O_3–SiO_2

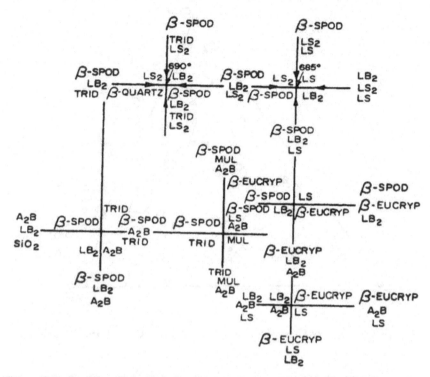

Figure 8.5. Invariant Point Relationships in the System $Li_2O-B_2O_3-A\ell_2O_3-SiO_2$

IV. SOLID MODELS OF REAL QUATERNARY SYSTEMS

A. Primary Phase Volumes

The $CaO-Fe_2O_3-A\ell_2O_3-SiO_2$ diagram of Lea and Parker (6), shown as Figure 8.6 (Figure 943, p. 305, 1964 collection) is one of the most elegant examples of the use of the perspective drawing to visualize a primary phase volume in a real system. The very small wedge of the calcium trisilicate phase volume becomes obvious, as well as other primary phase volumes important in cement chemistry.

The diagrams of Koch, Troemel and Heinz (7,8) and Schairer and Yoder (9) for the system $CaO-MgO-A\ell_2O_3-SiO_2$ (pages 254, 256, 257 and 260 of the 1981 edition of "Phase Diagrams for Ceramists") also show primary phase volumes, saturation surfaces and a relationship of quaternary invariant points.

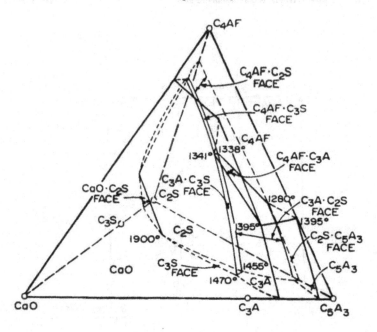

Figure 8.6. Primary Phase Volume in the System $CaO-Fe_2O_3-Al_2O_3-SiO_2$

B. Compatibility Tetrahedra

Two examples (10,11) of ceramic systems which demonstrate the usefulness of the compatibility tetrahedra are shown in Figures 8.7a,b and 8.8a,b.

Figures 8.7a and 8.7b are Figures 939 and 940, p. 304, 1964 collection.
Figures 8.8a and 8.8b are Figures 966 and 967, p. 312, 1964 collection.

The titanate-stannate system is important to electroceramics and the silica-zirconia system is related to refractories.

C. Nitride and "Sialon" Systems

One of the most significant new developments (1971) in ceramics is the Si_3N_4 solid solution data of Oyama and Kamigaito (12) and the so-called "Nitrogen Ceramics" of K. H. Jack (13). The entire development is based on the relatively low thermal expansion, good thermal conductivity, and high strength which can be achieved in β-Si_3N_4 by reaction bonding or hot pressing. The "Sialon" family of derivative compositions was typified by the addition of Al_2O_3 to Si_3N_4 to yield a series of $\beta'Si_3N_4$ solid solutions which have improved resistance to

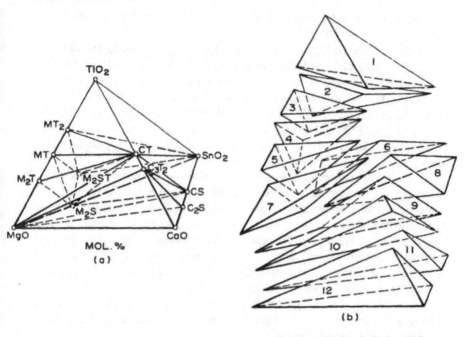

Figure 8.7a,b. Compatibility Tetrahedra in the System CaO–MgO–SnO$_2$–TiO$_2$

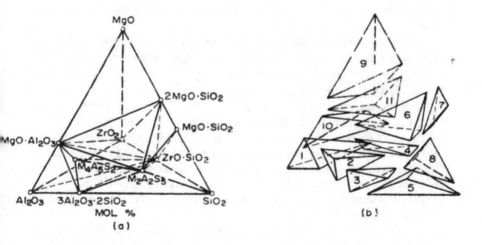

Figure 8.8a,b. Compatibility Tetrahedra in the System MgO–Al$_2$O$_3$–SiO$_2$–ZrO$_2$

oxidation relative to β-Si_3N_4, yet retain the relatively low expansion characteristics of the pure nitride phase.

Two additional facts are important. The Si_3N_4 phase easily develops an SiO_2 film on the surface of a fine particle 1) during firing in air or 2) when additives (Li_2O, BeO, MgO, Y_2O_3, TiO_2 or ZrO_2) are made to the fundamental Si_3N_4-$A\ell_2O_3$ mixture to improve certain properties. The fundamental system which takes into account the SiO_2 film is shown as the pseudoternary system Si_3N_4-$A\ell_2O_3$-SiO_2 and the systems with additives are shown as Li_2O-Si_3N_4-$A\ell_2O_3$, MgO-Si_3N_4-$A\ell_2O_3$, Be_2SiO_4-Si_3N_4-$A\ell_2O_3$, and Y_2O_3-Si_3N_4-$A\ell_2O_3$, for example, which neglect the presence of the SiO_2 film.

From the standpoint of fundamental phase equilibria, these systems are all based on the four-component system Si-Aℓ-N-O and require that the partial pressure of both nitrogen and oxygen be controlled during the investigation of the fields of stability for the β-Si_3N_4, the β' phase, Si_2N_2O, and other phases which appear in the system. This also applied to compatibility tetrahedra as well as primary fields (volumes) of stability in the four component system. If an additive is made to the fundamental system, it becomes a five component system with at least two gas phases, for example, the system Li-Si-Aℓ-O-N. During the practical heat treatments given to these systems, the atmosphere may be air, nitrogen or some undetermined combination which is developed during hot pressing or other forming methods.

From the standpoint of rigorous expression of the components and phases which exist in a chemical system, these systems are composed of solid, (sometimes liquid) and gaseous phases, demonstrating once again the danger of classifying ceramic systems as "condensed" systems containing only liquid and solid phases. The term "solid state" system is even more confining in cases such as these.

The general idea of substituting nitrogen for oxygen in many other types of compounds other than the "Sialons" is very significant and has already been demonstrated by Jack through the preparation of a "nitrogen" akermanite, $Y_2Si\,[Si_2(O,N)_7]$, the analogue of $Ca_2MgSi_2O_7$.

From the standpoint of crystal chemistry, it was unlikely that $A\ell_2O_3$ formed a single phase solid solution when added to Si_3N_4. In 1975, Gauckler, Lukas and Petzow (14) treated the system Si_3N_4-$A\ell N$-$A\ell_2O_3$-SiO_2 (Si-Aℓ-N-O) as a ternary reciprocal salt system and presented the 1780° isothermal relations as a square planar diagram (Figure 8.9a). It was proved that a 1:1 mixture of aluminum oxide and aluminum nitride ($A\ell N \cdot A\ell_2O_3$, $A\ell_3O_3N$, not a compound) was soluble to a large extent in Si_3N_4. Six compounds within the system were also discovered.

The basic discussions of reciprocal ternary systems and square planar diagrams are in Chapter XV (Reciprocal Ternary Systems, page 371) of the Ricci textbook and paragraphs 214-224 in the Ternary System chapter of the text

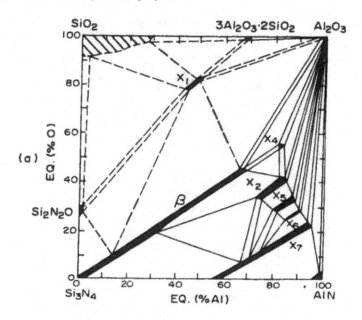

Figure 8.9a. The System Si_3N_4–SiO_2–AlN–Al_2O_3, $1780°C$ Isotherm

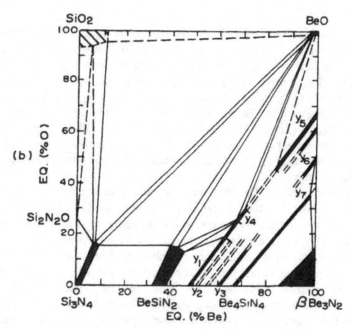

Figure 8.9b. The System Si_3N_4–SiO_2–Be_3N_2–BeO, $1780°C$ Isotherm

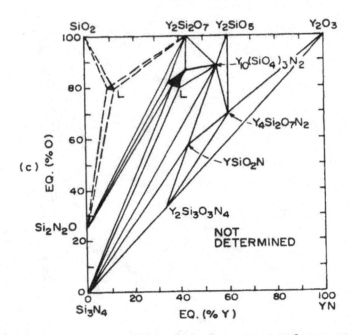

Figure 8.9c. The System $Si_3N_4-SiO_2-YN-Y_2O_3$, 1550°C Isotherm

"Silicate Melt Equilibria" by W. Eitel. The Ricci text discusses the fundamentals of reciprocal salt systems and the Eitel text relates the basic reciprocal salt systems to reciprocal oxide systems. Since 1975, the systems $Si_3N_4-SiO_2-BeO-Be_3N_2$ (15) (Figure 8.9b), $Si_3N_4-AlN-Mg_3N_2-MgO-Al_2O_3-SiO_2$ (16), $Si_3N_4-SiO_2-ZrN-ZrO_2$ (17), $Si_3N_4-YN-AlN-Al_2O_3-Y_2O_3-SiO_2$ (18), $Si_3N_4-SiO_2-Y_2O_3$ (19) (Figure 8.9c), $Si_3N_4-AlN-ZrN-Al_2O_3-SiO_2-ZrO_2$ (20), $Si_3N_4-ScN-SiO_2-Sc_2O_3$ (21) and $Mg_3N_2-AlN-MgO-Al_2O_3$ (22) have been studied isothermally between temperatures of 1500°C to 1800°C as four or five component oxynitride systems, and in several cases, as square planar diagrams of reciprocal oxynitride systems.

The basic Si-Al-N-O system was studied for solid-liquid relationships in 1978 (23) and more recently studies of oxynitride glasses (24,25,26,27,28) have been made. Jack (29) has made a review and analysis of the structure and phase equilibrium of oxynitride systems, compounds and solid solutions.

From the standpoint of fundamental phase equilibria, the oxynitride research has demonstrated that some four component systems may be represented as ternary isothermal, reciprocal, square planar diagrams. However, if oxynitride compounds melt congruently, incongruently or dissociate, enter into complete or partial solid solution, are involved in eutectic, peritectic, eutectoid or peritectoid reactions, become exsolved in the solid state or immiscible in the liquid state, and are dependent on oxygen or nitrogen pressure, one has to return to the basic temperature, pressure and compositional studies of four, five and multicomponent systems.

The same is true for oxyfluoride, oxysulfide, oxychloride, etc., types of ceramic systems. Once again, it demonstrates how many hundreds of phase equilibrium studies need to be done for three, four, five and multicomponent systems.

D. The B_2O_3-$A\ell_2O_3$-SiO_2-P_2O_5 System

The system B_2O_3-$A\ell_2O_3$-SiO_2-P_2O_5 may ordinarily be thought to be a four component system (30,31) composed of these oxides. However, the high vapor pressure and relative instability of B_2O_3 and P_2O_5 might require that the system in some cases be expressed in terms of the five components, B-Aℓ-Si-P-O.

This system is an example of a quaternary system where detailed phase equilibrium research is needed in binary and ternary subsystems before the final solid model of the quaternary system can be constructed. References 29 and 30 contain EXPLORATORY data on the quaternary system and its ternary subsystems, but considerably more data is necessary on the crystalline and glassy regions as follows:

i. Glass Regions

1. B_2O_3-SiO_2
2. $A\ell_2O_3$-SiO_2
3. P_2O_5-SiO_2
4. B_2O_3-$A\ell_2O_3$-SiO_2
5. B_2O_3-SiO_2-P_2O_5
6. $A\ell_2O_3$-SiO_2-P_2O_5
7. B_2O_3-$A\ell_2O_3$-SiO_2-P_2O_5

ii. Crystalline Regions

1. $A\ell_2O_3$-SiO_2
2. P_2O_5-SiO_2
3. B_2O_3-Al_2O_2 (9:2 compound)
4. Mullite-BPO_4 (B_2O_3, P_2O_5)
5. Mullite-$A\ell PO_4$ ($A\ell_2O_3$, P_2O_5)
6. 9:2-BPO_4
7. 9:2-$A\ell PO_4$
8. B_2O_3-$A\ell_2O_3$-SiO_2-P_2O_5

This type of attitude and analysis is needed for many other quaternary systems which are the basis for ceramic technology and ceramic products.

V. SOLID SOLUTION IN QUATERNARY SYSTEMS

A. General

The number of geometrical types of quaternary solid solution systems would be enormous, far greater than the types of ternary solid solution systems discussed in Chapter 7.

The reasons for the tremendous number of possibilities is as follows:

1. The boundary binary and ternary systems may be complete solid solution types. The binary systems may or may not have a maximum or minimum in the liquidus and the ternary systems may have many types of liquidus relations.
2. The bounding binary and ternary systems may have partial solid solution relationships.
3. The boundary binary and ternary systems may have eutectoid or peritectoid relationships due to polymorphism of compounds.
4. The liquid regions of the binary and ternary systems may have immiscibility regions.
5. The boundary binary and ternary systems may have various types of intermediate compounds (congruently melting, incongruently melting or dissociating types) as well as complete or partial solid solution.
6. The quaternary system itself may contain immiscible liquid regions, a great variety of liquidus relationships, complete or partial solid solution, polymorphic compounds which give rise to eutectoid or peritectoid relationships and intermediate compounds which melt congruently, incongruently or dissociate.

B. Diagrams Based on Two, Three, Four or Five Phase Equilibria

In the metallurgical texts of Rhines and Prince, quaternary solid solution is based on two, three, four and five phase equilibria, similar to the two, three and four phase equilibria in ternary solid solution systems. As far as a homogeneous liquid and solid solutions are concerned, the relationships are as follows:

1. $\ell \rightleftarrows \alpha$ (tie line)
2. $\ell \rightleftarrows \alpha + \beta$ (tie triangle)
3. $\ell \rightleftarrows \alpha + \beta + \gamma$ (tie tetrahedron)
4. $\ell \rightleftarrows \alpha + \beta + \gamma + \delta$

As stated in Section A, if the "condensed" quaternary system has two to four immiscible liquids, if the end member components or the binary, ternary or quaternary compounds have polymorphic inversions and if there are various types of complete or partial solid solution, there will be monotectics, peritectics, eutectoids, peritectoids and an enormous number of complex types of two, three, four and five phase equilibria, in addition to the very simple types of eutectic equilibria listed above (1–4).

It is very interesting at this point to make two quotes from Chapter 19 in Rhines:

a) "From the standpoint of metallurgical practice, phase diagrams of alloy systems involving four and more elements are needed more than are the diagrams of simpler systems, because many of the alloys of commerce contain more than three elements added deliberately, and if significant impurities are included in the count, the majority contain a still larger number of elements. About 40 quaternary systems (Rhines is a 1956 textbook) have been investigated in part, none completely, and portions of perhaps a half dozen quinary systems (five components) have been studied."

b) "Were it mechanically (geometrically) feasible to do so, a three-dimensional grid for reading the composition of alloys, analogous to that employed with ternary diagrams, might be constructed by passing regularly spaced planes through the tetrahedron parallel to each of its four faces. This is not practical, however, with a two-dimensional drawing of a space model. Consequently, there is little that can be done, short of making actual space models, to make four-component isotherms quantitative. Several schemes for reading composition in two-dimensional drawings of space figures have been proposed, but none has been found satisfactory. The principal usefulness of the three-dimensional isotherm is, therefore, in exhibiting the structure of the quaternary diagram in a qualitative way. Isopleths may then be used to advantage to give the temperature and compositions at points distributed upon the lines and surfaces depicted in the isotherms. This is a cumbersome procedure, indeed, but nothing better is at hand."

With regard to quotation "a", one could state that phase equilibria (or nonequilibria) has been studied in a relatively small number of quaternary CERAMIC systems (See Table I). Calculations of the number of binary (Chapter 3) and ternary (Chapter 5) ceramic systems based on 90 elements have been made using the combination equation:

$$C_{(n,r)} = \frac{n!}{r!(n-r)!}$$

In the case of four component systems, the number would be:

$$C_{(90,4)} = \frac{90!}{4!(90-4)!} = \frac{90!}{4!\times 86!}$$

$$= \frac{90\times 89\times 88\times 87}{24} = 2,555,190$$

Instead of using the number 90, assume that one would be interested in the four component systems involving the following common oxides:

Li_2O	MgO	MnO	B_2O_3	SiO_2	P_2O_5	MoO_3
Na_2O	CaO	FeO	Al_2O_3	TiO_2	V_2O_5	WO_3
K_2O	ZnO	CoO	Y_2O_3	GeO_2	Nb_2O_5	
	SrO	NiO	La_2O_3	ZrO_2	Ta_2O_5	
	BaO	CuO	Cr_2O_3	HfO_2		
	PbO					

The number of four component systems involving the thirty oxides would be:

$$C_{(30,4)} = \frac{30!}{4!(30-4)!} = \frac{30!}{4!\times 26!}$$

$$= \frac{30\times 29\times 28\times 27}{24} = 27,405$$

With respect to quotation "b", the conclusion is that space models of quaternary systems are so enormous, so complicated and so difficult to construct that an additional textbook is required.

VI. MULTICOMPONENT SYSTEMS

The number of significant five component systems contained in "Phase Diagrams for Ceramists are listed in Table 2.

At this point, one can proceed in one of two possible directions. One can think of multicomponent systems in mathematical terms, as illustrated by the text of Palatnik and Landau (32). One is lead into topology, Eulers theorem for polyhedra and other mathematical devices, which, although facinating as exercises, are not particularly useful to the practicing ceramist or material scientist or engineer.

The direction which has been taken by the practicing ceramist is reflected in books on glasses, glazes (33) and enamels (34) which deal with the art and science of manipulation of polycomponent systems toward the development of glass compositions with especially useful chemical and physical properties. Glasses, glazes and enamels frequently contain from five to ten components in

Table 8.2 Five Component Systems

	Page	Figure
1964 Edition		
$Na_2O-CaO-MgO-A\ell_2O_3-SiO_2$	314, 315	973-981
$CaO-MgO-A\ell_2O_3-Fe_2O_3-SiO_2$	317-319	991-1002
1969 Edition		
$Na_2O-CaO-MgO-A\ell_2O_3-SiO_2$	196-202	2679-2697
$CaO-MgO-A\ell_2O_3-Cr_2O_3-SiO_2$	202-205	2699-2713
1975 Edition		
$CaO-MgO-A\ell_2O_3-SiO_2-TiO_2$	288-291	4653, 4654
1981 Edition		
$K_2O-CaO-MgO-A\ell_2O_3-SiO_2$	272-274	5498, A-H
$Na_2O-CaO-MgO-A\ell_2O_3-SiO_2$	275, 276	5500, 5501

the Phase Rule sense and geometrical treatment becomes impossible. Modern ceramics has taken a turn toward compositionally simpler systems, but, on the other hand, the computer has taken over on the design of multicomponent glass, glaze and enamel compositions.

REFERENCES

1. J. F. Schairer, The System $K_2O-MgO-A\ell_2O_3-SiO_2$: 1, Results of Quenching Experiments on Four Joins in the Tetrahedron Cordierite-Forsterite-Leucite-Silica and on the Join Cordierite-Mullite-Potash Feldspar, J. Am. Ceramic Society, 37(11) 526, 1954.
2. K. H. Kim and F. A. Hummel, private communication, Dec. 20, 1961, PhD thesis of K. H. Kim, Phase Equilibria in the System $Li_2O-B_2O_3-A\ell_2O_3-SiO_2$, The Pennsylvania State University, August 1961.
3. M. D. Karkhanavala and F. A. Hummel, Reactions in the System $Li_2O-MgO-A\ell_2O_3-SiO_2$: I, The Cordierite-Spodumene Join, J. Amer. Cer. Soc. 36(12) 393-397, 1953.
4. T. I. Prokopowicz and F. A. Hummel, Reactions in the System $Li_2O-MgO-A\ell_2O_3-SiO_2$: II, Phase Equilibria in the High Silica Region, J. Amer. Cer. Soc. 266-278, 1956.
5. W. B. Badger, PhD thesis in Geochemistry, Phase Equilibria in the System $Li_2O-Na_2O-A\ell_2O_3-SiO_2$ with Special Emphasis on the Silica Polymorphs, The Pennsylvania State University, June, 1968.

6. F. M. Lea and T. W. Parker, Trans Roy Soc (London) Ser. A 234A(731) 16, 1934.

7. K. Koch, G. Troemel and G. Heinz, Tonind-Ztg, Keram. Rundsch., 99 (3) 57, 1975.

8. K. Koch, G. Troemel and G. Heinz, Arch. Eisenhuettenwes., 46 (3) 165, 1975.

9. J. F. Schairer and H. S. Yoder, Jr., Yearbook, Carnegie Inst., Washington, 68, 202, 1970.

10. L. W. Coughanour, R. S. Roth, S. Marzullo, and F. E. Sennett, Solid State Reactions and Dielectric Properties in System Magnesia-Lime-Tin Oxide-Titania, J. Res. Natl. Bur. Stds. 54(3) 149-162, 1955.

11. P. G. Herold and W. J. Smothers, Solid State Equilibrium Relations in the System $MgO-Al_2O_3-SiO_2-ZrO_2$, J. Amer. Cer. Soc. 37(8) 351-353, 1954.

12. Y. Oyama and O. Kamigaito, Solid Solubility of Some Oxides in Si_3N_4, Jap. J. Appl. Phys., 10(11), 1637, 1971.

13. K. H. Jack, Nitrogen Ceramics, XVII Mellor Memorial Lecture, Trans. and Jour. Brit. Cer. Soc. 72(8) 376-384, 1973.

14. L. J. Gauckler, H. L. Lukas and G. Petzow, Contribution to the Phase Diagram $Si_3N_4-AlN-Al_2O_3-SiO_2$, J. Am. Ceram. Soc., 58(7-8), 346, 1975.

15. Irvin C. Huseby, Hans L. Lukas and Günter Petzow, Phase Equilibria in the System $Si_3N_4-SiO_2-BeO-Be_3N_2$, J. Am. Ceram. Soc., 58(9-10), 377, 1975.

16. L. J. Glaucker, J. Weiss and T. Y. Tien, Insolubility of Mg in βSi_3N_4 in the System Al-Mg-Si-O-N, J. Am. Ceram. Soc., 61, (9-10), 397, 1978.

17. J. Weiss, L. J. Gauckler and T. Y. Tien, The System $Si_3N_4-SiO_2-ZrN-ZrO_2$, J. Am. Ceram. Soc., 62, (11-12), 632, 1979.

18. I. K. Naik and T. Y. Tien, Subsolidus Relations in Part of the System Si,Al,Y/N,O, J. Am. Ceram. Soc., 62, (11-12), 642, 1979.

19. L. J. Gauckler, H. Hohnke and T. Y. Tien, The System $Si_3N_4-SiO_2-Y_2O_3$, J. Am. Ceram. Soc., 63, (1-2), 35, 1980.

20. J. Weiss, L. J. Gauckler, H. L. Lukas, G. Petzow and T. Y. Tien, Determination of Phase Equilibria in the System Si-Al-Zr/N-O by Experiment and Thermodynamic Calculation, Jour. Mat. Science, 16, 2997, 1981.

21. P. E. D. Morgan, F. F. Lange, D. R. Clarke, and B. I. Davis, A New Si_3N_4 Material: Phase Relations in the System Si-Sc-O-N and Preliminary Property Studies, J. Am. Ceram. Soc., 64, 4, C-77, 1981.

22. J. Weiss, P. Greil and L. J. Gauckler, The System Al-Mg-O-N, J. Am. Ceram. Soc., 65, 5, C-68, 1982.

23. I. K. Naik, L. J. Gauckler and T. Y. Tien, Solid-Liquid Equilibria in the System $Si_3N_4-AlN-SiO_2-Al_2O_3$, J. Am. Ceram. Soc., 61, (7-8), 332, 1978.

24. Ronald E. Loehman, Preparation and Properties of Yttrium-Silicon-Aluminum Oxynitride Glasses, J. Am. Ceram. Soc., 62, (9-10), 491, 1979.

25. C. J. Leedecke and Ronald E. Loehman, Electrical Properties of Yttrium-Aluminum-Silicon Oxynitride Glasses, J. Am. Ceram. Soc., 63, (3-4), 190, 1980.

26. Paul E. Jankowski and Subhash H. Risbud, Synthesis and Characterization of an Si-Na-B-O-N Glass, J. Am. Ceram. Soc., 63, (5-6), 350, 1980.

27. C. J. Leedecke, Electrical Properties of Some Y-Si-Aℓ Oxynitride Glass-Ceramics, J. Am. Ceram. Soc., 63, (7-8), 479, 1980.

28. C. J. Brinker, Formation of Oxynitride Glasses by Ammonolysis of Gels, J. Am. Ceram. Soc., 65, (1), C-4, 1982.

29. Phase Diagrams; Materials Science and Technology, Volume 6-V, Chapter V, K. H. Jack, The Relationship of Phase Diagrams to Research and Development of Sialons, p. 242-283, Academic Press, 1978.

30. F. A. Hummel and W. F. Horn, The Quaternary System B_2O_3-$Aℓ_2O_3$-SiO_2-P_2O_5; Part I, Literature Review and Exploratory Data on the Ternary Subsystems, Jour. of the Australian Ceramic Society, 17, No. 1, 25-32, 1981.

31. W. F. Horn and F. A. Hummel, The Quaternary System B_2O_3-$Aℓ_2O_3$-SiO_2-P_2O_5; II, Exploratory Data, Jour. of the Australian Ceramic Society, 17, No. 2, 33-36, 1981.

32. L. S. Palatnik and A. I. Landau, Phase Equilibria in Multicomponent Systems, 1964, Holt, Rinehart and Winston, Inc., 383 Madison Avenue, New York, 10017.

33. C. G. Harman, Ceramic Glazes, Third Edition, (Revised from C. W. Parmelee) Cahners Books, 89 Franklin Street, Boston, Mass., 02110.

34. A. I. Andrews, Porcelain Enamels, Second Edition, The Garrard Press, Champaign, Illinois.

INDEX

Printed in the United States
by Baker & Taylor Publisher Services

Printed in the United States
by Baker & Taylor Publisher Services